AMORPHOUS MAGNETISM

AMORPHOUS MAGNETISM

Proceedings of the International Symposium on Amorphous Magnetism,
August 17-18, 1972, Detroit, Michigan

Edited by
Henry O. Hooper and Adriaan M. de Graaf

Department of Physics
Wayne State University
Detroit, Michigan

PLENUM PRESS • NEW YORK–LONDON • 1973

Library of Congress Catalog Card Number 72-96493
ISBN-13: 978-1-4613-4570-1   e-ISBN-13: 978-1-4613-4568-8
DOI: 10.1007/978-1-4613-4568-8

© 1973 Plenum Press
Softcover reprint of the hardcover 1st edition 1973

A Division of Plenum Publishing Corporation
227 West 17th Street. New York, N.Y. 10011

United Kingdom edition published by Plenum Press, London
A Division of Plenum Publishing Company, Ltd.
Davis House (4th Floor), 8 Scrubs Lane, Harlesden,
London NW10 6SE. England

PREFACE

The title of amorphous magnetism was conceived to encompass the study of the magnetic properties of dilute random substitutional alloys, amorphous metallic alloys, amorphous magnetic semiconductors and the conventional glasses, including chalcogenide, organic and inorganic glasses. These materials have heretofore been considered to be quite different because of the differences in their atomic structures. However, when the magnetic behavior of these materials is carefully examined it becomes clear that these materials exhibit certain similarities. For the first time at the International Symposium on Amorphous Magnetism held at Wayne State University in Detroit, Michigan on August 17 and 18, 1972, scientists working in these areas were assembled to examine the similarities of these systems. This volume contains a summary of the 9 invited talks, 23 delivered, contributed papers and an additional 12 contributed papers whcih were not presented because of insufficient time in the program. This volume presents for the first time a collection of the most current theoretical and experimental studies in the area of amorphous magnetism.

We wish to thank the authors of these papers for their outstanding cooperation in preparing their manuscripts and for editing our transcriptions from tape recordings of the discussions which followed their presentations. In addition, we wish to thank the participants at this conference whose enthusiasm and interest made the conference so successful.

We wish to acknowledge and deeply thank the Department of Physics, the Graduate Division, and the Wayne State Fund of the Alumni Association (through a Research Recognition Award presented to one of us, Henry O. Hooper) for their generous financial support.

We are indebted to B.R. Cooper, T.A. Kaplan, D.L. Weaire, and S.A. Werner for so effectively chairing the four sessions of the Symposium on Amorphous Magnetism.

We wish to acknowledge the assistance and express our deep appreciation to the members of our organizing committee; V. Cannella, J.T. Chen, G. Gerhart, R. Henderson, P.F. Kenealy, R. Kline, Y.W. Kim, T.J. Moran, R.L. Thomas and R. Verhelst and to the many other members of the Staff of the Department of Physics including Suzanne Pfunk, Barbara Collins, Mary Phillips, Patricia Watson, Bonnie Serman, Gail Davis, Virginia Blannon, and Bernice Winter. These people are the ones most responsible for the smooth operation of the Symposium.

We are particularly indebted to Mrs. Suzanne Pfunk for her outstanding secretarial assistance. Her enthusiastic and energetic effort enabled us to meet the numerous deadlines associated with the Symposium and these Proceedings.

November 1, 1972   Henry O. Hooper
                                                 and
                                          Adriaan M. de Graaf

LIST OF PARTICIPANTS

The following list includes the names of those participants who registered at the Symposium on Amorphous Magnetism, August 17 and 18, 1972, Wayne State University. The asterisk indicates the names of the speakers.

K. Aaland
V.P. Agrwal
H. Alloul
H.A. Alperin
*P.W. Anderson
G.H. Azarbayejani
*B.G. Bagley
N.K. Batra
*P.A. Beck
W. Beres
L.H. Bieman
S.G. Bishop
H.V. Bohm
S.P. Bowen
S.M. Bose
J.R. Bright
T. Brun
R.R. Bukrey
T.J. Burch
*W.J.L. Buyers
M.C. Cadeville
*V. Cannella
*G.S. Cargill
R.W. Catchings
R.E. Chase
J.T. Chen
R. Cochrane
*B.R. Coles
M.E. Collins
M.M. Collver

J.F. Cooke
B.R. Cooper
J. Cowen
J. Cusick
M. Cyrot
M.D. Daybell
A.M. de Graaf
J.E. Dickman
K.J. Duff
D.M. Esterling
A.J. Fedro
G.A. Ferguson
A. Fert
J. Flouquet
N.E. Frankel
M.A. Frenkel
R.E. Frenkel
E.J. Friebele
R.J. Gambino
L.M. Geppert
G.R. Gerhart
C.D. Graham, Jr.
D.L. Griscom
*J.E. Gubernatis
E. Gurmen
D.R. Gustafson
*R. Hasegawa
W.M. Hartmann
L.G. Hayler
R.G. Henderson

*H.O. Hooper
P.M. Horn
S.J. Hudgens
G.P. Huffman
I.S. Jacobs
G.O. Johnson
T.A. Kaplan
P.F. Kenealy
*D.J. Kim
Y.W. Kim
R.W. Kline
*E.S. Kirkpatrick
W.C. Koehler
*J.A. Krumhansl
M. Kwan
K.C. Lee
*P. Lederer
H. LeGall
*D.C. Licciardello
R.L. Lintvedt
L.L. Liu
N. Lurie
*R.K. MacCrone
*D.E. MacLaughlin
S.D. Mahanti
J.R. Marko
*G.W. Mather, Jr.
R.E. Mills
T. Mizoguchi
C.G. Montgomery
R. Moon
T.J. Moran, Jr.
S.C. Moss
W.B. Muir
*L.N. Mulay
*J.A. Mydosh
B.G. Nickel
K.H. Oh
S. Ovshinsky
M.A. Paesler
D. Pan
D. Paul
C.H. Perry
R.A. Reck
C.W. Rector

*J.J. Rhyne
*N. Rivier
H. Rockstad
I. Rosenstein
R.P. Santoro
K. Sapru
M.P. Sarachik
H. Sato
*A.I. Schindler
F.C. Schwerer
*D.J. Sellmyer
S.M. Shapiro
C.H. Sie
E.J. Siegel
*H.C. Siegmann
L.R. Sill
Major W.C. Simmons
*A.W. Simpson
M.M. Sokoloski
J. Souletie
M.B. Stearns
M.G. Stewart
*R.M. Stubbs
W. Stutius
C.C. Sung
*R. Tahir-Kheli
*J. Tauc
R.L. Thomas
D. Thoulouze
R.J. Todd
R. Tournier
D.J. Treacy
C.C. Tsuei
R.D. Turoff
H.L. Van Camp
R.A. Verhelst
T.G. Verlinde
R. Viswanathan
S. von Molnar
*P. Wachter
D.L. Weaire
S.A. Werner
*L.K. Wilson
*B. Window
J. Wong

M. Yessik
D.R. Zrudsky

CONTENTS

*These papers were not delivered at the Symposium on Amor-
phous Magnetism due to insufficient time.

TOPICS IN SPIN GLASSES*

P. W. Anderson

Bell Laboratories, Murray Hill, New Jersey,

07974, and Cavendish Laboratory, Cambridge,

England

ABSTRACT

A discussion of the general theory of spin glasses and specifically, how the ideas of Anderson, Halperin and Varma on low-energy excitations in real glasses may be made applicable to random spin arrangements, will be presented. Attention will be called to the phonon-spin wave analogy. The work of Kok and Anderson on accounting in the simplest possible fashion for the properties of various alloys, such as amorphous solvent metals and giant susceptibility systems, will be discussed.

INTRODUCTION

I would like to use this talk as an opportunity to expand on, and to some extent publicize, work which various co-workers and I have published in this field over the past few years. I won't be saying, then, very much that is really new to science, but I will be trying to bring together a group of ideas which make up what seems to me to be rather unified point of view on this subject. In particular, in my

*The portion of work done at Cavendish Laboratory was sponsored by the Air Force Office of Scientific Research under contract No. F44620-71-C-0108.

title I have used the key phrase "spin glass," and it is
the "spin glass" point of view to which I refer.

First of all, what is a spin glass? This is a very apt
term which was invented, as far as I know, by Brian Coles,
to describe the entire class of magnetic alloys of moderate
dilution in which the magnetic atoms are far enough apart
so that the magnetic structure no longer resembles that of
the pure metal, but close enough so that their exchange
interactions dominate other energies such as the Kondo
effect and other free electron interactions. The canonical
example is Mn in Cu, in the range of 1/2-10% concentration,
but there are many others - La-Gd, Au and Ag-Mn, etc. I
will talk about various of the properties later, but the
most characteristic is a susceptibility maximum with no
genuine evidence either of antiferro- or ferro-magnetism.

The physical model that has come to be accepted for
these substances is that the exchange interactions follow
the R-K-Y theory, oscillating and falling off as $1/r^3$, and
that aside from mediating the R-K-Y interaction the free
electrons don't play much of a role. The very important
property of R-K-Y which gives spin glasses their unique
properties is the oscillation, because the randomly located
spins then have interactions of essentially random sign, so
that no ferromagnetic, or (with less confidence) long-range
ordered antiferromagnetic state is particularly favored
energetically. The fact that R-K-Y falls off as $1/r^3$ is
also confirmed by the reasonably universal tendency for
temperatures to scale with concentration: $E \sim T \propto 1/r^3 \sim n$.

The basic idea of the spin glass goes back to Herring,
Marshall,[1] and Klein-Brout:[2] that each individual spin sits
in its own random internal field H_i, and that the material
is characterized essentially by the distribution of random
fields $p(H_i)$. I pointed out recently[3] that this essentially local view of the individual spins is an indication that
their behavior must be localized in the sense of localization theory; but that is merely to say that at least at
high temperatures the standard theory should be more or
less correct.

The first bit of work we will discuss here was carried
out by my student at Cambridge, Miss Kok.[4] The point of
this was to clear up some of the mysterious-looking

phenomena in these materials and to show that they were compatible with the standard model and the spin glass idea. The
first was a set of measurements by Hilsch's group[5] who showed experimentally that amorphous films, evaporated at He
temperatures, behaved somewhat differently from the same
films allowed to warm up in room temperature and crystallize.
In particular, it was quite noticeable that both in the case
of Mn in Cu, and of Gd in La, there was a marked shift in
the Curie-Weiss θ for the high-temperature susceptibility:

$$\chi = \frac{C}{T-\theta} \qquad \text{or} \qquad \frac{1}{\chi} = \frac{T-\theta}{C}$$

and it was quite striking that $\theta_{amorphous} \simeq 0$ in all cases,
whether θ_{cryst} was greater or less than zero (see Figs. 1
and 2).

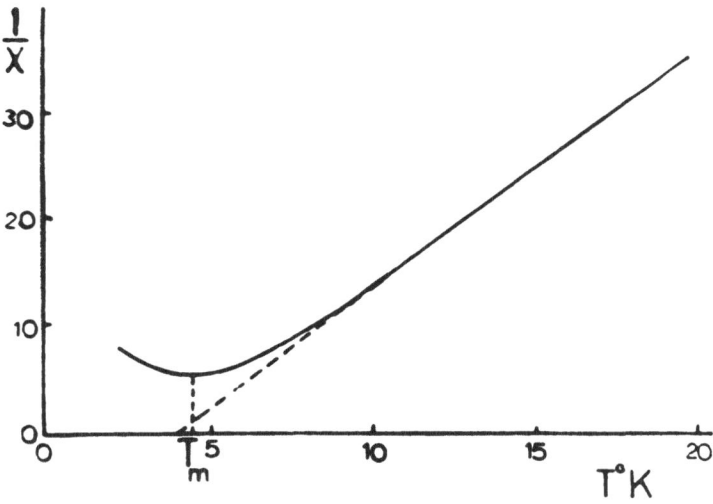

Figure 1

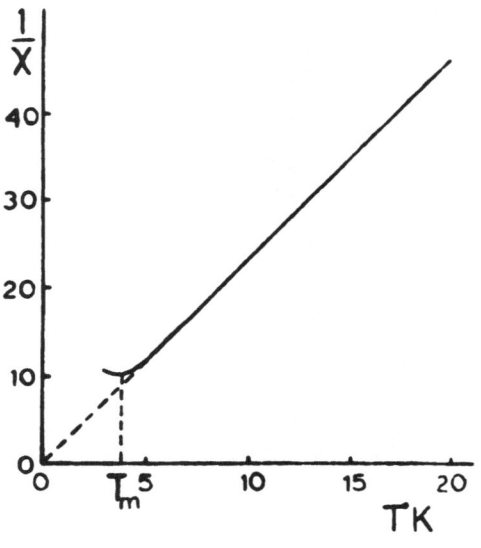

Figure 2

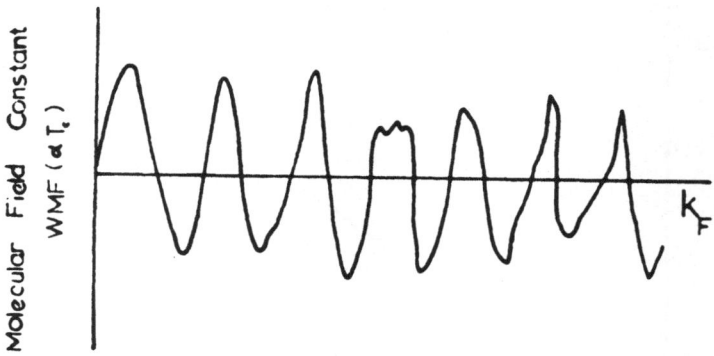

Figure 3

The same generalization is true of liquid metals.[6] The explanation of this is quite obvious if you remember a figure from Mattis' book,[7] a rough sketch of which is shown in Fig. 3; this is the Curie-Weiss θ as predicted by R-K-Y theory as a function of the size of the Fermi surface, i.e. of the ratio $k_F a$ of lattice constant to Fermi surface wavelength. The reason for this rather rapid oscillation is that the average θ is zero, because of the alternating R-K-Y interactions. Only because the crystalline nature of the material puts the neighbor atoms at specific points on the R-K-Y oscillation do we get any θ at all. In the amorphous case, we have no fixed a, so we can think of these films as <u>averaged</u> over several cycles of Mattis' curve; it's not really quite as easy as that because of the nonuniform radial distribution but that's what the story amounts to. Even if the films are not truly amorphous, θ depends only on the radial distribution function, which is experimentally measured, not on more complex measures of disorder.

A somewhat less wholly satisfactory study[8] was made of the phenomenon of "giant susceptibilities" in some of these materials.[9] Here again, the idea was to treat the problem as simply as possible. The giant susceptibilities when plotted as $1/\chi$ vs. T looked really no different from ordinary Cu-Mn, except that the θ-values are positive rather than negative. Thus if we could produce a theory of the χ of CuMn, it ought to fit La-Gd reasonably well too. Our theory was just a slight extension of molecular field theory in which we allowed a random distribution of molecular fields in very much the spirit of the Marshall-Klein-Brout work, with some reasonable further improvements. There was no difficulty indeed in giving a description of the susceptibility at and above the maximum, but the very rapid non-linear decrease of χ with H was a little hard to manage. I feel that there are appreciable, but not controlling, effects of the cooperative motion of rather large clusters of spins in these materials, even at and near the susceptibility maximum, which are important in certain effects such as this, but do not basically invalidate the theory.

All of this work looks at the spin-glass problem from the high-temperature side. The starting approximation is molecular field theory:

$$\overline{M_i} = \frac{C}{T} H_i^{eff} = \frac{C}{T} \sum J_{ij} \overline{M_j} \; .$$

The scheme of Miss Kok's work was to attempt to improve this by estimating higher, nonlinear terms.

The idea I worked out in cooperation with Halperin and Varma[10] goes at the spin glass from the opposite end: we are concerned with the very low temperature end, and with exploiting the idea of the spin glass as a specific analogy to the real, ordinary glass, also at low T.

For over a decade one of the most puzzling facts about the Cu-Mn system has been the large linear specific heat at low temperatures, more or less independent of spin concentration.[11] This specific heat is several times that caused by the free electrons, and is independent of concentration. The total entropy represented by this specific heat is a reasonable fraction of the total entropy of the spins; this is proportional to n because it cuts off at about T_m, with $T_m \propto n$:

$$S = \int^{T_m} \frac{Csp}{T} \; dT$$

$$= \int^{T_m} const \; dT \propto n.$$

At first sight, actually, this is just the result we would expect from the simple Marshall-Klein-Brout theory, because it is what one would get from a probability distribution $p(H_i)$ of local fields H with a finite probability of zero fields. But although this explanation appears in the literature, it seems likely that it is not completely correct, at least in the way that it is stated, as has been emphasized by Herring and various others. The difficulty is that although it is reasonable to expect the field in, say, the z direction to have a finite probability of being zero, of the same order as the probability that it has any other reasonable value (since the fields due to any pair of neighbors are as likely to cancel as to add) it is not reasonable to expect the fields in <u>all</u> directions to cancel (if one is thinking of the system classically).

Another way to say it which is quantum mechanically correct is to pick an axis of quantization z and to write

$$H = \sum_{ij} J_{ij} S_i \cdot S_j$$

$$= \sum_{ij} J_{ij} S_{zi} S_{zj} + \sum_{ij} J_{ij} [S_+^i S_-^j + S_+^j S_-^i] = H_o + H'.$$

We can expect that the eigenvalues of H_o are just like what we have described: each spin has a best direction, + or -, which determines the various S_z^j's and each spin i will experience a field $H_i = \sum_{ij} J_{ij} S_j^z$ which has the given kind of probability distribution. But the terms $S_+S_- + S_-S_+$ provide <u>off-diagonal</u> <u>matrix</u> <u>elements</u> between the states of H_o, which will couple then together. Thus when $H_i \to 0$ so that two energy levels m and n are very close together,

$$E_m - E_n \to 0$$

there will be a matrix element V_{mn} which will <u>mix</u> the two levels and cause them to repel each other. The effective Hamiltonian for these few levels will be like

$$H_{eff} = \begin{pmatrix} E_m & V_{mn} \\ V_{mn} & E_n \end{pmatrix}$$

with eigenvalues

$$E = \frac{E_n + E_m}{2} \pm \frac{1}{2} \sqrt{(E_m - E_n)^2 + |V|^2}$$

which will have a minimum separation

$$\Delta E = \left| V_{mn} \right| .$$

Thus V_{mn} will act just like the x and y components of the
random field, preventing an actual zero value of the energy
level splitting unless $V_{mn} \to 0$ also. Since V_{mn} is <u>two</u> random variables, (it has a real and an imaginary part) the
probability that it also goes to zero appears to be $\propto VdV$
so that

$$p(\Delta E)_{\Delta E \to 0} \propto (\Delta E)^2 .$$

Actually, this is known to be a rather general property of
random matrices: there is a <u>hole</u> in the distribution of
$E_m - E_n$.

Various attempts have been made to get around this
argument, or to find a different source for the linear specific heat, but so far none of them have appeared very successful to me. Kondo tried to ascribe it to the free electrons,
but in the absence of a Kondo effect such a large change in
density of states is a bit implausible.

Recently it has been discovered that there is also an
unexpected linear specific heat observed in <u>real</u> glasses,[12]
with a coefficient which, while not as big as that for the
spin glasses, is not negligible compared to typical metals
either. Halperin, Varma and I have proposed an explanation
for <u>that</u> anomaly which we believe may also by analogy be
the answer to the problem in the spin glasses, and if so, it
suggests a rather new way of looking at the spin glasses
which may be of more general validity.

At first sight in the case of ordinary glasses one may
make just the same kind of argument, and have it break down
in just the same way. Namely, since the configuration of
the atoms is random, there must be very many places (of order
N) where there are two possible positions of very similar
energy (see Fig. 4). Again, E_1 and E_2 are independent random variables so $p(E_1 - E_2 \to 0)$ is finite. But again also we

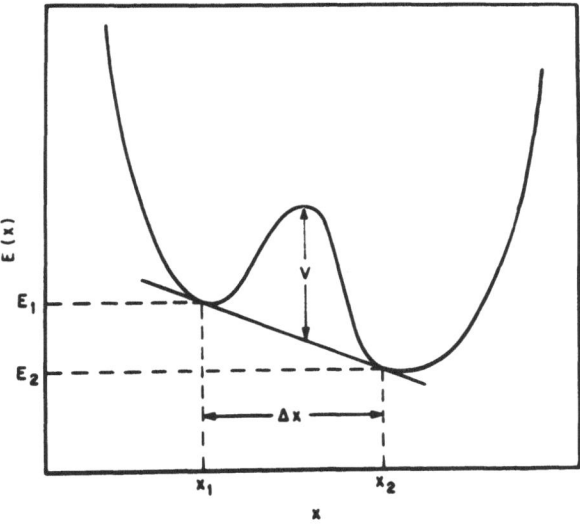

Figure 4

have a problem – if it is possible to tunnel between the
two energy minima with a tunneling matrix element M_{12}, the
energy level separation will be at least $\Delta E > M_{12}$, so that
we **cannot have** any separations of exactly zero energy.

Now in the glass situation it is pretty clear that the
tunneling matrix element

$$M_{12} = h\omega_o e^{\int \sqrt{mV(x)dx}}$$

is not randomly distributed with a uniform probability, but
actually has a high probability of being near zero: it is
the barier height $V(x)$ and Δx where are nearer to being
uniformly distributed random variables. But there is one
other difficulty: that V or Δx must not be too big, or the
minima E_1 and E_2 will not be accessible to each other; there
will be a relaxation time, quantum or thermal, like
$\tau \propto e^{+V/kT}$ or $\tau \propto e^{+\int \sqrt{V}dx}$ which may become so large that
thermal equilibrium cannot be reached. Those pairs of
levels which are not connected practically because τ is too

large contribute to the zero-point entropy of the glass,
which is indeed known experimentally to be finite. Thus we
made three remarks about ordinary glasses:

(1)   The linear specific heat comes from $E_1 - E_2 \to 0$ and
      <u>moderate</u> $V\Delta x$.

(2)   The same kinds of centers with <u>small</u> $V\Delta x$ contribute a
      scattering mechanism for phonons which explains anoma-
      lous low-temperature thermal resistivities.[12]

(3)   We expect pairs with <u>large</u> $V\Delta x$ to contribute to a whole
      spectrum of slow-relaxation and creep-like phenomena in
      glasses, especially to slowly-relaxing thermal pheno-
      mena but also to creep under stress, etc.

Now what about the corresponding problem in the spin
glasses? We suggest that it may also in this case be a
valid view to consider the spins as classical dynamical
objects with an energy surface in an N-dimensional configuration space which also has very many directions along
which there are different energy minima. When we concentrate
on cases where $\Delta E_z = 0$ for a single spin, we are concentrat-
ing only on flat planes where the intermediate V is zero;
but such a place has a very large tunneling matrix element
and can't lead to zero energy splitting. There are, however,
undoubtedly energy minima of nearly the same energy which
require several spins to turn over to get from one to the
other, and can only be connected by an activated process or
by a quantum-mechanical tunneling. To write down the full
dynamics or even to have a clear understanding of the multi-
dimensional geometry of this system is not the easiest thing
in the world, but I would essentially make the following
point: that in the dynamics of antiferromagnets it has always been possible to make the so-called "Heller-Kramers"
semi-classical approximation, in which the energy is expanded
about any initial configuration $(S_z^i)_0$ (or S_x^i or S_y^i) of the
spin vectors, and the other two components then act almost
exactly as would the conjugate position and momentum variables of mass particles. Both the mass (kinetic) energy and
the potential energy must come from the exchange, which is
a difference from the case of ordinary glasses and may lead
to interesting dynamical scaling effects; but otherwise at
least nearby configurations should behave in complete analogy
to the dynamics of a crystal lattice (as e.g. the analogy of
phonons and antiferromagnetic spin waves).

In conclusion of these remarks, let me make some points about experimental effects in spin glasses, the study of which is suggested by analogy.

(1)   Most straightforward: we again suspect the presence of very slow relaxation as well as fast ones; can we study magnetic as well as thermal creep phenomena in these materials? I suspect that many of the phenomena observed by Beck and Kouvel are related to this.

(2)   As in the glasses, may I make a heartfelt plea for measurements of that elusive quantity the zero-point entropy,[13] (if any)? In this case it should be somewhat easier, since we can if all else fails cool in an enormous magnetic field and measure the zero of entropy on a saturated specimen.

(3)   A very interesting problem is that of scattering of spin waves, and of electrons. Clearly we expect phenomena like those seen in regular glasses: resonant scattering by extremely narrow resonances.

REFERENCES

1.   W. Marshall, Phys. Rev. 118, 1519 (1960).
2.   M. Klein and R. Brout, Phys. Rev. 132, 2412 (1963).
3.   P.W. Anderson, Materials Research Bulletin 5, 549 (1970).
4.   W. C. Kok and P.W. Anderson, Phil. Mag. 24, 1141 (1971).
5.   R. Hilsch and D. Korn, Kurznachr. Akad. Wiss. Göttingen 13, 61 (1965); D. Korn, Z. Phys. 187, 463 (1965); 214, 136 (1968).
6.   S. Tamaki and S. Takeuchi, J. Phys. Soc. Jap. 22, 1042 (1967).
7.   D.C. Mattis, Theory of Magnetism, Ch. 7, (Harper and Row, New York, 1965).
8.   W.C. Kok, Phil. Mag. (to be published).
9.   R. P. Guertin, J. E. Crow, and R. D. Parks, Phys. Rev. Letters 16, 1095 (1965); D. K. Finnemore, L. J. Williams, F. H. Spedding, and D. C. Hopkins, Phys. Rev. 176, 712 (1968).

10. P. W. Anderson, B. I. Halperin, and C. M. Varma, Phil.
     Mag. 25, 1 (1972).
11. J. E. Zimmerman and F. E. Hoare, J. Phys. Chem. Solids
     17, 52 (1960).
12. R. C. Zeller and R. O. Pohl, Phys. Rev. B4, 2029 (1971).
13. I. Gutzow, Z. Phys. Chem. 221, 153 (1936); F. Simon
     and F. Lange, Z. Phys. 38, 227 (1926).

DISCUSSION

Question: Are there any consequences for the resistivity of
your results?

P.W. Anderson: I haven't really thought about what the consequences are but I am sure there are. I guess perhaps the
scattering from a very narrow resonance is rather different
from the scattering from a spin which is in exactly zero
field and one might expect some rather funny resonance
scattering phenomena.

N. Rivier: Two points about the objections you mentioned
against the p(H) explanation for the specific heat: 1) If
you allow for a Heisenberg RKKY interaction, i.e., for local
fields in all directions, the relevant coefficient to the
linear specific heat is $p(H_z) = \int dH_\perp \, H_\perp \, p(\vec{H})$, not $p(|\vec{H}|)$.
$p(H_z=0)$ is finite. The reason is that the magnetic energy
per spin is $\vec{\mu} \cdot \vec{H} = \mu H_z$, i.e., samples not $|\vec{H}|$, but $H_z =$
$|H| \cos\theta$. (I shall discuss this point in my talk). 2) The
"universality" of the coefficient of the linear specific
heat which, in spin glasses, correspond to a concentration
independent coefficient, is also fundamental. Does your
theory give such a concentration independence?

P.W. Anderson: 1) Confusion on this point is widespread.
I believe Herring's work is correct and the above comment is
just incorrect; it is quite clear that $\vec{\mu} \cdot \vec{H} = \mu |H|$. To show
the result in another way I also did it quantum mechanically
in the talk, in which case the terms corresponding to H_x
and H_y are the off-diagonal matrix elements. 2) Yes, this
is merely the statement that energies scale as c.

B.R. Coles: If one were to start from a totally different point of view and regard the spin glasses and real glasses as not too bad a deviation from a long range ferromagnet or antiferromagnetic and a long range lattice, and say what we have are excitations or damped spin waves on the one hand and damped phonons on the other which would distort our spin wave specific heat, could you possibly get out the linear specific heat? I mean, it always seemed to me that what bounded you was that p(H) left collective excitations out of the system.

P.W. Anderson: That is the nice thing about the ordinary glasses. You know that doesn't work because there the phonon specific heat is visible and it is T^3. On the other hand, one of the most facinating things about the ordinary glasses is that the coefficient of T^3 is wrong, relative to sound velocity. Which as yet is not explained. Chandra Varma and I have some ideas.

B.R. Coles: Doesn't that mean that part of the quasi phonon like excitations are no longer going as T^3 but as T?

P.W. Anderson: Well, the point is you see the phonons, in fact, you see more than you expect from the phonons already and that is something else again. That is really there as T^3. There are no theorems, but I have a feeling (I think Brenig has made this point also) that there are in some sense two entirely separate branches to the spectrum. There is a long wave length collective branch, but there is at the same time something else at very short wave lengths that is very local.

J. Wong: You use the term "glass" for amorphous alloys like Cu-Mn. Are these amorphous alloys truly glasses in the sense that they exhibit a glass transition phenomenon. If so, are there any drastic changes in magnetic properties as these amorphous alloys go through the glass transition?

P.W. Anderson: There are two questions here, and I am not sure which is being asked. In the one case of Miss Kok's theory of θ in amorphous alloys, it is a good question whether even depositing at helium temperatures can prevent some crystallinity. But θ depends only on the 2-particle correlation function, which is what is measured directly in the X-ray pattern, so microcrystallinity is irrelevant.

The second question is whether in the case of "spin glasses" any fuller analogy to a true glass exists, e.g., is there a T_g. I do not know and would be eager to learn and to see better experiments. In one sense, the randomness of the alloys (which for most measurements are atomically of a <u>crystalline</u> nature but of random <u>composition</u>) is an extraneous source of randomness for the spin structure, so that there simply may not <u>be</u> an ordered ground state; but still, whatever T_g is, one expects that there must be some point at which the spin glass settles down and chooses a specific configuration. This is one of the reasons I called for thermal measurements and particularly for zero-point entropy: to investigate the analogy further. Are the sharp phenomena seen in some experiments a T_g?

IT'S A RANDOM WORLD*

J. A. Krumhansl

Laboratory of Atomic and Solid State Physics

Cornell University

Ithaca, New York 14850

ABSTRACT

Random semiconductors, alloys, and magnetic materials have recently received considerable attention by solid state physicists, and a variety of theoretical approaches have been developed. However, these developments are only part of an extensive history of randomness in physical situations. This paper is a report on some of the related history in other fields where randomness has been an ingredient; a basic bibliography is provided.

I.

It is a random world; in almost any physical, chemical or biological system - indeed even geographical and sociological interactions - one can find that randomness in either the parameters, or in the structure, is an essential element of important situations.

Physicists have recently devoted considerable attention to the properties of random semiconductors,[1] alloys,[2] and magnetic[3] materials - primarily from the quantum mechanical point of view. The properties of liquid metals have been discussed theoretically.[4]

*This work has received support by the Atomic Energy Commission, and by the Advanced Research Projects Agency.

However, the list of statistical applications which one finds in the literature is truly impressive. Several major reviews have recently been written; the book "Statistical Continuum Theories", by M.J. Beran,[5] and the review article, "Wave Propagation in Random Media", by U. Frisch[6] will give the reader extensive bibliographical compilations of studies of the properties and dynamics in random systems, while the review "Some Fundamental Ideas in Topology and Their Application to Problems in Metallography", by Lida K. Barrett and C.S. Yust[7] will introduce the reader to an extensive literature on structural questions in random arrays.

I have not yet found any simple way to neatly categorize the structure of the subject of randomness, although a few regularly recurring themes will be discussed in the next sections. But to give some idea of the range of problems which have been looked at here are some examples:* In geophysical applications – the thermal, mechanical, and electrical properties of randomly polycrystalline rocks and minerals; the pore and fissure structure; the seismological characteristics of waves in geological matter; wave motion in the sea, and its power spectrum; underwater sound propagation in water with random density and temperature variations; turbulence in the sea, atmosphere, and ionosphere; the twinkling of stars; radio wave propagation in the above random media; radiative and convective transfer in astrophysical applications. In materials science – the structure and properties of heterogeneous materials, such as polycrystalline commercial alloys or sintered ceramics; flow through randomly porous media; the properties of reinforced composite materials; wood, polymers, paper. In biological science I have not made any extensive survey, but at the very least the structure of plant and animal tissue immediately provides the same kinds of situations encountered above in materials and geophysical situations; in addition, the biochemical and genetic processing in such structures must inevitably reflect consequences of randomness; similarly, while it is one thing to model biological or human population situations with averaged para-

---

*Detailed references are not given except where references are not to be found in the reviews cited above.

meters, since many of those descriptions are non-linear, the introduction of randomness can lead to entirely new qualitative behavior.[8] In modern communications technology, stochastic problems abound; the response of linear and non-linear electrical networks to random noise signals is a central issue, which has led to an extensive mathematical development; of course fading and the like in radio wave propagation has long been recognized as having essential stochastic characteristics. In reactor technology, neutron transport is a stochastic diffusion process and is therefore a subject which has produced an extensive literature on transport in stochastic systems. Partial coherence in either acoustic or optical wave applications (holography, etc.) is of considerable current interest. And then there is the last subject of equilibrium and non-equilibrium. Undoubtedly this list is far from complete, but it should give some idea of the ever widening spiral which the writer has been led in by an early summer's fancy that it might be a good idea to browse the world of randomness.

II.

There does seem to be one methodology which has been useful in a wide variety of instances, and which is remarkable in regard to the number of independent rediscoveries of essentially the same approach. In solid state physics, the most recent examples[2] have been "Coherent Potential Approximation" (CPA) or "self consistent" methods in alloys. Previous or parallel developments in other contexts of this method are the subject of this section.

It is frequently possible to divide the physical problem in a random system into two parts - an average behavior, plus the fluctuations from it. An exact treatment must keep both, and when the latter becomes comparable to the former, the separation is meaningless. Fortunately, there are many situations in which the separation is both useful and meaningful. In this case the "self consistent local field theory" is quite useful for describing the average field; however, to complete sum rules (energy conservation, for example) the fluctuations must be included in principle.

Mean field theories have usually been invented as needed for the application at hand. A. Einstein[9] was concerned with the properties of a suspension of small particles in his classic Browian motion studies. Lord Rayleigh[10] and many others[11] have since addressed the question of the average permativity of a heterogeneous medium, both in the static and wave propagation regime.

The essence of the standard problem is relatively simply stated. The field (electromagnetic, elastic, Schrödinger, etc.) denoted by u, obeys a linear equation of motion characterized by an operator L, which in turn is parameterized by specification of the properties of the medium. In a homogeneous medium, L depends only on a few parameters, while in a random medium these parameters (conductivity, potential, index) vary from place to place according to some probability distribution; for brevity we denote the random parameter as ε in $L(\varepsilon)$. The equation of motion is

$$L(\varepsilon) \ u = 0 \qquad . \tag{1}$$

The operator L may also depend on other important parameters which are not random (e.g., E in H - E for Schrödinger waves, ω^2 in acoustic and electromagnetic waves. The essence of the averaging methods is to define $L = < L > + L_1$, and attempt to calculate $< u >$ or other quantities such as $< \varepsilon u >$, and from them effective parameters $\varepsilon^* = [< \varepsilon u >/< u >]$. This may be done in various approximations, both self-consistent or not. When stated in this way a variety of methods[6] from the literature of perturbation theory in applied mathematics, as well as from field theoretical methods in physics may be applied. It is feasible to obtain the "self consistent local field theory" as the best low order local approximation to a resummed perturbation series; it also can be obtained by choosing a local field such that scattering vanishes to first order in the "concentration" of defects, i.e., all repeated scatterings by L_1 at a particular position. One may them compute an average ε^*, as above, to obtain an effective medium parameter.

Neither time nor space allow a detailed comparison of various applications to be made here, but we may illustrate by the calculation of the dielectric constant of a composite medium consisting of small spheres of dielectric ε_2 imbedded in ε_1 which are assumed to be randomly located, and non-overlapping. The "equation of motion" is

$$\{ [\vec{\nabla} \cdot \varepsilon(\vec{r})] \ \vec{E}(\vec{r}) \} = 0 \quad , \tag{2}$$

in the absence of free charges. The field $u \rightarrow \vec{E}(\vec{r})$ and $L \rightarrow \vec{\nabla} \cdot \varepsilon(\vec{r})$. If an applied charge $\rho(r)$ is imposed then it is useful in solving the problem to have the Green's function $G = [\vec{\nabla} \cdot \varepsilon(\vec{r})]^{-1}$ which satisfies $LG = \delta(\vec{r} - \vec{r}')$. Indeed, a systematic method for solving may now be based on writing $L = \ <L> + L_1 = L_0 + L_1$, whence the operator relation holds that

$$G = G_o + G_o L_1 G \quad . \tag{3}$$

Further there is an exact scattering representation for fields

$$\vec{E} = \vec{E}_o - G L_1 \vec{E}_o = \vec{E}_o + \vec{E}_{scatt.} \tag{4}$$

where $\vec{E}_o$ is any field which satisfies $<L> \vec{E}_o = 0$. Equation 4 can be used to define an ε^*. In the example at hand $L_1 = \varepsilon_2$ within the spheres, ε_1 is the host; thus $L_1 = \Sigma_i (\varepsilon_2 - \varepsilon_1)$, i running over spheres.

The specific practical problem for a dielectric is to compute the average dielectric function for some direction, say x: $\varepsilon^*_{xx} = [<\varepsilon E_x>/<E_x>]$. Here, the averaging is volume averaging.

The basic scattering problem is to compute the scattering around a sphere in the presence of some local field. In fact, it is tedious but not difficult to solve this problem using a Green's function G_0; on the other hand, the exact fields around a single sphere or ellipsoid can easily be calculated using spherical harmonics. When this is done, if we take spheres to scatter independently except for each to contribute to a volume average field, then

$$\frac{E_x}{E_{xo}} = 1 + \frac{\varepsilon_1 - \varepsilon_2}{2\,\varepsilon_1 + \varepsilon_2} \ \frac{a^2}{r^3} \ (1 - 3\cos^2\theta), \ \text{outside a sphere.}$$

$$= 1 + \frac{\varepsilon_1 - \varepsilon_2}{2\,\varepsilon_1 + \varepsilon_2} \quad , \ \text{inside a sphere.} \qquad (5)$$

$$= 1 + (E_{scatt.}/E_{ox}).$$

If a volume average of $<\varepsilon\, E_x>/<E_x>$ is taken, and f_2 is the volume fraction occupied by ε_2, the integrations yield

$$\varepsilon^*_{xx} = \frac{<\varepsilon> + f_2\,\varepsilon_2\big(\dfrac{\varepsilon_1 - \varepsilon_2}{2\,\varepsilon_1 + \varepsilon_2}\big)}{1 + f_1\,\big(\dfrac{\varepsilon_1 - \varepsilon_2}{2\,\varepsilon_1 + \varepsilon_2}\big)} \quad . \qquad (6)$$

This expression is not self consistent, since the field incident on the spheres was assumed to be E_0 in a medium of $\varepsilon = \varepsilon_1$. For the moment, however, expanding (6) to lowest order of $\varepsilon_2 - \varepsilon_1$ which would apply to small dielectric fluctuations one obtains an expression obtained by various means:[11,12]

$$\varepsilon^*_{xx} \simeq <\varepsilon> - \frac{f_2\,(\varepsilon_1 - <\varepsilon>)^2}{3 <\varepsilon>} \quad . \qquad (7)$$

A study of the methods shows that (6) is equivalent to the "average T matrix approximation, ATA", used in random alloys,[2] while (7) is a low order correction to the virtual crystal approximation.

In order to make the method totally symmetric in ε_1 and ε_2, as well as self consistent, we take the medium to be describable by an average ε^* and then when in medium 1 scatter by $(\varepsilon_1 - \varepsilon^*)$, when in 2 by $(\varepsilon_2 - \varepsilon^*)$. With the assumptions of statistical independence and volume averaging one obtains the implicit equation for ε^*_{xx} (after much algebra):

$$1 = \sum_i \frac{3 \, f_i}{2 + (\epsilon_i / \epsilon^*_{xx})} \tag{8}$$

On the other hand, taking volume average x-components only, the scattering problem (neglecting interparticle scatterings) in an effective medium ϵ^*_{xx} looks like:

$$< E_x > = < E_{ox} > + (\sum_i \frac{\epsilon^*_{xx} - \epsilon_i}{2\epsilon^* + \epsilon_i} \, f_i) < E_{ox} > + \ldots \tag{9}$$

If we make what at first sounds like a physically different condition from the previous, which is that the first order scattering shall vanish on the average

$$\sum_i (\frac{\epsilon^*_{xx} - \epsilon_i}{2\epsilon^* + \epsilon_i}) \, f_i = 0. \tag{10}$$

But since $\Sigma_i \, f_i = 1$ it may be found that (10) and (8) are exactly the same. Thus, the self consistent local field is equivalent to the choice of an effective medium such that the averaged single site scattering vanishes. This is identical to the CPA method in alloys,[2] and we now pass on to cite a number of historically parallel developments in different contexts. However, to summarize, the essence of the method is to solve for the exact field around a representative element of the medium, which is taken to be imbedded in an effective medium determined in turn by requiring that the mean of the scattering by the random elements taken as statistically independent shall vanish.

Here are a number of examples where the same philosophy has been developed in one formalism or another: Yonezawa[13] and Leath[14] provide a diagrammatic basis for the self-consistent method (CPA) as applied to the quantum mechanics of random alloys.

Landauer,[15] in 1952, analyzed the conductivity of a random mixture of different materials, and derived an expression which is simply Eq. 10 above, with conductivities σ substituted for ϵ. Comparisons were made with experiment.

The thermal properties of random composites have
received considerable theoretical and experimental attention. Important contributions to the subject were made
by Kerner,[16] and more recently by Budiansky.[17] A brief
review of experimental and theoretical work has been
given.[18]

The understanding of elastic properties of a random
(polycrystalline) aggregate is a long standing subject of
concern in many subjects - geology, agronomy, civil engineering, continuum mechanics, and ceramics, to mention
a few. Here again the "self-consistent" method developed,
with original contributions to the subject by Budiansky,[17]
by Hill,[18] and by Kröner;[19] indeed the similarity of
Kröner's formalism to that developed for studying electronic properties of alloys is remarkable.

One should, of course, recognize that the elastic
fields are tensor quantities so the problem is considerably more complicated than that of Schrödinger scalor
waves in random systems, and thus the idealizations of
the self-consistent method may be more serious. None
the less Thomson[20] has recently applied Kröner's method
to the elasticity of polycrystals and rocks, with some
success in comparison with experiment.

It might also be mentioned in passing that an entirely different approach to the properties of heterogeneous
materials uses variational methods[21] to set bounds on
effective parameters; to the extent that I have checked
several cases the self-consistent results fall between
the expected upper and lower bounds (e.g., $< \epsilon >$ vs.
$< \epsilon^{-1} >^{-1}$ for ϵ^* in the case of dielectrics).

Another area in which the same philosophy has developed is that of determining an effective propagation constant
for the average wave in a random medium. In the context
of both acoustic wave propagation and electromagnetic wave
propagation a tremendous amount has been done, and Ref. 6
contains a bibliography with 154 entries up to 1968.
Noteworthy in physics, the work of Foldy (1945)[22] and Lax
(1951, 1953)[23] set out the approach which has since evolved
into self consistent methods in various wave propagation
applications. While at first it may not seem that the
mean static parameters discussed above are closely related

to wave propagation, once the wave equation is Fourier transformed, the technical aspects of the formalism are essentially the same, i.e., self consistent local field theory.

It is difficult to know how to conclude this section since it is certain that this list of applications is not nearly complete. Perhaps the best message to the reader is that he also go out and shop, with the above list as a beginning.

III.

In this last section I wish to make a few comments on special topics in the theory of random systems.

First of all, the self-consistent local field theory is suited to random systems only in circumstances where the average (sometimes called "coherent") field is a useful, meaningful quantity. On the other hand, sometimes we have "localized" resonant phenomena, or effects due to clustering, in which case the approach of a self-consistent field has limited value. It is possible to extend the idea to 2, 3, . . . n, site self consistent local fields? Perhaps, but there has not yet developed a clear basis on which to proceed (although several recent attempts have been made). So I am taking the course of not commenting on them in this writing. The essence of the problem is that for strong scattering there is no suitable parameter to use in collecting the various terms in approximate series expressions. It seems clear that beyond the self consistent local field approximation, higher order theories have not yet emerged in suitably effective form.

What about the more general problem of the complete description of the behavior of random systems? One sobering fundamental view of the outlook for progress may serve as useful perspective in closing this paper. The model problem discussed above is so easily stated, and indeed a stochastic operator L which depends linearly on random parameters is not all that complicated statistically, so what is the trouble? Probably many workers have recognized the simplistic nature of this view, but Kraichnan[24] and

Beran[5] particularly draw attention to the perfectly apparent fact that although the equation of motion may depend only linearly on the random parameters, the solutions are almost invariably highly non-linear functionals of the stochastic quantities - for whose description there does not exist straightforward methodology. Indeed the general problem shares the same intractabilities as appear in strong turbulence, strong interaction field theory, and many non-linear systems. Thus, it is likely that we must be content with solutions of special examples for guidance when we wish to go much beyond discussing average fields in random systems.

As I began, let me repeat in closing that its a random physical world, which on the one hand presents similar problems in widely different context, but on the other hand offers challenges beyond our common mathematical and conceptual tools. To paraphrase a common expression - that's good news, or bad news, depending on your taste.

REFERENCES

1. Many Aspects of Random Semiconductors are Regularly Discussed in J. Non Cryst. Solids.
2. P. Soven, Phys. Rev. 156, 809 (1967); ibid. 178, 1136 (1969). S. Kirkpatrick, B. Velický, and H. Ehrenreich, Phys. Rev. B1, 3250 (1970).
3. Proc. International Symposium on Amorphous Semiconductors, to be published, J. Non Cryst. Solids (1973).
4. L. Schwartz and H. Ehrenreich, Annals of Physics 64, 100 (1971).
5. M.J. Beran, Statistical Continuum Theories (Interscience, New York, 1968); also, Phys. Status Solidi a6, 365 (1971).
6. U. Frisch, Wave Propagation in Random Media, in Probabilistic Methods in Applied Mathematics, edited by A.T. Bharucha-Reid (Academic Press, New York, 1968).
7. Lida K. Barrett and C.S. Yust, Metallography 3, 1 (1970).

8. N.S. Goel, S.C. Maitra, and E.W. Montroll, Rev. Mod.
    Phys. 43, 231 (1971).
9. A. Einstein, Investigations on the Theory of Brownian
    Motion (Dover Reprint, New York, 1956).
10. Lord Rayleigh, Phil. Mag., Ser. 5, 34, 481 (1892);
    J.B. Keller, J. Math. Phys. 5, 548 (1964).
11. W.F. Brown, J. Chem. Phys. 23, 1514, (1955), reviews
    much previous work; see also Ref. 5.
12. L.D. Landau and E.M. Lifshitz, Electrodynamics of
    Continuous Media (Addison Wesley, New York), p. 45.
13. F. Yonezawa, Prog. Theor. Phys. 40, 734 (1968).
14. P.L. Leath, Phys. Rev. B2, 3078 (1970).
15. R. Landauer, J. Appl. Phys. 23, 779 (1952).
16. E.H. Kerner, Proc. Phys. Soc. 69B, 808 (1956).
17. B. Budiansky, J. Composite Materials, 4, 286 (1970).
18. R. Hill, J. Mech. Phys. Solids 15, 79 (1967).
19. E. Kröner, J. Mech. Phys. Solids 15, 319 (1967).
20. L. Thomsen, J. Geophys. Res. 77, 315 (1972).
21. Z. Hashin, Appl. Mech. Rev. 17, 1 (1964).
22. L.L. Foldy, Phys. Rev. 67, 107 (1945).
23. M. Lax, Rev. Mod. Phys. 23, 287 (1951); Phys.
    Rev. 85, 621 (1952).
24. R.H. Kraichnan, Proc. Symp. Appl. Math. (American
    Math. Soc.) 13, 199 (1962).

AMORPHOUS ANTIFERROMAGNETISM IN SOME TRANSITION ELEMENT-PHOSPHORUS PENTOXIDE GLASSES

T. Egami*, O. A. Sacli, A. W. Simpson and A. L. Terry

University of Sussex, Falmer, Brighton, Sussex

and F. A. Wedgwood

Atomic Energy Research Establishment, Harwell

ABSTRACT

Phosphorus pentoxide glasses containing high concentrations of Mn^{2+}, Fe^{3+}, Ni^{2+}, Co^{2+} and Cu^{2+} have been prepared and measurements made of their magnetic susceptibility and low temperature specific heat. Also neutron diffraction measurements were made on two of the samples. All the materials measured showed antiferromagnetic interactions but only one of the families of glasses, those containing Fe^{3+}, showed evidence of an antiferromagnetic transition (Néel temperature). The absence of magnetic order in the bulk of the glasses is attributed to a consequence of enhanced zero point energy effects in the amorphous state.

INTRODUCTION

Experimental data indicating the existence of strong antiferromagnetic interactions in random structures have been available for at least thirty years in the form of susceptibility measurements on various solutions containing magnetic ions. For example, in 1937, Nicolau (1) made some accurate measurements of the susceptibility

---

* Now at the Max-Planck-Institut für Metallforschung, Stuttgart.

of a concentrated aqueous solution of nickel sulphate which, if replotted as the reciprocal susceptibility as a function of absolute temperature, gives a straight line indicating a paramagnetic Néel temperature of roughly -20°K.   More recently, in 1965, Schinkel and Rathenau (2) have investigated the susceptibility of borate glasses containing high concentrations of manganese ions.   These glasses also showed large paramagnetic Néel temperatures (-85°K) and an abnormal downwards curvature of the reciprocal susceptibility-temperature relationship at low temperatures.   Independently, Simpson and Lucas (3,4) observed a similar reciprocal suscepti-bility behaviour in amorphous rare-earth oxide-ferric oxide materials with garnet ferrite, and other stoichiometries, prepared by the anodization of suitable homogeneous alloys.

The anodization technique has the severe experimental disadvantage that only very small samples can be produced which essentially restricts their assessment to susceptibility measure-ments only.   The manganese borate glasses also appeared to have preparation problems and therefore alternative methods of making amorphous antiferromagnets were considered.   On the basis of the ease of preparation, and the large solubility range for additional ions, phosphorus pentoxide glasses were chosen and a series of iron group transition element-phosphorus pentoxide antiferro-magnetic glasses were produced.   (A preliminary report on these glasses was presented at the Symposium on Electrotechnical Glasses in London in 1970 (5), and more recently the properties of just two of the samples, a cobalt and an iron glass, were described by the present authors (6)).

All the glasses were prepared by the conventional technique of melting the reagents in a platinum crucible and pouring the resulting liquid on to a polished copper plate (6).   For each comp-osition a right cylinder of 3.0 mm diameter and roughly 3 mm high was ultrasonically cut from the glass for the susceptibility and specific heat measurements.

MAGNETIC MEASUREMENTS

The magnetization as a function of applied magnetic field was measured in fields up to 10,000 oersteds for temperatures between liquid helium temperature and room temperature using a vibration magnetometer of the Foner type (7). For all the samples discussed here, the magnetization–field relationship was an accurate straight line through the origin at all temperatures indicating no spontaneous magnetization or superparamagnetism. For some of the glasses the susceptibility was measured below $4.2^{\circ}K$ using a mutual inductance method (8).

The reciprocal susceptibility–temperature relationship for one of the divalent cobalt glasses, $(CoO)_{1.4} \cdot P_2O_5$, is shown in figure 1, and compared with crystalline anhydrous cobaltous

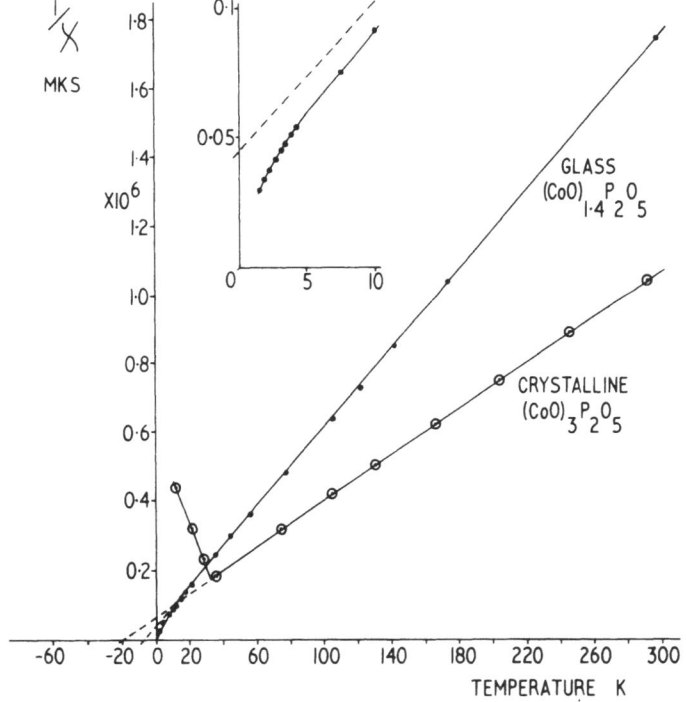

Figure 1   The temperature variation of the reciprocal susceptibility for crystalline cobalt orthophosphate and a cobalt glass. (The insert shows the low temperature region of the cobalt glass.)

orthophosphate, $(CoO)_3 . P_2O_5$; a crystalline material with the closest chemical formula to that of glass, prepared by heating the hydrated orthophosphate $(CoO)_3 . P_2O_5 . 8H_2O$ to $800^{\circ}C$ for one hour. It is seen that whereas the crystalline orthophosphate shows a reciprocal susceptibility–temperature behaviour characteristic of a normal antiferromagnet with a Néel temperature at $31.5^{\circ}K$, the cobalt glass shows no evidence for any critical temperature down to $1.5^{\circ}K$. Also, the insert in figure 1, which is an expanded view of the low temperature region, shows a similar downwards curvature of the reciprocal susceptibility of the glass to that previously observed in the manganese borate glass (2) and the rare earth oxide-iron oxide amorphous antiferromagnets (3,4).

The magnetic moments of the cobalt ions, calculated from the high temperature slope of figure 1, were 4.51 and 4.81 bohr magnetons for the glass and crystalline materials respectively; these values are in agreement with that normally found in crystalline solids (9).

The absence of clear magnetic evidence for any magnetic ordering temperature, or Néel temperature, is consistent with the molecular field models of the amorphous antiferromagnet (10–12) which do not predict any sharp discontinuities in the slope of the reciprocal susceptibility. It must also be emphasised that a molecular field model interpretation of the data of figure 1, supposing a large reduction of the Néel temperature to below $1.5^{\circ}K$, is not consistent with the observed paramagnetic Néel temperatures. The decrease in cobalt content in the glass compared to the crystalline material, by roughly a factor of two, causes a reduction of the superexchange interaction between cobalt ions, as indicated by the reduction in paramagnetic Néel temperature from -20 to -8°K, by a factor of 2.5, whereas the Néel temperature appears to have fallen from 31.5 to below 1.5°K; a factor of greater than twenty times. For open magnetic amorphous structures where the magnetic atoms may be regarded as still lying on two interpenetrating antiparallel sublattices, as treated by Kobe and Handrich (11) and Hasegawa (12), the molecular field model predicts a small increase in the Néel temperature when a material becomes amorphous, and for effectively close packed structures as treated by Simpson (10), a reduction by at most 30%.

The low temperature reciprocal susceptibility–temperature
data for the cobalt glass is roughly as predicted by the molecular
field treatment (10, 4) with the downwards curvature arising as a
consequence of a proportion of isolated paramagnetic atoms.
Since the number of interacting near neighbours is only roughly
three or four (see the discussion section) the probability of a
particular cobalt atom being surrounded entirely by phosphorus
atoms and hence not experiencing any significant exchange inter-·
action is quite large.   These essentially isolated cobalt atoms
have a very high atomic susceptibility at low temperatures which
swamps the contributions of the interacting atoms.

In figure 2 the temperature variation of the reciprocal susceptibility of a divalent manganese phosphate glass, $(MnO)_{1.30} \cdot P_2O_5$,
is compared to that observed by Fowlis and Stager (13) for a
crystalline material with a similar stoichiometry, $(MnO)_2 \cdot P_2O_5$
or $Mn_2P_2O_7$.   (L.K. Wilson et al have also reported a similar

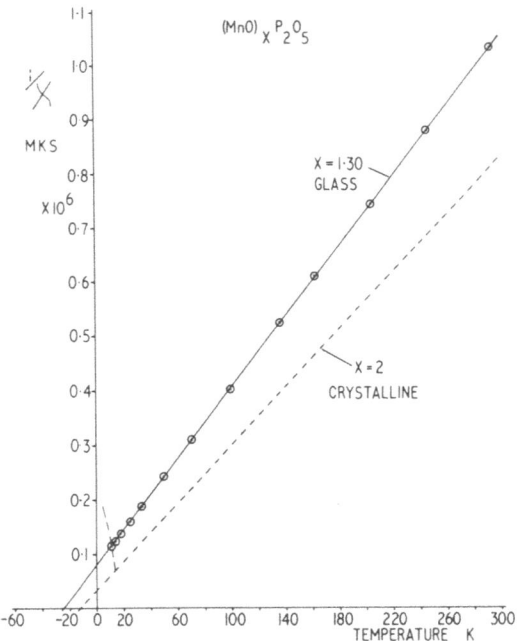

Figure 2   The temperature variation of the reciprocal susceptibility
for a manganese glass.   The dotted line shows the data of Fowlis
and Stager (13) for crystalline $(MnO)_2 \cdot P_2O_5$.

behaviour in a manganese phosphate glass and have observed a
downward curvature of the reciprocal susceptibility at low
temperature (L.K. Wilson, these Proceedings). For this glass the
paramagnetic Néel temperature (-26^OK) is just twice that for
the crystalline material (-13^OK) which has a higher concentration
of manganese ions. This implies an increase in the superexchange interaction between the manganese ions by roughly
two times. This effect may be interpreted by considering the
structure of the two materials. Collins et al (14) have shown
that the manganese ions in crystalline $Mn_2P_2O_7$ are roughly
arrayed in planes with the magnetization in alternate planes aligned
antiparallel. The near neighbouring cations to any manganese
ion are all phosphorus ions. In the glass, however, the manganese
and phosphorus ions will presumably randomly occupy the available sites in the amorphous oxygen anion lattice, and therefore now
roughly 40% of the near neighbouring cations to any manganese ion
will be manganese. Since the superexchange mechanisms
increase rapidly with decreasing distance these near neighbours
will make a disproportionately larger contribution to the total
superexchange than they would at the greater distances in the
crystalline material. (A similar analysis cannot be made for the
cobalt glasses because the structure of cobalt orthophosphate is
not known.)

Although accurate measurements of the susceptibility were
not obtained below 11^OK, cruder measurements indicated no
evidence for a sharp minimum in the reciprocal susceptibility
down to 4.6^OK. Again, the atomic magnetic moment of the
glass was typical of crystalline materials, being 5.97 bohr magnetons.

The results for two of the trivalent iron phosphate glasses
$(Fe_2O_3)_{0.79}.P_2O_5$ and $(Fe_2O_3)_{0.63}.P_2O_5$, together with Welo's
data (18) for crystalline anhydrous ferric orthophosphate,
$Fe_2O_3.P_2O_5$ or $FePO_4$, are shown in figure 3. The results for
the high iron content glass $(Fe_2O_3)_{0.79}.P_2O_5$ are in sharp contrast
to those of the cobalt and manganese glasses in that this glass
shows a clear Néel temperature at 7^OK. A sharp transition of
this type would not be expected on the basis of the molecular field
treatments of the amorphous antiferromagnets (10-12) and seems
to indicate the possibility that this iron glass may contain microcrystalline precipitates, probably of ferric orthophosphate.

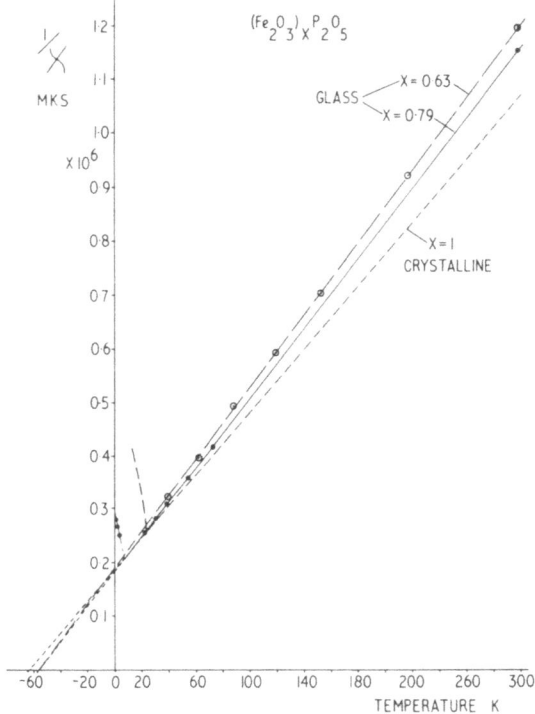

Figure 3 The temperature variation of the reciprocal susceptibility for two of the iron glasses.   The dotted line shows the data of Welo (18) for crystalline ferric orthophosphate.

However, the X-ray and neutron diffraction measurements give very little evidence for such a phase, and therefore if a crystalline phase does exist the particle size must be very small.   (The neutron diffraction data, see next section, indicates a coherence length of 20 to 30 angstroms).   A very small microcrystalline antiferromagnetic precipitate of ferric orthophosphate is also consistent with the reduction of the Néel temperature in the glass to 7°K from 25°K for the bulk material, since a high proportion of the interacting atoms will lie on or near the surface and will experience a low superexchange interaction.

Another unique function of the iron glasses is that the paramagnetic Néel temperature does not vary with the concentration of iron ions in the glass.   This is shown for the two glasses in figure 3 where both have a paramagnetic Néel temperature of roughly -58°K. This behaviour is to be expected if the glass contains microcrystalline precipitates. The atomic moments of the two glasses shown in figure 3 are 6.01 and 5.85 bohr magnetons for the $(Fe_2O_3)_{0.63} \cdot P_2O_5$ and $(Fe_2O_3)_{0.79} \cdot P_2O_5$ glasses respectively.

NEUTRON DIFFRACTION MEASUREMENTS

As was previously reported (6), neutron diffraction data was taken on an iron glass $(Fe_2O_3)_{0.79} \cdot P_2O_5$ and a cobalt glass $(CoO)_{1.4} \cdot P_2O_5$ using the conventional powder diffraction technique (19) at 300°K, 77°K and 4.2°K for both glasses and also 1.8°K for the cobalt glass. The variation of count rate with diffracted angle for the iron glass at 77°K and 4.2°K is shown in the lower two curves of figure 4. The room temperature data was identical

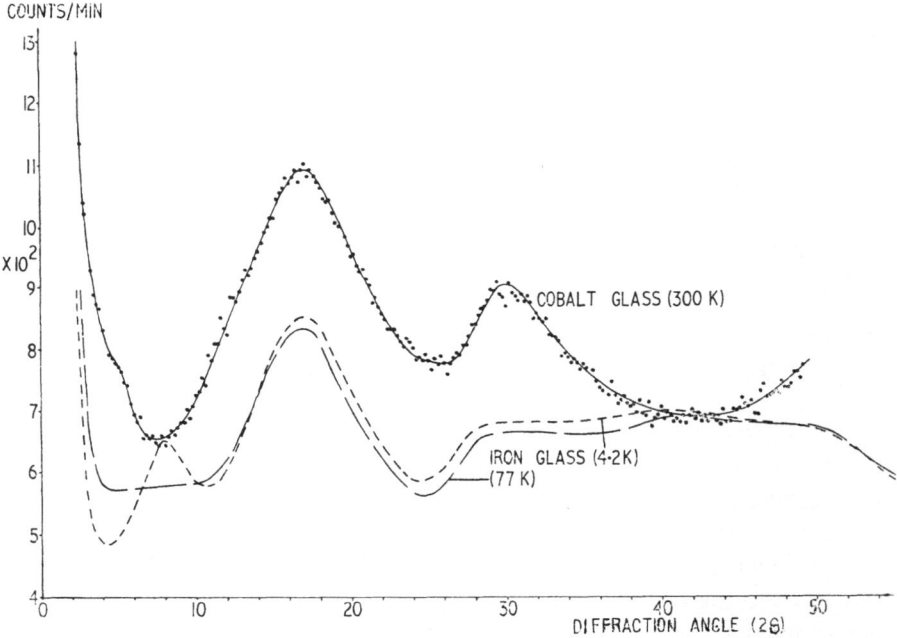

Figure 4 The variation of neutron count rate with diffracted angle (2 θ) for the $(Fe_2O_3)_{0.79} \cdot P_2O_5$ glass at 4.2 and 77°K and for the $(CoO)_{1.4} \cdot P_2O_5$ glass at 300°K. (For clarity the data points are omitted for the iron glass).

to 77°K results after corrections were made for thermal expansion.   It is seen that an additional large broad peak occurs in
the liquid helium data at a 2θ value of 17 degrees, indicating the
presence of some form of antiferromagnetic order.   The whole
spectrum may be interpreted as being associated with a
coherence length of 20 to 30 angstroms.

The temperature variation of the height of the antiferromagnetic peak is shown in figure 5.   The initial rapid fall in the
peak height presumably corresponds to the fall in long range
antiferromagnetic order, whereas the more gradual change is
associated with the variation in short range order.   The intercept
of the linear extrapolation of these two regions gives a value for
the Néel temperature of 6.5^{O}K, in excellent agreement with the
value found from the reciprocal susceptibility graph of figure 3.

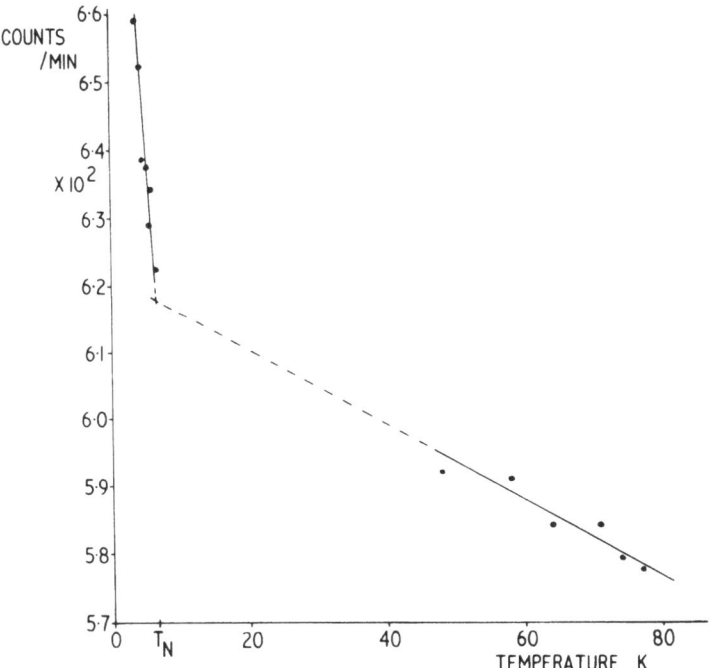

Figure 5   The temperature variation of the antiferromagnetic
neutron diffraction peak height (2θ = 8) for the iron glass.

The neutron diffraction data for the cobalt glass is in sharp
contrast to that for the iron composition as was the susceptibility
data.   After corrections for thermal expansion, all the spectra
were identical, within the experimental error, down to 1.8°K,
indicating no evidence for any magnetic transition.   The upper
curve in figure 4 shows the room temperature results for the
cobalt glass.   This spectrum has less structure than that for
the iron glass, indicating a shorter coherence length.

SPECIFIC HEAT MEASUREMENTS

The specific heat of several of the samples was measured in a
conventional He^3 cryostat between 0.4°K and 4.2°K using a quasi-
adiabatic technique where a mechanical heat switch was used to
cool the calorimeter.   Details of the cryostat were described
elsewhere (20).   Figure 6 shows the variation of specific heat with
temperature for the cobalt glass $(CoO)_{1.4} \cdot P_2O_5$, clearly indicating
a linear variation at low temperatures.   This variation is in
contrast to the phonon contribution for normal glasses which roughly

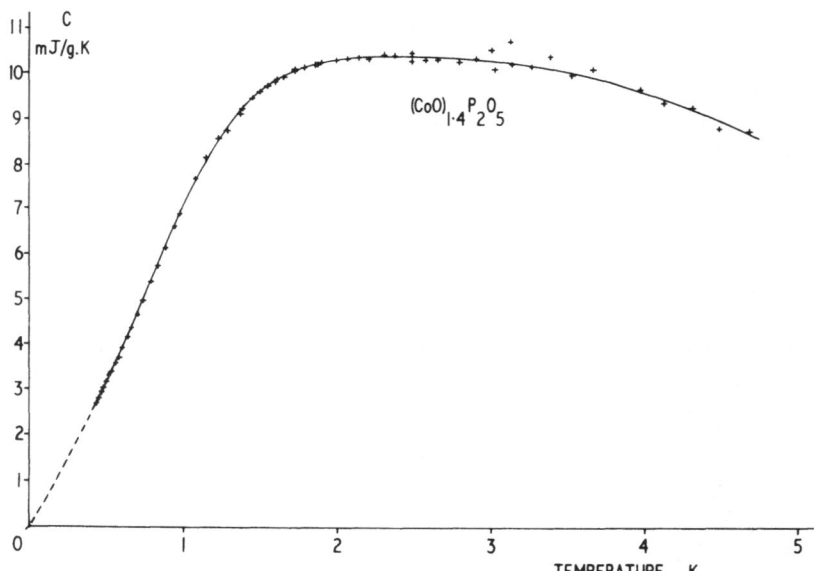

Figure 6   The temperature variation of the specific heat for the
$(CoO)_{1.4} \cdot P_2O_5$ glass.

obeys a T^3 law and, over this temperature range, is much smaller with a typical value rising to less than 5 mJ/g^{O}K at 5°K (17, 21). Therefore this anomalous specific heat must be entirely attributed to the magnetic contribution associated with the amorphous anti- ferromagnet. It should be noted that anomalous specific heat terms proportional to T, or with other variations, have been found in several non-magnetic glasses (17, 21-24); however these effects are some 10^4 times smaller than that shown in figure 6. A similar anomalous linear specific heat term is also observed in some dilute alloys at low temperatures (25-29).

The specific heat data for the cobalt glass reduces still further the temperature above which no magnetic transition occurs, since the curve shows no evidence of any peak that should be associated with a long range order-disorder transition. A possible inter- pretation of this data is to assume that the very broad peak in the specific heat at a temperature of about 2.5^{O}K is associated with a magnetic transition; however, if this were the case, the neutron diffraction or the susceptibility measurements would have detected the ordering.

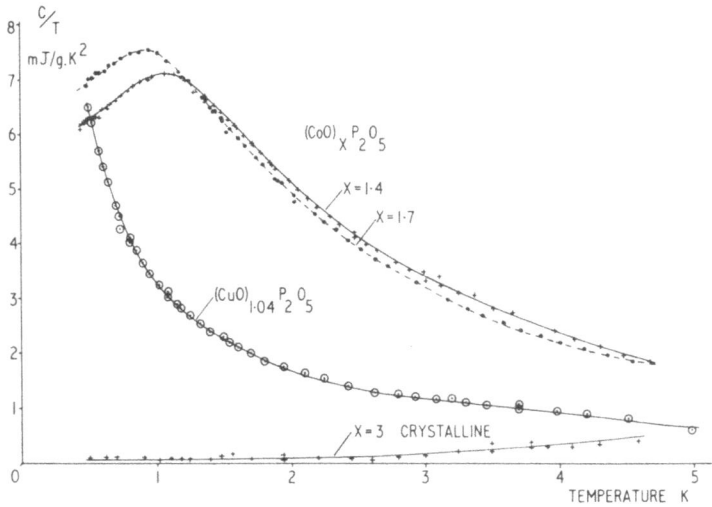

Figure 7   The specific heat divided by the temperature as a function of temperature for two cobalt and one copper glass and crystalline cobalt orthophosphate.

In figure 7 the specific heat divided by the absolute temperature (C/T) is plotted against temperature for two cobalt glasses, $(CoO)_{1.4} \cdot P_2O_5$ and $(CoO)_{1.7} \cdot P_2O_5$, and also a copper glass $(CuO)_{1.04} \cdot P_2O_5$. Measurements for the crystalline anhydrous cobaltous orthophosphate $(CoO)_3 \cdot P_2O_5$ are also included for comparison; it is seen that this compound obeys the normal T^3 relationship with a much smaller contribution over the temperature range considered. By making reasonable extrapolations for these results, in both the high and low temperature regions, an estimate of the area under the graph, and hence the entropy, may be made. The magnetic contribution to the entropy should be simply k log (2S + 1), that is k log 4 and k log 2 for the cobalt and copper glasses respectively. However for all three glasses the estimated entropy is roughly half that expected. (Of course the extrapolations in the case of the copper glass are very unreliable). A similar situation is also found for the dilute magnetic alloys (25-28), and presumably a very high peak occurs in the entropy at very low temperatures.

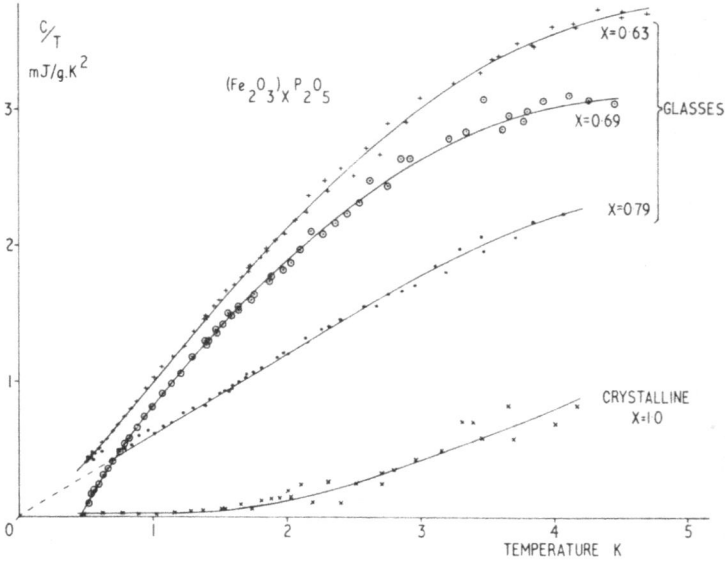

Figure 8   The specific heat divided by the temperature as a function of temperature for three iron glasses and crystalline ferric orthophosphate.

The specific heat data for the iron glass, shown in figure 8, is again in sharp contrast to that for the cobalt and copper materials as shown in figure 7. As with the crystalline cobalt orthophosphate, the crystalline iron orthophosphate, $Fe_2O_3 \cdot P_2O_5$ behaves in a normal manner, with a T^3 relationship; however the three glasses shown all roughly obey a T^2 law at low temperatures. Such behaviour is again consistent with the microcrystalline precipitate interpretation if it is supposed that surface antiferromagnetic spin waves can be excited on the individual particles.

The entropy estimates for the iron glasses over the temperature range shown in figure 8 are about twenty times smaller than would be expected from a classical order–disorder transition associated with the Néel temperature for a material with S = 5/2. (Crude specific heat estimates at higher temperatures for the $(Fe_2O_3)_{0.79} \cdot P_2O_5$ glass indicated a sharp peak at roughly the observed Néel temperature).

DISCUSSION

The experimental results indicate that the glasses may be divided into two groups; the iron glasses in one group and the rest in the other. The iron glasses have paramagnetic Neel temperatures independent of the magnetic ion concentration, neutron diffraction evidence for a magnetic transition, a T^2 term in their low temperature specific heat, and clear magnetic evidence for a Néel temperature for at least one glass. (A Néel temperature may be inferred for the other glasses from the specific heat data). The glasses containing cobalt, copper or manganese show a variation of paramagnetic Néel temperature with magnetic ion concentration, no temperature variation of neutron diffraction spectra (for one cobalt sample investigated), a large linear term in the low temperature specific heat, no magnetic evidence for a magnetic transition, and a downward curvature of the reciprocal susceptibility at low temperatures. To summarize, the cobalt groups of glasses resemble the dilute magnetic alloys in their magnetic susceptibility and specific heat, with no evidence for ordering. On the other hand, the iron glasses show antiferromagnetic order, and their susceptibility-temperature variations are similar to the crystalline antiferromagnets.

It appears to be possible to attribute such differences between the two groups of glasses to structural reasons, or differences in the "degree of amorphousness"; the cobalt group of glasses are highly amorphous, whereas the iron glasses tend to be microcrystalline. This presumption is supported by the fact that the paramagnetic Néel temperature of the former is proportional to the magnetic ion concentration, while that of the latter is independent of it, and the neutron diffraction patterns for the iron glasses are sharper than for the cobalt glasses. Also, the specific heat data of some iron glasses indicate the presence of a spin wave energy gap. If the solid is genuinely amorphous, there should be no spin wave gap (30). Furthermore, the T^2 specific heat relationship in the iron glasses indicates that most of the spin waves in those glasses are 'surface modes', which seems to be compatible with a very small particle microcrystalline system. Therefore we consider that the cobalt group of glasses are more appropriate for the study of the intrinsic properties of amorphous antiferromagnets.

The most conspicuous property of the amorphous antiferromagnets studied here is that in spite of the antiferromagnetic interaction among the spins they do not seem to show static long range spin correlations down to temperatures lower than 1/10 of the paramagnetic Néel temperature. Such an absence of the long range order has been known for some time for the substitutionally random magnets (dilute magnetic alloys) when the concentration of magnetic atoms is low, and many theoretical and experimental studies have been made on the critical concentration of the incipience of the long range order (31-35). The conclusion drawn so far is that the classical percolation limit (31-35) represents the critical concentration fairly well, that is, long range order sets in when the average number of interacting neighbours exceeds 1.5 or 2. In truly amorphous solids, not only the number of interacting neighbours but also the exchange constant varies randomly, therefore the problem is far more complicated. However, in our glasses the magnetic interactions arise as a consequence of superexchange via the oxygen atoms and consequently are short range. Therefore it seems reasonable to apply the percolation theory with little modification, since the number of interacting neighbours can be unambiguously determined, unlike the case for metals where the interactions are long range.

The Bernal model for the structure of amorphous solids (36) reveals that the coordination number is high, around 12 to 14. If we take for example the cobalt glass with the composition $(CoO)_{1.4} \cdot P_2O_5$, the number of neighbouring cations around a cobalt ion is, then, presumably 10 or more, so that the number of interacting cobalt neighbours is 3 or 4 as a conservative estimate. The superexchange interaction is most effective when the magnetic ion–oxygen ion–magnetic ion are collinear, so that the effective exchange constant in amorphous solids may be lower than in a corresponding crystalline solid. This is indicated by the relatively low paramagnetic Néel temperatures of the cobalt glass compared to $(CoO)_3 \cdot P_2O_5$ when the corrections are made, assuming that the Néel temperature is proportional to the magnetic concentration. However, it is very unlikely that the super-exchange is completely suppressed when the atomic alignment is not ideal. Therefore we may conclude that the magnetic concentrations of all the glasses studied here are well above the percolation limit.

Thus, the existing theories of the substitutional random magnets do not seem to explain the apparent absence of the long range order in the cobalt group of glasses. Any theory which is applicable to amorphous solids must fully take into account the randomness in the interaction energy, or the 'off-diagonal randomness'. The development of such theories has only recently commenced (37, 38), and has not yet been applied to systems with antiferromagnetic interactions. So that at this stage we have to begin with some qualitative discussion as we have published elsewhere (30). The first thing that we can say with little reservation is that in amorphous antiferromagnets the spin waves tend to be well localized, presumably more than in any other magnetic system. Then the density of the lower energy excitations can be high, as is indicated by the large low temperature specific heats. This arises because the low energy spin waves are not only long waves but are also the localized excitations around weakly coupled spins or the spins with low molecular fields due to the compensation of the neighbouring spins. These low energy localized spin waves are important in determining the thermal properties of the system with a low number of interacting neighbours, because the excitation of such waves can destroy the long range order, leaving uncorrelated spin clusters inside which spins are strongly correlated. Therefore we think that the ordering temperatures of amorphous antiferromagnets are generally significantly lower than the corresponding crystalline antiferromagnets.

One further speculation is that the zero point effect of such low energy localized modes can be large for small spin systems. If the spin defect of certain spins are large enough so that those spins are in the 'many-body singlet' state in the time scale of the observation, long range order cannot set in. Then the incipience of the long range order depends not only on the average number of interacting neighbours but also on the spin value. It is quite tempting to apply this argument to our glasses, since for iron (Fe^{3+}) S is 5/2, while it is 3/2 for cobalt (Co^{2+}) and 1/2 for copper (Cu^{+}), therefore the quantum effect will be much less for an iron ion since it is nearer the classical limit.

ACKNOWLEDGEMENTS

The authors would like to thank Professor Sir Nevill Mott and Drs. A.J. Leggett and M.A. Moore for useful discussions, Dr. C.B.P. Finn and Mr. R.J. Commander for measuring the susceptibility below 4⁰K, and Mr. D. Price for help with the glass preparation. We would also like to thank Telcon Metals, Crawley, U.K., for their financial support for one of us (TE).

REFERENCES

1. A. Nicolau, Comptes Rendus, 205 (1937) 557.
2. C.J. Schinkel and G.W. Rathenau, Physics of Non-Crystalline Solids (North-Holland, 1965).
3. A.W. Simpson and J.M. Lucas, Proc. Brit. Ceram. Soc., 18 (1970) 117.
4. A.W. Simpson and J.M. Lucas, J. App. Phys., 42 (1971) 2187.
5. A.W. Simpson and A.L. Terry, Symposium on Electrotechnical Glasses. Society of Glass Technology (Sheffield, 1970).
6. T. Egami, O.A. Sacli, A.W. Simpson, A.L. Terry and F.A. Wedgwood, J. Phys. C., 5 (1972)
7. S. Foner, Rev. Sci. Instr., 30, no. 7 (1959) 548.
8. F.R. McKin and W.P. Wolf, J. Sci. Instr., 34 (1957) 64.

9. C. Kittel, Introduction to Solid State Physics, Vol. 3, p. 438, (Wiley, New York, 1967).

10. A. W. Simpson, Phys. Stat. Sol., 40 (1970) 207.

11. S. Kobe and K. Handrich, Phys. Stat. Sol., 42 (1970) K69.

12. R. Hasegawa, Phys. Stat. Sol. (b), 44 (1971) 613.

13. D. C. Fowlis and C. V. Stager, Canadian J. Phys., 47 (1969) 371.

14. M. F. Collins, G. S. Gill and C. V. Stager, Canadian J. Phys., 49 (1971) 979.

15,16. Omitted.

17. R. C. Zeller and R. O. Pohl, Phys. Rev., B4 (1971) 2029.

18. L. A. Welo, Phil. Mag., 6 (1928) 481.

19. G. E. Bacon, Neutron Diffraction (Clarendon Press, 1962), p. 95.

20. D. R. Howe, D. Phil. Thesis, University of Sussex (1967).

21. W. T. Berg, J. Appl. Phys., 39 (1968) 2154.

22. P. W. Anderson, B. E. Halperin and C. M. Varma, Phil. Mag., 25 (1972) 1.

23. W. Reese, J. App. Phys., 37 (1966) 3959.

24. O. L. Anderson, J. Phys. Chem. Solids, 12 (1959) 41.

25. J. P. Frank, F. D. Manchester and D. L. Martin, Proc. Roy. Soc. (London), A263 (1961) 494.

26. B. W. Veal and J. A. Rayne, Phys. Rev., 135 (1964) A442.

27. O. A. Sacli and D. F. Brewer, to be published.

28. A. J. Heeger, Solid State Physics, 23 (1969) 283.

29. B. R. Coles et al., Proceedings of the Thirteenth Conference on Low Temperature Physics (August 1972), to be published.

30. T. Egami and A. W. Simpson, Phys. Stat. Sol., 53 (1972), to be published.

31. R. J. Elliott and B. R. Heap, Proc. Roy. Soc. (London), A265 (1964) 264.

32. S. F. Edwards and R. C. Jones, J. Phys. C: Solid St. Phys., 4 (1971) 2109.

33. G. A. Murray, Proc. Phys. Soc., 89 (1966) 87, B. J. Last, to be published.

34. T. Kaneyoshi, Prog. Theor. Phys., 42 (1969) 477.

35. M. F. Sykes and J. W. Essam, Phys. Rev., 133 (1964) A310.

36. J. D. Bernal, Nature, 183 (1959) 141.

37. Y. Izyumov, Proc. Phys. Soc. (London), 87 (1966) 505.

38. J. A. Blackman, D. M. Esterling and N. F. Berk, Phys. Rev., B4 (1971) 2412.

DISCUSSION

S.C. Moss: A comment on the neutron scattering: it seems
that if you get long range ordering of the spins on a lattice
in which the positional disorder is quite limited, one is
left, in the ferromagnetic case for example, with all the
spins aligned parallel but positionally disordered. This
means that the pair correlation function is quite short
ranged, although the spins are all lined up. A possibility
therefore exists in neutron scattering of lining all spins
parallel to the diffraction vector and wiping out that contribution and then preserving that contribution and getting
just the iron-iron pair correlation function, and that is
interesting.

A.W. Simpson: Yes, I agree. However, I think one can obtain the iron-iron pair correlation directly from the type
of experiment described here. This may be done by analyzing the difference between the diffraction data above and
below the Néel temperature to give the scattering due to
just the magnetic atom, and hence an iron-iron pair correlation function. We have done this and it gives a near
neighbor distance for iron atoms in the iron phosphate
glass (Fig. 4) essentially the same as that for the crystalline orthophosphate.

H. Sato: I have some conceptual difficulty understanding
amorphous antiferromagnetism. The ferromagnetic case, of
course, is (just like you said) conceivable, but I can't
imagine what kind of structure you can think of in the
antiferromagnetic case.

A.W. Simpson: On a very simple level, for the case of magnetic atoms with spin one half, and bearing in mind that
there is no anisotropy in an amorphous material, one may
regard the amorphous antiferromagnet as two interpenetrating amorphous 'sublattices' with the spins on one sublattice
antiparallel to the other.

P.W. Anderson: A remark to the same point. I think your
interpretation of the absence of an antiferromagnetic transition should include some kind of degree of misfit. I'm
not sure whether amorphous materials don't have an infinite
degree of misfit, but even in regular lattices, antiferromagnetism can be disfavored by having misfit structures, so
that when nearest neighbors are nearest neighbors of each

other, this is well known to lower the transition temperature sometimes to zero.

A.W. Simpson: Yes, this may be so. For a face-centered cubic lattice with nearest neighbor interactions only where, of course, the nearest neighbors are nearest neighbors of each other, there is no transition temperature.

H. Sato: Are you sure?

A.W. Simpson: I'm not sure, but a paper by Ziman (Proc. Phys. Soc. (London), A66, 89-94, 1953) predicts no transition in a face-centered cubic lattice.

MAGNETIC ORDER IN ALKALIBORATE AND ALUMINOSILICATE GLASSES

CONTAINING LARGE CONCENTRATIONS OF IRON-GROUP IONS[†]

H.O. Hooper, G.B. Beard, R.M. Catchings,*

R.R. Bukrey,** M. Forrest, P.F. Kenealy,

R.W. Kline, T.J. Moran, Jr., J.G. O'Keefe,***

R.L. Thomas, and R.A. Verhelst

Department of Physics, Wayne State University

Detroit, Michigan 48202

INTRODUCTION

In this paper is presented a brief review of our studies of magnetic interactions and atomic structure of borate and silicate base inorganic glasses containing large amounts of iron group ions. Several experimental techniques have been employed in this laboratory to study the atomic structure and magnetic properties of alkaliborate glasses containing large amounts of iron and aluminosilicate glasses containing large percentages of cobalt and manganese ions. This work includes extensive Mössbauer studies, low field magnetic susceptibility measurements, x-ray diffraction studies, magnetic resonance measurements and ultrasonic sound velocity measurements. Various magnetic behavior; paramagnetic, superparamagnetic, and antiferromagnetic; are exhibited by these glasses, depending on glass composition and the glass preparation procedure.

[†]Research Supported in part by the Air Force Office of Scientific Research under grant AFOSR-72-2002.

*Present Address, Department of Physics, Howard University Washington, D.C.

**Present Address, Department of Physics, Loyola University Chicago, Illinois.

***Present Address, Department of Physics, Rhode Island College, Providence, Rhode Island.

47

In some glasses there is definite evidence of the presence of small particles of crystalline iron oxide which
are responsible, in part, for the magnetic behavior of
these samples. However, in numerous other glasses it does
appear that the observed magnetic behavior is not due to
crystalline material. Samples of the glass system,
$CoO \cdot Al_2O_3 \cdot SiO_2$, exhibit an antiferromagnetic transition
with the transition temperature depending on glass composition. A general review of the results of the experimental
studies of the magnetic behavior of these glass samples
will be given.

$Fe_2O_3 \cdot Na_2O \cdot B_2O_3$ GLASS SYSTEM

Our initial investigations[1] of magnetic interactions
in an oxide glass matrix began by examining a series of
glass samples of sodium borate glass into which iron oxide
was doped. The base glass composition consisted of 1 part
Na_2O and 2 parts B_2O_3. To this was added varying amounts
of Fe_2O_3 to form samples containing up to about 40 wt.%
Fe_2O_3 ($\sim$22 mole %). The structure of alkaliborate glasses
has been rather clearly determined, through the use of
numerous experimental techniques including nuclear magnetic
resonance[2] (NMR), to be a random network formed of planar
BO_3 groups and negatively charged BO_4^- tetrahedra with the
alkali ion (Na^+) residing in holes in the network. When
iron ions in this case Fe^{3+} are introduced into the glass
there is the possibility of the iron residing within holes
in the network or bonding in the lattice as a network
former. The initial studies of these glasses were limited
to the room temperature NMR of the ^{11}B nuclei and to x-band
electron spin paramagnetic resonance (EPR) study of the
glass samples. The ^{11}B NMR results were somewhat surprising in that the ^{11}B line width increased only about 30%
from samples with no Fe_2O_3 up to samples containing as much
as 25 wt% Fe_2O_3. This would indicate a rather fast relaxation time for the iron, which is consistent with Mössbauer
data. In addition, the NMR data on the ^{11}B nuclei enabled
a measurement of the fraction of the boron atoms which are
four-coordinated. When Na_2O is added to pure B_2O_3, where
only planar BO_3 units exist, two boron atoms become four-
coordinated. The sodium is called a modifier and exists in
the glass as a Na^+ ion presumably near a BO_4^- tetrahedron
to maintain charge neutrality. As Fe_2O_3 is added to the

1/3 molar $Na_2O \cdot 2/3$ molar B_2O_3 the number of four-coordinated boron atoms drops to about 1/2 the value found in the glass with no iron by the time 1.5 wt% (0.65 mole %) Fe_2O_3 has been added. As more iron is added up to 25 wt% Fe_2O_3 the fraction of four-coordinated borons remains constant. It would appear, therefore, that only when very small amounts (less than about 0.65 mole%) of Fe_2O_3 are added to the glass does the iron possibly act as a modifier, and for amounts of iron exceeding a fraction of a mole percent the iron goes into the network, most likely bonding to four oxygen atoms.

An examination of the EPR and Mössbauer data produces a conclusion which is consistent with that drawn from the ^{11}B NMR results as to site of the Fe^{3+} ions in these borate base glasses. The EPR data are shown in Fig. 1. These data and Mössbauer data[3,4] lead to the conclusion that the

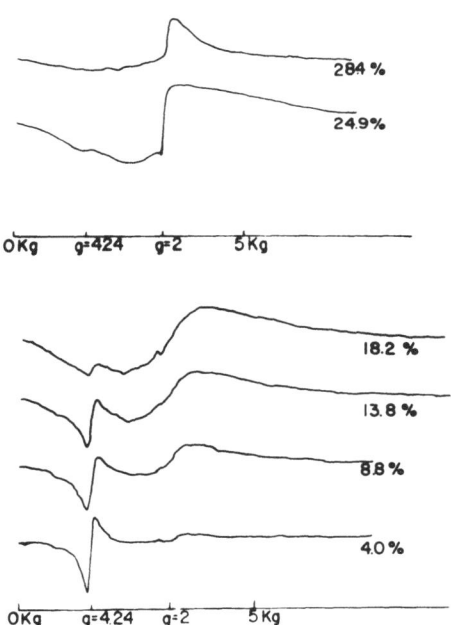

Figure 1.   X-Band EPR spectra for iron containing sodium borate glasses. Spectra are identified by the wt% Fe_2O_3 present in each sample.

iron is in the Fe^{3+} state and situated predominately in a
distorted tetrahedral coordination. The EPR (Fig. 1) spectra are characterized by the appearance of a rather sharp
resonance at g = 4.2 for low iron content, which is replaced by a very broad, intense resonance at g = 2 for the samples with high iron content. This type of resonance behavior
has been observed in a number of crystalline and glassy
systems containing iron[1],[5-9]. An early interpretation[7] of
this type of EPR spectrum associated the two resonances
(g = 2 and g = 4.2) with two distinct sites in the glass,
glass modifier (interstitial) and glass former. However,
more recent interpretations,[5,6,8,9] agree that the appearance of the g = 2 resonance is not due to the formations
of a new (interstitial) site that was not present in the
glass at lower concentrations. Rather, they agree that the
strong g = 2 resonance is due to spin-spin interaction of
neighboring iron atoms, whose separation decreases as the
iron concentration increases. The importance of the EPR
data is that the appearance of the g = 2 resonance is due
simply to the increased iron interaction rather than to the
formation of a new iron site. Furthermore, it has been the
usual interpretation[6],[9] to attribute the g = 4.2 resonance
to ferric iron situated exclusively in distorted tetrahedral coordination, which agrees with the interpretation
of the Mössbauer doublet[3],[4] spectrum observed in these
glasses. There is still, however, a question as to whether
these observations can be interpreted exclusively as due to
Fe^{3+} in tetrahedral coordination. For example, Kurkjian
and Sigety[8] on the basis of optical absorption studies on
silicate and phosphate glasses showed that both tetrahedral
and octrahedral coordination of the ferric ion in these
glasses gave an EPR resonance at g = 4.2 for low iron concentration. There is also the observation that the
Mössbauer doublet observed in these glasses remains unchanged throughout the region of iron concentration in
which the transformation in the EPR spectra was observed.
However, in light of the NMR, EPR, and Mössbauer data for
these iron containing, alkali borate glasses the assignment
of the ferric iron to a distorted tetrahedral coordination
appears consistent and reasonable.

The susceptibility of these iron-containing sodium
borate samples has been measured employing a low field ac
bridge. In Table I are shown the room temperature susceptibility of some of the samples prepared.

TABLE I

AC Susceptibility of $Fe_2O_3 \cdot Na_2O \cdot B_2O_3$ Samples

wt% Fe_2O_3	$\chi \times 10^6$ emu/gm	wt% Fe_2O_3	$\chi \times 10^6$ emu/gm
18	18	30	778
25	25	30	1536
26	53	35	2450
28	136		
29	173		
30	213		
35	496		
37	422		

Samples containing less than 26 wt% Fe_2O_3 yield a negative paramagnetic Néel temperature and $1/\chi$ vs. T is a straight line. Above this concentration of iron two types of behavior appear. First, some of the glasses are paramagnetic while for the other samples (for example, the three samples listed at the right in Table I) which have a much greater room temperature susceptibility have a complex behavior which may be due to the presence of fine particles of crystalline material.

In Fig. 2 the temperature dependence of χ and $1/\chi$ for a paramagnetic glass containing 26 wt% Fe_2O_3 is shown. In Fig. 3 the temperature dependence of χ and $1/\chi$ is shown for the glass sample containing 35 wt% Fe_2O_3 and having a room temperature susceptibility of 496 x 10⁻6 emu/gm. This sample appears to be glassy and presumably the shape of $1/\chi$ is due to magnetic clustering. It should be noted that for the paramagnetic glasses with iron concentration up to 30 wt% Fe_2O_3, the effective magnetic moment is essentially constant and has a value of 3.7 ± 0.4 μ_B. The magnetic moment for the 35 wt% Fe_2O_3 glass for which data are shown in Fig. 3 has a higher effective moment, the exact value of which depends on the temperature region used to evaluate the magnetic moment.

Rather interesting data were obtained for several samples (for example, the last three samples in Table I) whose room temperature susceptibility was very large. Data for two such samples, one of 30 wt% Fe_2O_3 and one of 35 wt% Fe_2O_3 are shown in Fig. 4. The 35 wt% Fe_2O_3 sample (data indicated by triangles in Fig. 4) shows a drop in its sus-

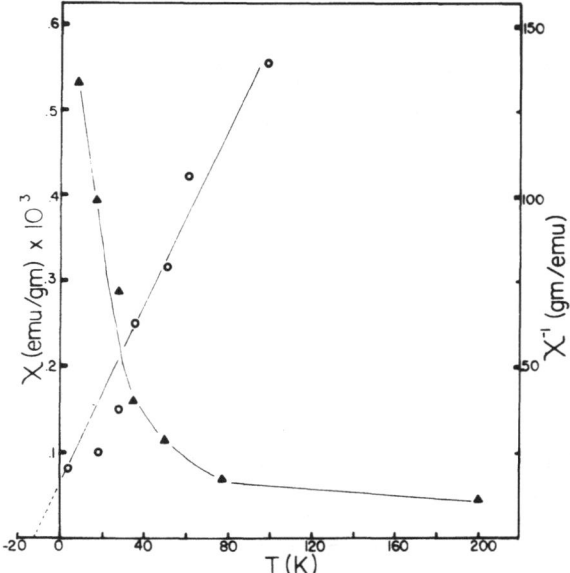

Figure 2.   Low field a.c. susceptibility (▲) vs. temperature
            and $1/\chi$ (O) vs. temperature for a sodium borate
            glass containing 26 wt% Fe_2O_3.

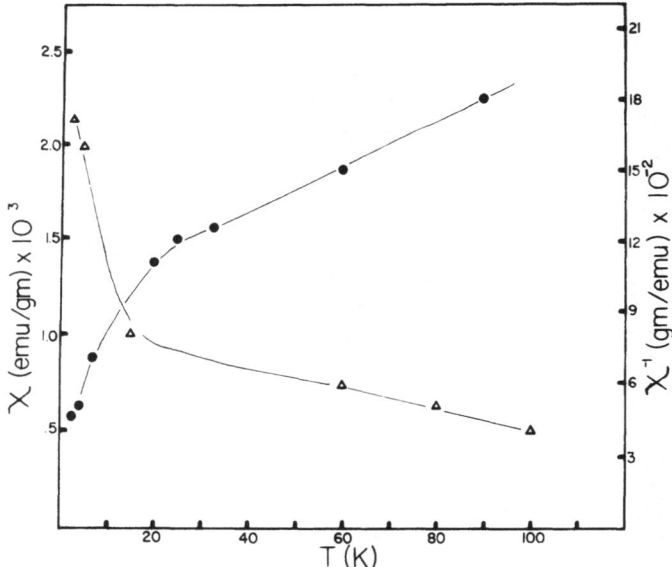

Figure 3.   Low field a.c. susceptibility (Δ) vs. temperature
            and $1/\chi$ (O) vs. temperature for a sodium borate
            glass containing 35 wt% Fe_2O_3.

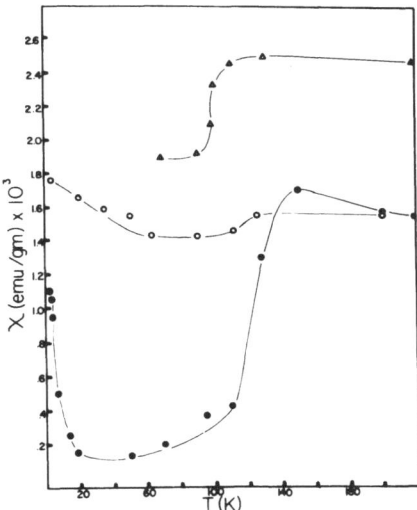

Figure 4.   Low field a.c. susceptibility vs. temperature for
sodium borate samples containing (a) 35 wt%
Fe_2O_3 (A); (b) 30 wt% Fe_2O_3 (0); (c) same 30 wt%
Fe_2O_3 sample in an applied field of 2kOe (0).

ceptibility as the temperature is lowered at about 113°K.
The 30 wt% Fe_2O_3 sample (data indicated by circles in Fig.
4) shows a similar drop in susceptibility at about 126°K.
Below 60°K the susceptibility of the 30 wt% Fe_2O_3 sample
for which data are shown in Fig. 4 behaves almost identically to another 30 wt% Fe_2O_3 sample whose room temperature
susceptibility is 0.21 x 10^{-3} emu/gm. That is, at low temperatures both samples show the same increase in χ as the
temperature is lowered which is very close to paramagnetic
behavior, the deviation from paramagnetic behavior being
similar to that shown in Fig. 3. A third temperature dependence shown in Fig. 4 by the open circles illustrates
the variation in χ for this same 30 wt% Fe_2O_3 sample when
the sample is kept in a field of 2 kOe. It appears that
this change in susceptibility is due to the presence of
fine particles of magnetite in the samples. Bulk magnetite
is known to exhibit a transition at about 113°K. The susceptibility observed in the sample would be a combination
of that due to fine particles of crystalline material and
of that due to a highly paramagnetic glass matrix. The
ordered crystalline material and an external field would
both interact with the iron at the paramagnetic regions of
the glass and conceivably could account for the strange
behavior of χ as a function of temperature. Such data

point out the importance of determining whether or not a
given sample is amorphous.

$$Fe_2O_3 \cdot Li_2O \cdot Na_2O \cdot B_2O_3 \text{ GLASS SYSTEM}$$

In order to study more carefully the structure and
magnetic properties of alkali borate glasses containing
large amounts of iron ions, an extensive study employing
Mössbauer and x-ray techniques was undertaken. The most
systematic study has involved samples of the system
$Li_2O \cdot Na_2O \cdot Fe_2O_3 \cdot B_2O_3$. The amount of iron oxide varied from
15 wt% (6.1 mole% Fe_2O_3) to 49 wt% (27.1 mole% Fe_2O_3). The
molar concentration of alkali oxide was either equal to that
of the B_2O_3 or one half of it, while the molar ratio of
Li_2O to Na_2O was usually 1. The magnetic behavior seemed
not to depend u p o n   the amount of alkali present except
that the amount of iron which could be incorporated into the
glass did depend upon the Na_2O concentration presumeably
because of its function as a flux in the melting process.

The iron in these glasses exhibits three modes of magnetic behavior depending upon the iron concentration: long
relaxation-time paramagnetism, ordinary paramagnetism, and
superparamagnetism, respectively, in order of increasing
iron concentration. Typical Mössbauer spectra are shown in
Fig. 5 for the samples containing the lower amounts of
Fe_2O_3. Below about 30 wt% Fe_2O_3 the Mössbauer spectra consist of a broadened, quadrupolar-split doublet while for
iron oxide concentrations of 30 wt% and above the spectra
consist of a variety of complex hyperfine patterns. There
is evidence of some broadened hyperfine lines in the 30 wt%
Fe_2O_3 sample as shown in Fig. 5.

The Mössbauer spectra of samples containing greater than
than 30 wt% Fe_2O_3 are shown in Fig. 6. Here the spectra
are arranged from top to bottom in decreasing amounts of
Fe_2O_3, and it can be seen that the intensity of the magnetic hyperfine interaction increases as the iron concentration increases. The major hyperfine lines can be identified as due to the presence of crystalline iron oxides
γ-Fe_2O_3, α-Fe_2O_3 and Fe_3O_4. However, there is an additional
12 line spectrum which appears to arise from a garnet type
structure in which a boron replaces the yttrium.

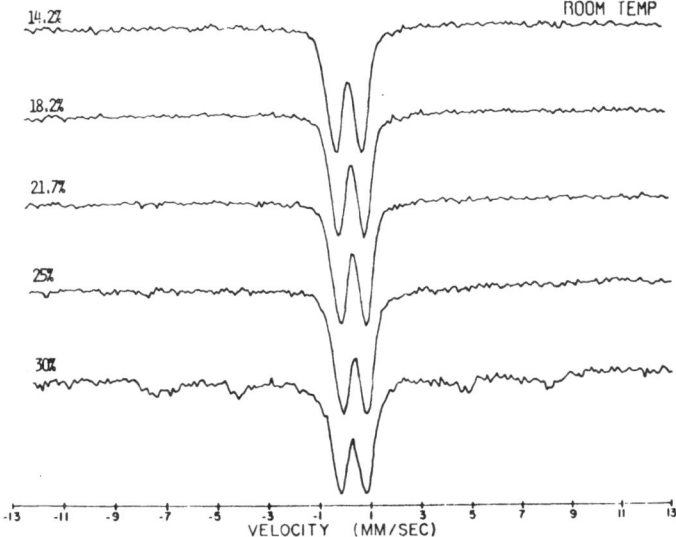

Figure 5. Room temperature Mössbauer spectra for sodium-
lithium oxide boron oxide glasses containing
14.2 to 30 wt% Fe_2O_3.

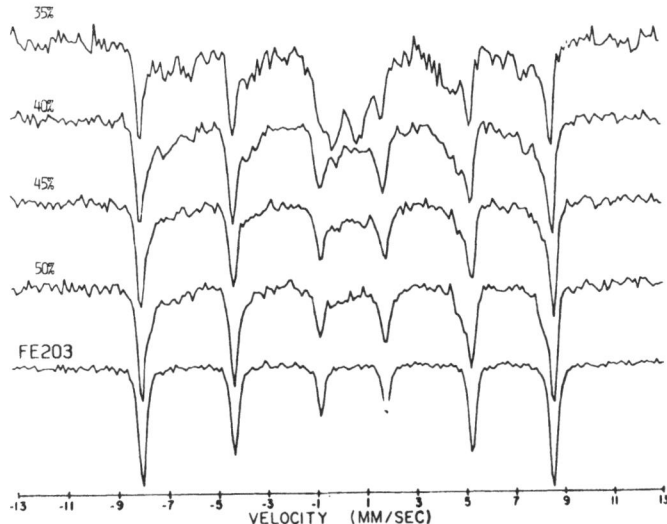

Figure 6. Room temperature Mössbauer spectra for sodium-
lithium oxide boron oxide samples containing
35 to 50 wt% Fe_2O_3.

In Fig. 7 are shown the x-ray spectra of each of these samples. The sharpness of the x-ray lines indicates that a as the α-Fe$_2$O$_3$ content increases above 40 wt% the sample contains a large amount of crystalline α-Fe$_2$O$_3$. The x-ray spectra of the 29.5 and 25 wt% samples show diffuse peaks characteristic of microcrystalline material or clusters, while the Mössbauer data for these samples are essentially doublets. Therefore the appearance of a simple doublet spectrum at room temperature does not preclude the existence of microcrystalline material or clustering in the sample.

The room temperature Mössbauer spectra of the glasses can be characterized by three main features:

1.   For iron concentrations below 30 wt% Fe$_2$O$_3$ the spectra consists of a single quadrupole-split doublet with isomer shift $\sim$0.29 $\pm$ .01 mm/sec., quadrupole splitting -.51 $\pm$ .01 mm/sec., and width $\sim$.6 mm/sec. Note that these values of isomer shift and quadrupole splitting remain the same at 4.2°K but the lines broaden.

2.   At the highest concentrations of iron, the doublet is replaced by hyperfine patterns characteristic of several simple iron oxides.

3.   In the neighborhood of 30 - 35 wt% a complex hyperfine pattern appears which, in addition to the eight lines attributed to α-Fe$_2$O$_3$ and glass doublet, consists of 18 lines of uncertain origin.

If the discussion is limited to the glasses which essentially contain a quadrupole doublet, interesting conclusions can be reached concerning the interaction between the iron ions in these glass samples. The susceptibility of these glasses increases three orders of magnitude as the concentration changes from 14 wt% Fe$_2$O$_3$ to 30 wt% Fe$_2$O$_3$. The values for the room temperature susceptibility in emu/gram x 10^6 are as follows: 14.2 wt% Fe$_2$O$_3$, 18.3; 18.2 wt% Fe$_2$O$_3$, 115; 21.7 wt% Fe$_2$O$_3$, 354; 25 wt% Fe$_2$O$_3$, 2770; and 30 wt% Fe$_2$O$_3$, 3660.

This large susceptibility, when combined with the doublet Mössbauer spectra (which indicate a lack of magnetic order) and the weak, broad x-ray patterns (indicating the presence of microcrystals or clustering) is strongly suggestive of the presence of superparamagnetism in these

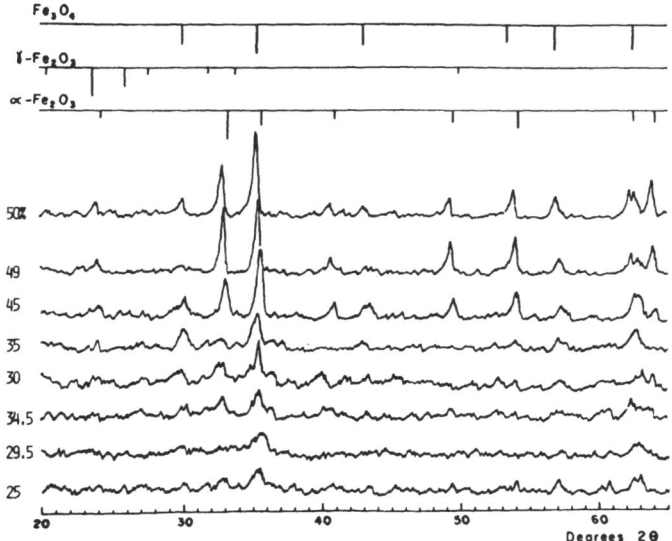

Figure 7. X-ray diffraction patterns for selected samples.
Spectra are identified by their wt% Fe_2O_3.

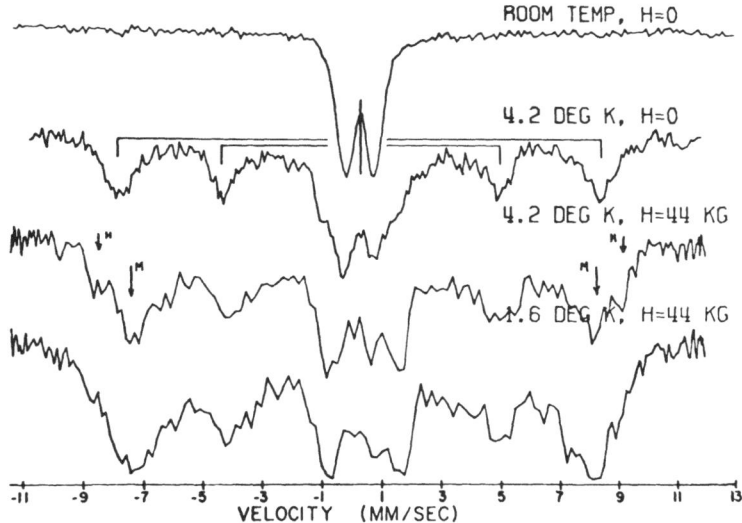

Figure 8. Mössbauer spectra for a sodium-lithium oxide,
boron oxide glass containing 25 wt% Fe_2O_3 at
various temperatures and applied fields.

samples for as a rule of thumb, the susceptibility of superparamagnetic material is on the order of 1000 x 10^{-6}
emu/gm[11]- comparable to the values obtained for the 25 and
30 wt% Fe_2O_3 samples.

The Mössbauer spectra for these samples, 14.2 wt%
Fe_2O_3 to 30 wt% Fe_2O_3, was examined at low temperatures
and in a magnetic field of 44 kOe parallel to the gamma
ray beam. The effect of lowering the temperature is to
resolve hyperfine lines; the higher the iron concentration
the better the hyperfine lines are resolved at a particular
temperature. The effect of the large magnetic field is to
split the quadrupole doublet into two or more lines and in
addition, each spectrum has superimposed on it a broad
diffuse magnetic hyperfine pattern. Figure 8 illustrates
how the spectrum from the 25 wt% Fe_2O_3 sample changes as a
function of temperature and magnetic field. These data and
similar data taken for each sample containing less iron indicate a relaxation spectra characteristic of collective
magnetic behavior. An analysis[4] of these spectra using the
model of Van de Woude and Dekker[12] indicates these Mössbauer
spectra are indeed due to changes in relaxation time and
changes in the degree of magnetic order as the temperature
is lowered. The details of this analysis will soon be
published.

ALUMINOSILICATE GLASS SAMPLES

Samples of the glass system $XO \cdot Al_2O_3 \cdot SiO_2$ have been
prepared in air in an arc-image furnace constructed from
a 5 ft. diameter search light. To date good glasses with
metal oxide (XO) content (batch composition) from ∿40 mole%
to ∿85 mole% have been prepared where X is Co or Mn. These
samples show no visible signs of devitrification and appear
to be good glasses. Limited attempts to make NiO, and FeO
glasses have not been successful to data although it is
presumably possible to make such glasses.[13]

The samples have been examined employing low-field
magnetic susceptibility measurements and ultrasonic sound
velocity measurements (ν = 10 MHz). Using both techniques
an antiferromagnetic-type transition has been found in these
glass samples. The initial experimental results on the
$CoO \cdot Al_2O_3 \cdot SiO_2$ system are described briefly here.

In Fig. 9 is shown a plot of χ versus temperature for two samples of CoO Al_2O_3 SiO_2 glasses. The rather surprisingly sharp transition appears in all of the samples, but it is broader and less easily distinguished in samples containing larger amounts of CoO. The transition temperatures determined by the peak in the susceptibility vs. temperature curve are shown in Table II.

TABLE II

CoO Concentration (Molar %)	Susceptibility Peak (°K)	Sound Velocity Minimum (°K)
40	2.5 $\pm$ 0.1	2.75 $\pm$ 0.5
50	A. None observed to 1.6°K	
	B. 2.3 $\pm$ 0.1	B. 2.5 $\pm$ 0.5
	C. 2.5 $\pm$ 0.1	
53	5.25 $\pm$ 0.1	6.25 $\pm$ 1.0
60	4.75 $\pm$ 0.1	6.5 $\pm$ 1.0
71.2	7.25 $\pm$ 0.1	6.25 $\pm$ 0.75
83	7.8 $\pm$ 0.1	Too Broad to Resolve

The multiple values for the transition temperatures for the samples with 50 molar % CoO are due to the fact that three samples were cut from different regions of the same piece of glass. The samples are approximately the same in size and were cut from a piece of glass which was about 3 cm in diameter with a thickness varying from about 1 mm at the edge to about 1 cm near the center. This variation in transition temperature is probably due to a composition gradient across the glass sample, or it is possible variations in the internal strains produce this effect. Annealing studies and chemical analysis are now being carried out in an attempt to resolve this question.

Graphs of the susceptibility versus temperature and the inverse susceptibility versus temperature are shown in Fig. 10 for one of the 50 molar % CoO samples. This figure is

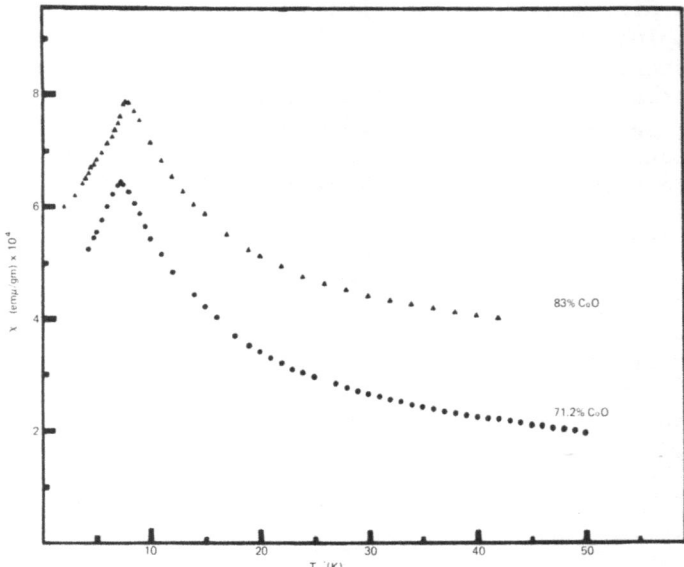

Figure 9. A.C. susceptibility vs. temperature for two
           glass samples of the system $CoO \cdot Al_2O_3 \cdot SiO_2$.

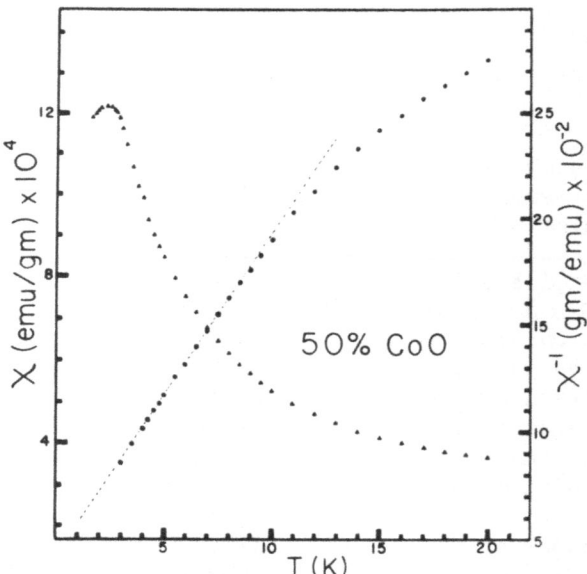

Figure 10. A.C. susceptibility and $1/\chi$ vs. temperature
            for a 50% $CoO \cdot Al_2O_3 \cdot SiO_2$ glass.

typical of the results found for all of the CoO glasses.
The inverse susceptibility deviates from linearity above
about 8 - 10°K but has a linear region both at high temperatures above about 15°K and at low temperatures.

In addition to magnetic susceptibility studies, the
change in ultrasonic sound velocity in the same glass samples was measured as a function of temperature. These data
are shown in Fig. 11 where the critical change in sound
velocity $\Delta v/v_0$ is plotted as a function of temperature.
The temperatures at which the minimum change occurs in
sound velocity are shown in Table II. The minimum in the
sound velocity appears at a slightly higher temperature
than that at which the peak in the susceptibility occurs.
The susceptibility peaks appear slightly more narrow. The
sound velocity measurements indicate that extensive strain
exists in these glass samples and a variation in composition
may exist within the sample. The annealing studies should
enable a determination of more information on the internal structure of these glasses.

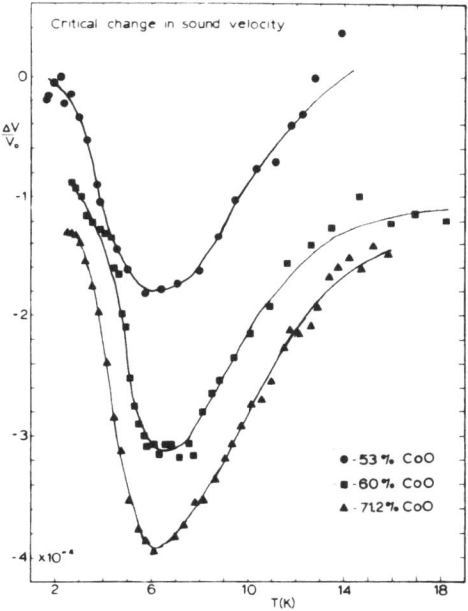

Figure 11.   Critical change in sound velocity vs. temperature for three glass samples of the system
$CoO \cdot Al_2O_3 \cdot SiO_2$.

In addition to these measurements, x-ray diffraction and and electron microscopic studies of each sample are being performed to determine if there is any crystalline material present. Initial x-ray data indicated that there were no crystalline particles present in these samples.

The susceptibility data are now being carefully examined and analyzed in terms of recent theoretical models to see if the nature of the magnetic interactions can be more clearly determined.

REFERENCES

1. H.O. Hooper and R.M. Catchings, Bull. Am. Phys. Soc. II 11, 188 (1966); R.M. Catchings, M.S. Thesis, Wayne State University 1966. R.M. Catchings, Ph.D. Dissertation, Wayne State University, 1970.
2. P.J. Bray and A.H. Silver, Modern Aspects of the Vitreous State, Vol. 1, Chap. 5, Butterworths, London (1960); P.J. Bray and J.G. O'Keefe, Phys. Chem. Glasses 4, 37 (1963).
3. R.R. Bukrey, P.F. Kenealy, G.B. Beard and H.O. Hooper, J. Appl. Phys. 40, 4289 (1969); R.R. Bukrey, R.M. Catchings, H.O. Hooper, and P.F. Kenealy, Bull. Am. Phys. Soc. 15, 108 (1970).
4. R.R. Bukrey, Ph.D. Dissertation, Wayne State University (1972).
5. T. Birchall, N.N. Greenwood and A.F. Ried, J. Chem. Soc. (A), 2382 (1969).
6. A.F. Ried, H.K. Perkins and M.J. Sienko, Inorganic Chemistry 1, 119 (1968).
7. R.F. Tucker, Advances in Glass Technology, Plenum Press, New York (1962).
8. C.R. Kurkjian and E.A. Sigety, Phys. Chem. Glasses 9, 73 (1968).
9. E.J. Friebele, L.K. Wilson, A.W. Dozier, and D.L. Kinser, P. Status Solidi, 45, 323 (1971).
10. C.A. Domenicali, Phys. Rev. 78, 458 (1950).
11. D.W. Collins, J.T. Dehn, and L.N. Mulay, Mössbauer Effect Methodology, Vol. 3 Plenum Press, New York (19 (1967).
12. F. Van der Woude and A.J. Dekker, P. Status Solidi 9, 775 (1965).
13. P.W. McMillan, Advances in Glass Technology, Plenum Press, New York (1962).

DISCUSSION

R.K. MacCrone: In terms of your odd ball glasses and your crystallized phase, I would just like to make one comment. We have made quenched glasses containing much less iron then you have made that give every evidence of being good glasses, no small angle x-ray scattering, no diffraction peaks, and yet the replica electron microscopy shows very well developed microstructure present in very well quenched specimens.   So one has to be very careful here.

H.O. Hooper: I think the lesson to learn here is that one should use every technique available to him to determine the atomic arrangement in the samples.   We have made arrangements to use electron microscopy to examine the cobalt aluminosilicate glasses.

J. Wong: Do you have any idea as to where the region of phase separation is in the pure sodium borate glass?

H.O. Hooper: As long as one stays below about 40 mole % Na_2O you are in a good glass forming region.   Above that you have problems with crystallization.

ANTIFERROMAGNETISM IN THE VANADIUM, MANGANESE AND IRON
PHOSPHATE GLASS SYSTEMS

L. K. Wilson, E. J. Friebele and D. L. Kinser

Vanderbilt University

Nashville, Tennessee 37235

ABSTRACT

Variable temperature magnetic susceptibility measurements have been made on a series of concentrated vanadium,
manganese and iron phosphate glasses. The high temperature
magnetic susceptibility of all glasses studied obeys a
Curie-Weiss law with an antiferromagnetic Curie temperature.
The magnetic properties of these glasses are a function of
composition and approach the properties of related crystalline compounds at the limit of glass formation. The low
temperature susceptibility measurements of the glasses show
anomalous behavior and are discussed in terms of recent
theoretical models for amorphous antiferromagnets.

INTRODUCTION

In 1964 Shinkel and Rathenau (1) reported susceptibility measurements on a series of manganese borate glasses.
The temperature dependance of the magnetic susceptibility
of these glasses at high temperature indicated antiferromagnetic behavior similar to that ordinarily found in crystalline oxides. Since the publication of this work, antiferromagnetic coupling has been reported in chromium phosphate glass (2), iron phosphate glass (3), vanadium phosphate glass (4), manganese phosphate glass (5) and in an
amorphous yttrium-iron compound (6).

Recently, theoretical calculations by Simpson (7) have shown that negative exchange interactions can occur in amorphous systems. The effect of the statistical distribution of near neighboring distances was incorporated by assuming a distribution of effective field coefficients between interacting ions. Simpson's model predicts a precise ordering temperature proportional to the square root of the average number of interacting near neighbors. This temperature is somewhat lower than the Neél temperature of the analogous crystalline material in which the ordering temperature is proportional to the first power of the number of interacting neighbors. Another characteristic predicted is that the projected intercept of the reciprocal susceptibility versus temperature plot is essentially the same in both the amorphous and crystalline cases. Another model of amorphous antiferromagnetism was developed by Kobe and Handrich (8), who included structure fluctuations in the calculation of susceptibility. In contrast to the Simpson model, the Kobe and Handrich model predicted a Neél temperature above that of the crystalline compound and an increasing inverse susceptibility below the transition. Hasegawa (9) reported studies of a model which took into account both first and second nearest neighbor interactions. The Hasegawa model predicts a Neél temperature which may be either above or below the crystalline Neél temperature, depending upon the relative distribution widths of the exchange integral distributions of the first and second nearest neighbor interactions. In addition, the Hasegawa model predicted an increasing inverse susceptibility below the Neél temperature.

The purpose of this paper is to report the results of an investigation of the bulk magnetic properties of some transition metal phosphate glasses and to discuss these properties in view of recent theoretical analyses of amorphous antiferromagnets.

EXPERIMENTAL

The glasses of the present study were prepared by melting a mixture of the constituent transition metal oxide and P_2O_5 in air for one hour at 1300°C. The glasses were quenched to room temperature on copper blocks but were not annealed.

Finely powdered samples were examined for crystallinity in a vacuum Guinier DeWolff x-ray camera. This technique permits detection and identification of as little as 0.1 wt. % crystal for crystals larger than about 0.05μm. No evidence of crystallinity was found by x-ray techniques in any of the as-cast glasses. The glasses were also examined using conventional replica electron microscopy techniques on both freshly fractured and etched surfaces. None of the glasses examined exhibited any crystalline features, although liquid-liquid immiscibility was noted in the vanadium and manganese glasses.

RESULTS AND DISCUSSION

The manganese phosphate, vanadium phosphate, and iron phosphate glass systems were studied. Typical high temperature susceptibility data for the glasses of these three systems are shown in Figures 1 and 2. It may be seen that

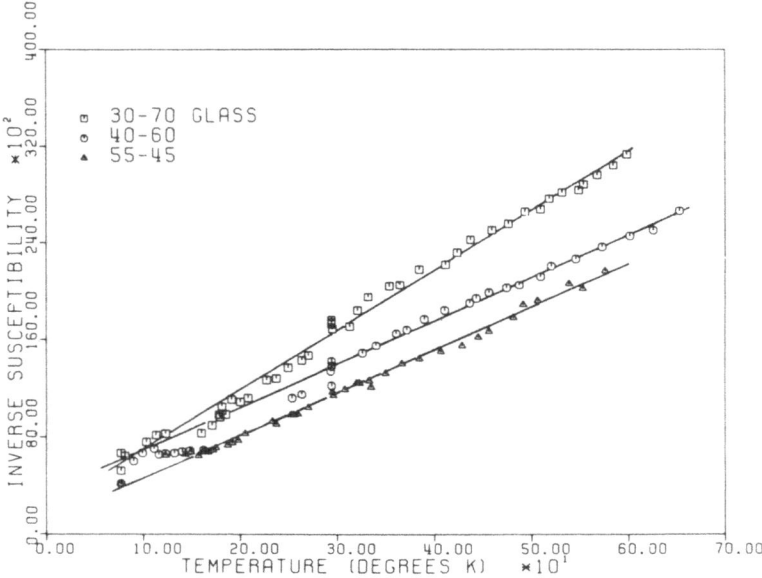

Figure 1. The inverse susceptibility as a function of temperature for a series of manganese oxide phosphate glasses [xMnO-(1-x) P_2O_5].

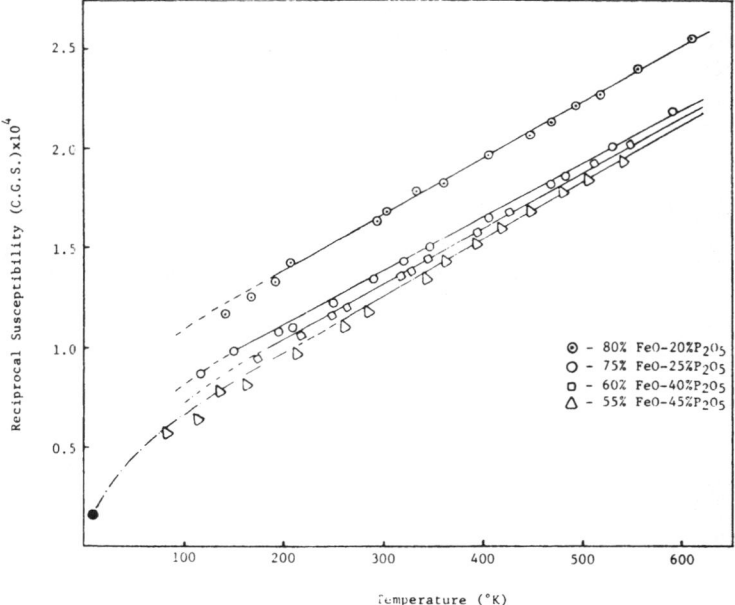

Figure 2. The inverse susceptibility as a function of temperature for a series of iron oxide phosphate glasses.

the high temperature inverse susceptibility obeys a Curie-Weiss law with an antiferromagnetic Curie temperature. These results reveal the universal existence of antiferromagnetic coupling between transition metal ions in the glasses studied. In the manganese phosphate system, the manganese ions exist primarily in the divalent state so that the antiferromagnetic coupling is between Mn^{2+} ions (5). In the vanadium phosphate system, the pentavalent vanadium ions are diamagnetic with S=0, and coupling in this system occurs between V^{4+} ions (4). However, in the iron phosphate system, both ferrous and ferric ions are magnetic, and since both valence states exist in the glasses (10), coupling occurs between divalent, between trivalent, and between divalent and trivalent ions. It has been shown previously that extensive coupling occurs between aliovalent iron ions (3) in contrast to the coupling between like valence states which occurs in other transition metal glass systems.

The high temperature susceptibility data is summarized in Table I.

TABLE I

Magnetic Susceptibility Data of Some Amorphous Antiferro-magnets and Related Compounds

Composition (Mol %)	Effective Paramagnetic Curie Temp. θ (°K)	Curie Constant C (CGS Units)
30 MnO 70 P_2O_5	-48	.0205
40 MnO 60 P_2O_5	-64	.0260
50 MnO 50 P_2O_5	-90	.033
55 MnO 45 P_2O_5	-31	.028
Crystalline $Mn_2P_2O_7$	-13	.0255
60 V_2O_5 40 P_2O_5	-117	.0015
65 V_2O_5 35 P_2O_5	-82	.0013
75 V_2O_5 25 P_2O_5	-274	.0009
Crystalline $FePO_4$	-60	--
55 FeO 45 P_2O_5	-105	.0288
60 FeO 40 P_2O_5	-109	.0298
65 FeO 35 P_2O_5	-126	.0327
70 FeO 30 P_2O_5	-137	.0333
75 FeO 25 P_2O_5	-184	.0352
80 FeO 20 P_2O_5	-289	.0357
Crystalline $\alpha\text{-}Fe_2O_3$	-2000	--

In general, the exchange integral increases as the concentration of the transition metal ion is increased in the glass. As a result, the magnitude of the antiferromagnetic Curie temperature increases. Likewise, the Curie constant increases as the concentration increases. The decrease in C and θ values of the 55-45 manganese phosphate glass may be interpreted by comparison with the C and θ values of manganese pyrophosphate ($Mn_2P_2O_7$ = 67-33 $MnO-P_2O_5$). It is apparent that the magnetic properties of this glass are approaching that of the crystal, whose properties have been reported by Fowlis and Stager (11). Likewise, the magnetic behavior of the iron phosphate glass system approaches the behavior of $FePO_4$ and Fe_2O_3 at the end points of the compositions studied (12).

The low temperature behavior of the 50-50 manganese phosphate glass is shown in Figures 3 and 4. As previously mentioned, the downward curvature of the inverse susceptibility plot at low temperature has been theoretically explained by Simpson. The Simpson model assumes that the random nature of the glass matrix allows a fraction of the transition metal ions to be "shielded" from exchange interaction with another metal ion. These shielded ions give rise to paramagnetic behavior, which tends to dominate the susceptibility at low temperatures. As shown in Figure 5, it is also possible to obtain the downward curvature of the inverse susceptibility plot using a model based on the low

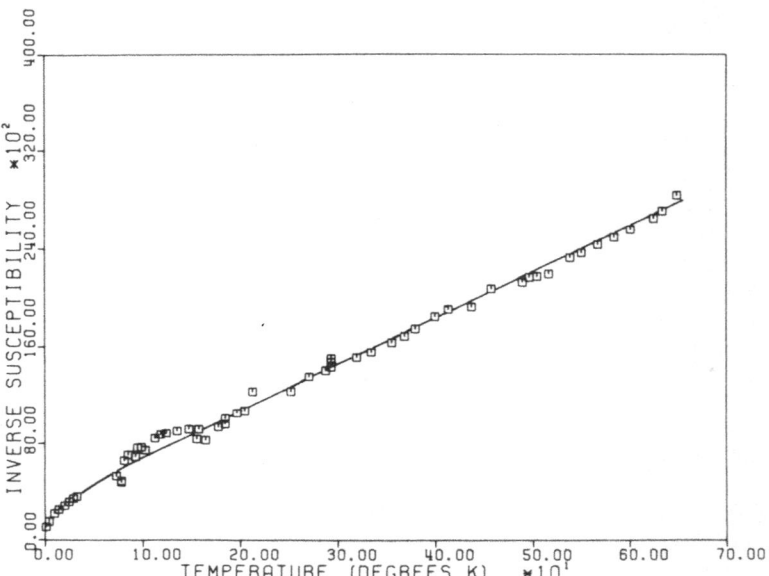

Figure 3.   The inverse molar susceptibility of a 50 mol % MnO 50 mol % P_2O_5 glass.

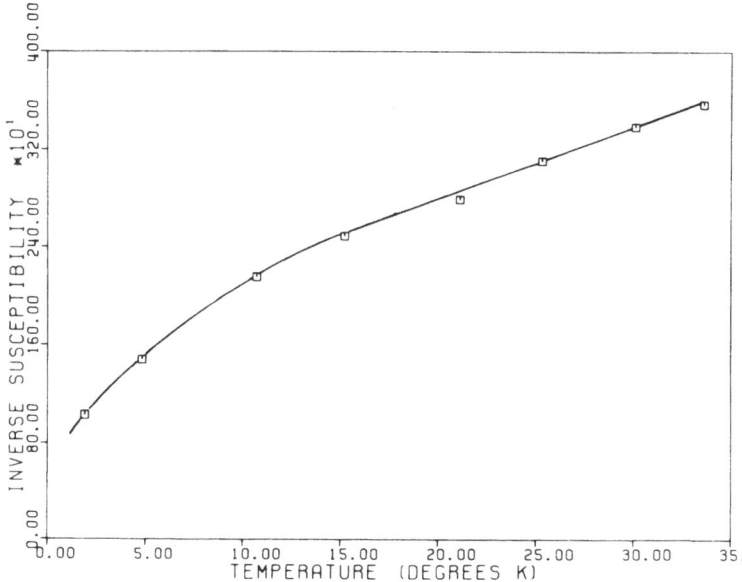

Figure 4.   The low temperature inverse susceptibility of a
50 MnO 50 P_2O_5 glass.

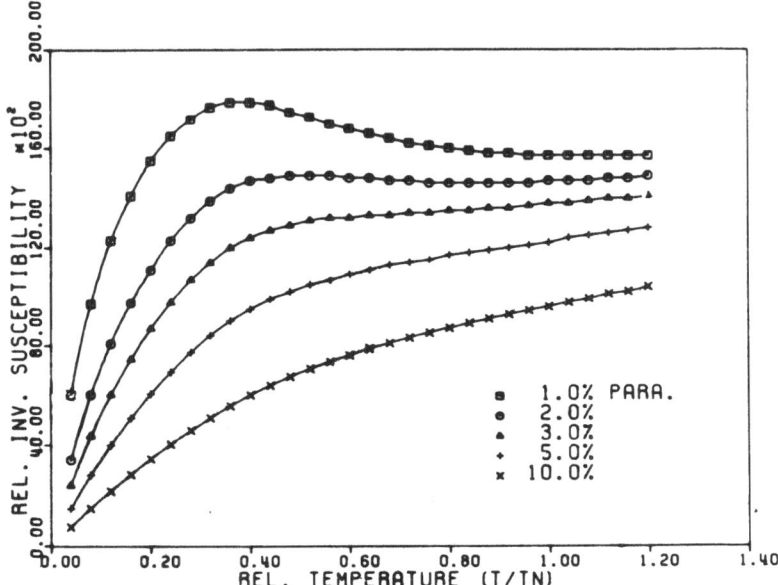

Figure 5.   The effect of paramagnetic ions on the powder
susceptibility for a polycrystalline antiferromagnet.

temperature powder susceptibility of a crystalline antiferromagnet with a small fraction of paramagnetic ions present.
For this case, the total magnetic susceptibility is given by

$$\chi_T = f\frac{C_p}{T} + (1-f)\ [1/3\ \chi_{||} + 2/3\ \chi_{\perp}]$$

where f is the fraction of paramagnetic ions present.

The existence of isolated ions in a glassy matrix has
been previously established by observation of the crystal
field electron spin resonance absorption of "isolated" Fe^{3+}
in a 55-45 iron phosphate glass (3). Furthermore, if structural inhomogeneities such as liquid-liquid phase separation
or devitrification occur in a glass, compositional segregation
will occur leaving one of the phases with a proportionally
lower transition metal ion concentration. The probability of
isolated ions in this phase is substantially greater, and the
low temperature susceptibility behavior of such materials is
affected. Liquid-liquid separation has been observed in the
vanadium phosphate (4, 13) and manganese phosphate systems
(5) and the downward curvature of the manganese phosphate
data is, in part, related to this separation.

The low temperature inverse susceptibility behavior of
the iron phosphate glass is shown in Figure 6, and is seen to
behave differently from the manganese glass. A possible explanation of this behavior may be found in the models of Kobe
and Handrich (8) and Hasegawa (9). However, these models are
based upon a totally exchange coupled amorphous system. Since
isolated ions have been detected in this glass, another explanation for this anomalous low temperature effect is the
existence of extensive antiferromagnetic coupling between Fe^{3+}
and Fe^{2+} ions. It is also possible that microcrystals smaller
than the minimum detectable size of the x-ray technique may be
present in the glass. Further investigation of this anomalous
behavior is in progress.

In conclusion, it has been shown that antiferromagnetic
coupling exists between transition metal ions in the vanadium,
manganese and iron phosphate glass systems. The magnetic
properties of these glasses are functions of the composition
of the glass and approach the properties of the crystalline compounds at the limits of glass formation. The

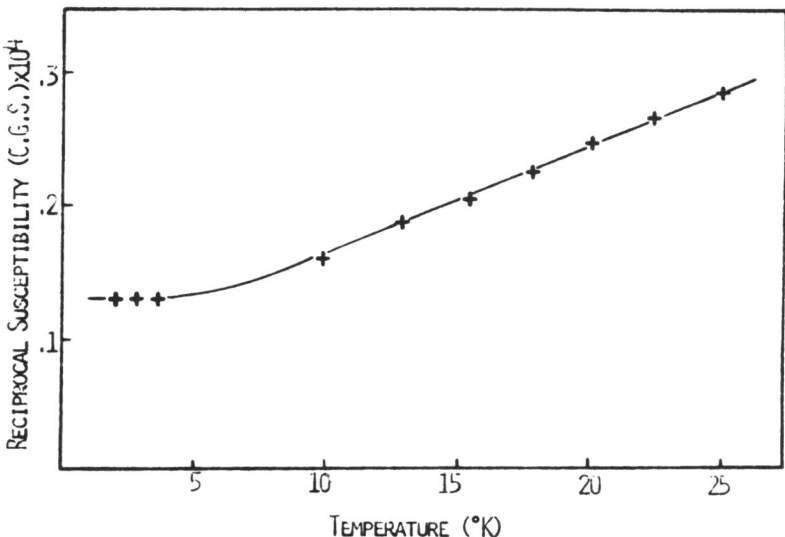

Figure 6. Low temperature inverse susceptibility of a 55 FeO 45 P_2O_5 glass.

high temperature inverse susceptibility of these glasses obeys a Curie-Weiss law. The low temperature inverse susceptibility of the manganese phosphate glass system exhibits a downward curvature which may be explained in terms of uncoupled metal ions behaving paramagnetically in the glassy matrix. At low temperatures, the susceptibility of the iron phosphate system becomes constant, perhaps as a result of a superposition of paramagnetic behavior of the isolated Fe^{3+} ions and antiferromagnetic behavior of the Fe^{3+} -Fe^{2+} ion pairs.

ACKNOWLEDGEMENTS

The authors would like to thank Mr. John Dayani and Mr. Charles Perry for their technical assistance and Dr. Israel S. Jacobs and Dr. N. F. Koon for their helpful discussions. The support of the Army Research Office – Durham is gratefully acknowledged.

REFERENCES

1. C.J. Schinkel and G.W. Rathenau, Physics of Non-Crystalline Solids, (North Holland Publishing Co., Amsterdam, 1965), pp. 215-219.
2. R.J. Landry, J.T. Fournier and C.G. Young, J. Chem. Phys. 46, 1285 (1967).
3. E.J. Friebele, L.K. Wilson, A.W. Dozier and D.L. Kinser, Phys. Stat. Sol. (b) 45, 323 (1971).
4. E.J. Friebele, L.K. Wilson and D.L. Kinser, J. Am. Cer. Soc. 55, 164 (1972).
5. E.J. Friebele, L.K. Wilson and D.L. Kinser, Am. Cer. Soc. Bull. 51, 352 (1972).
6. A.W. Simpson and J.M. Lucas, J. Appl. Phys. 42, 2181 (1971).
7. A.W. Simpson, Phys. Stat. Sol. 40, 207 (1970).
8. S. Kobe and K. Handrich, Phys. Stat. Sol. 42, K69 (1970).
9. R. Hasegawa, Phys. Stat. Sol. (b) 44, 613 (1971).
10. A.W. Dozier, L.K. Wilson, E.J. Friebele and D.L. Kinser, J. Am. Cer. Soc. 55, 373 (1972).
11. D.C. Fowlis and C.V. Stager, Can. J. of Phys. 47, 371 (1969).
12. J.H. Dayani, L.K. Wilson and D.L. Kinser, Am. Cer. Soc. Bull. 51, 365 (1972).
13. D.L. Kinser, E.J. Friebele, A.W. Dozier and L.K. Wilson, "Liquid-Liquid Immiscibility in $V_2O_5 - P_2O_5$ Glasses", to be published.

DISCUSSION

A.W. Simpson: I am delighted to see us in excellent agreement. There are ways of thinking that we disagree on Fe glass. But ours is trivalent whereas yours seems to be divalent. It is possible that this different behavior is due to the divalent iron.

L.K. Wilson: Our glasses were, in fact, multivalent glasses. By chemical analysis and later Mössbauer Effect Spectroscopy, we determined that we had both Fe^{2+} and Fe^{3+} ions present in the glass. Our glasses were made by melting Fe_2O_3 and P_2O_5 in air, but the glass compositions are based on an FeO equivalent as in Hansen's work [J. Electrochem. Soc. 112, 10, 1965]. The 55-45 glass, as cast, turned out to be 75% Fe^{3+} and 25% Fe^{2+}.

A.W. Simpson: I would like to ask you in the manganese glasses, was there evidence for recrystallization for compositions greater than 50% where your curve of the paramagnetic Néel temperature started bending down?

L.K. Wilson: We found the limit of glass formation to be 55 mole % MnO. All our glasses were analyzed by Guinier - de Wolff X-ray techniques and no evidence of crystallinity was found.

R. Hasegawa: Do you have a slide of crystalline compounds corresponding to the glasses you have shown in your last slide [Fig. 6].

L.K. Wilson: The crystalline phase which precipitates in our glasses most readily is $FePO_4$, although other crystalline phases have been detected. The Néel temperature of $FePO_4$ is 25°K and the Curie temperature is -60°K. Our data seems to approach these numbers [see Table 1]. However, I do not have a slide of magnetic data for this compound.

E.J. Friebele: I think it is important to point out it is no simple matter to determine which crystalline compound is most analogous to the 55-45 iron glass. On intentional devitrification one obtains a small amount of Fe_3O_4 as well as $FePO_4$. Further devitrification produces even more complex iron phosphate complexes, such as Fe_3PO_7 and $Fe_4(P_2O_7)_3$. So, it is difficult to make a direct comparison between any one particular iron compound and the 55-45 glass.

J. Wong: Ferric metaphosphate offers a good example in which a direct comparison of the magnetic properties can be made between the crystalline and homogenous glassy states. In fact, the Mössbauer spectra of both crystalline and glassy $Fe(PO_3)_3$ has been studied by Kurkjian and Buchanan, [Phys. Chem. Glasses, 5, 63 (1964)].

J.R. Marko: With respect to that last slide (Figure 6), Kobe and Handrich made the calculation that you are referring to. They wound up predicting a Néel temperature which was greater than the Curie temperature that you would get from a plot of the inverse susceptibility. What was the slope of that curve immediately above where you got that break? Did that curve go through the origin? Would it have gone through the origin? What is the relationship of your 10°K to the Curie temperature in that glass?

L.K. Wilson: The Curie temperature for the 55-45 iron phosphate glass based on the high temperature data is -105° [See Table 1]. As can be seen from Fig. 2, the inverse susceptibility in the low temperature region is bending toward the origin, as is the case for the $MnO-P_2O_5$ glass. The particular anomaly in the iron glass at 10°K might be related to $FePO_4$ which has a Néel temperature at 25°K. If we use Simpson's theoretical results we could expect an "amorphous" Néel temperature associated with $FePO_4$ at the Néel temperature divided by the square root of the coordination number of the iron ions. If we use a coordination number of 8, then the expected Néel temperature is around 9°K. Of course there may be microcrystals in the glass. [As has been stated in the paper, this anomaly may not be related to $FePO_4$ at all, but rather to a more complex relation between aliovalent iron ions.]

J.R. Marko: It was one of the features, I thought, of the Kobe and Handrich calculation that you would get this break at a higher temperature than you would expect from extrapolations the high temperature reciprocal susceptibility.

MAGNETIC INHOMOGENEITIES IN BaO B_2O_3 - Fe_2O_3 OXIDE GLASSES

R.K. MacCrone

Division of Materials Engineering, Rensselaer

Polytechnic Institute, Troy, New York

INTRODUCTION

Metastable oxide glasses, rich in iron oxide, are known[1-7] to develop interesting magnetic properties as crystalline ferritic phases precipitate on suitable heat treatment. While some of the investigations were primarily concerned with identifying the crystalline phases formed and others studied mainly the magnetic properties, both groups implicitly assumed that the precipitates were homogeneous. In this paper we describe an investigation in which the magnetic properties and the microstructure of an inhomogeneous glass of this type produced by heat treatment were simultaneously measured. This investigation showed that the precipitates were not only bimodally distributed in size but in addition were not homogeneous in structure. Some interesting magnetic behavior due specifically to the inhomogeneity of the precipitate is described.

MAGNETIC PROPERTIES

(a) Specimens Heat Treated at 590°C

The magnetization of specimens progressively heat treated at 590°C for the times indicated are shown in Fig. 1. There are several features which are of interest: Firstly, the H/T superposition of measurements made at 300 and 77°K. This is to be interpreted as evidence for the super-

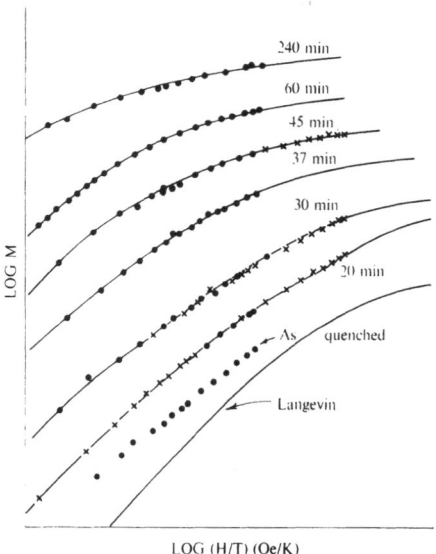

Figure 1. Magnetization as a function of H/T for specimens heat treated at 590°C for various times indicated; • 293°K, x 77°K.

paramagnetic behavior of single domain particles with a blocking temperature below liquid nitrogen temperatures. Secondly, the magnetization at a given magnetic field increases with increasing time of heat treatment. This indicates that the magnetic moments and/or number of particles are increasing with increasing time of heat treatment. Thirdly, the good fit to the experimental curves in all cases by the superposition of two Langevin functions, i.e.,

$$M = N_1\mu_1 \, L(\mu_1 H/kT) + N_2\mu_2 \, L(\mu_2 H/kT).$$

(van der Giessen[10] has shown how a log-log plot, as Fig. 1, is particularly useful in analyzing such data.) This is to be interpreted that the magnetic properties are due to two groups of precipitates with different magnetic moment sizes. The magnitude of each magnetic moment and their number may be determined from the magnetic measurements. These values are given in Table I.

The morphology as revealed by replica electron microscopy is shown in Fig. 2. The micrograph shows that two

TABLE I

Time of Heat Treatment Mins.	20	30	45
$N_1 \times 10^{-16}$	4.51	3.18	2.41
$\mu_1 \times 10^{16}$	0.094	0.205	0.648
$N_2 \times 10^{-17}$	9.62	1.9	3.37
$\mu_2 \times 10^{17}$	0.135	0.293	0.926
r_1 Å	16.8	21.6	31.9
r_2 Å	8.75	11.3	16.7

particles sizes are indeed present whose size and number closely reflect the two magnetic moment sizes and numbers as deduced from the magnetic measurements.

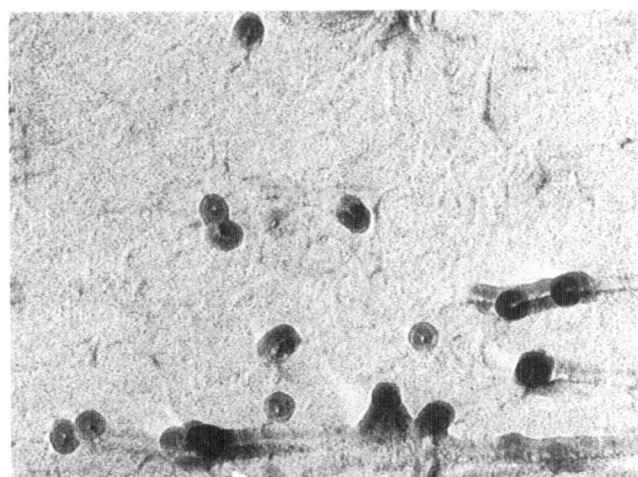

Figure 2. Replica Electron Micrograph of Specimen Heat Treated at 590°C.

A discrepancy however arises when the magnetic moment sizes deduced from the morphology $\mu = I_s V$ (I_s = saturation magnetization of the ferrite, V = volume of precipitate determined from the photomicrograph) are compared with those deduced from the actual magnetization measurements, μ_m. The former moments are too large. This discrepancy is most satisfactorily removed by assuming that the precipitates are themselves inhomogeneous, and that the two phase structures implied by the photomicrograph is indeed real. In this case, the inner volume of each precipitate may be estimated assuming this to be the ferrimagnetic phase. These values are also listed in Table I. An assumed skin of ferrite, the other possibility, leads to skin thicknesses unreasonably small. The work of Tanigawa and Tanako[6] implies that Fe_3O_3 and Fe_3O_4 are the only precipitated phases here, and indeed, only Fe_2O_3 was detected by wide angle x-ray diffraction in these specimens.

(b)   Specimens Heat Treated at 618°C

The magnetization of specimens heat treated at 618°C is shown in Fig. 3.   It is to be noticed that the magneti-

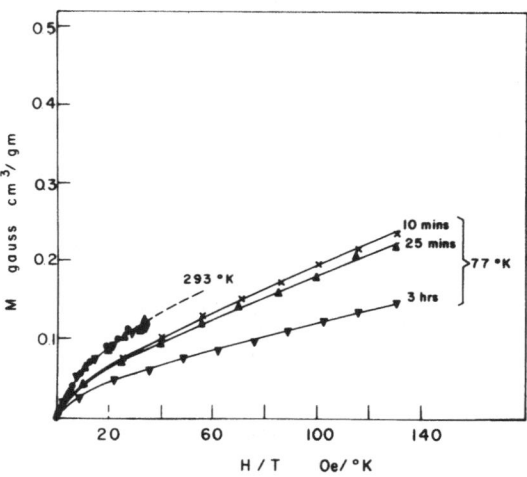

Figure 3. Magnetization as a function of H/T for specimens heat treated at 618°C for the times indicated.

zation is much smaller in this case and that no superparamagnetic behavior is observed. This decrease in magnetization is due to the oxidation of any Fe_3O_4 to Fe_2O_3 which
occurs rapidly at 618°C, less rapidly at 590°C (A specimen
heat treated at 590°C, for four hours for example, has a
magnetic moment far below that of a specimen heat treated
for one hour, equal about to that of a quenched specimen.)
This increased rate of oxidation is due to the increased
rate of oxygen diffusion and oxygen activity with temperature.

The magnetic behavior, however, is not that expected
from simple Fe_2O_3 precipitate, as may be seen in Fig. 4.
At 77°C, a specimen cooled from room temperature in zero

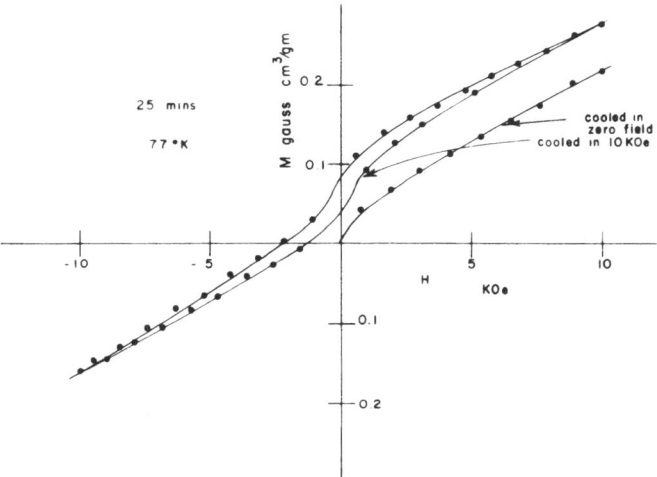

Figure 4. Displaced magnetization curve at 77°K of specimen (heat treated for 25 mins.) cooled in positive field
from room temperature to 77°K.

magnetic field displays a symmetrical magnetization curve.
In contrast, the same specimen cooled in a small positive
field displays a magnetization curve displaced also in the
positive direction. Experiments (about which space pre-
cludes discussion) show that this displaced magnetization
is not due to simply cooling below the blocking temperature
in a magnetic field which could give similar displacement,
but rather due to a mechanism infinitely connected with

the Morin transition. At the Morin transition temperature the antiferromagnetic direction in Fe_2O_3 changes from the a direction to the c direction, and it is at this temperature that the thermo-remanent magnetization, Fig. 4, finally disappears.

Displaced hysteresis curves, such as this in Fig. 4, are known to result from exchange anisotropy, an effect in which the magnetization of one phase (Fe_3O_4 say) is coupled to the magnetization of another phase (Fe_2O_3 say) by an exchange interaction across the boundary between the two phases. Appealing to the model for the inhomogeneous precipitate particles in Section (a), we consider the consequences of very small inclusions of Fe_3O_4 within the Fe_2O_3. As pointed out by Neel,[11] good epitaxy is obtained if the [111] axis of the Fe_3O_4 is parallel to the [001] axis of the Fe_2O_3. Below the Morin transition, the magnetic axis of Fe_2O_3 is along the c axis, and presumably, the Fe_3O_4 magnetic axis is also parallel to this direction, an energetically favorable situation, since the [111] in Fe_3O_4 is a direction of low magneto-crystalline energy. We assume that there is an exchange coupling between the two magnetizations, so that the Fe_3O_4 magnetization is fixed in direction once it is formed, giving rise to the displaced loop of Fig. 4 when cooled in a magnetic field. Above the Morin temperature, the Fe_2O_3 magnetic axis is perpendicular to the [111] easy direction of magnetization in Fe_3O_4, and should exchange coupling to the Fe_2O_3 force this direction, the magnetic energy of the Fe_3O_4 will be unfavorably large. To be consistent with our experimental observations, we must suppose that in this latter case ($M \perp [111]$) the unfavorable magnetic energy of the Fe_3O_4 is sufficiently large to preclude the usual magnetic ordering, and that the small inclusions of Fe_3O_4 are thus paramagnetic above the Morin temperature. Further experiments are in progress to test the validity of this model.

DISCUSSION

The bimodal size distribution that is observed in specimens heat treated at 590°C and 618°C is a result of two effective heat treatments, one unintended during the quench, which cannot be avoided without splat cooling quenching rates, and the other during the intended anneal.

During the quench, the specimen apparently passes through a temperature range where the nucleation rate is low but growth rate rapid and this results in the formation of the large particles. On the other hand, during the intended heat treatment, the nucleation rate is rapid but the growth rate slow and this heat treatment produces the small particles.

It is not clear from these experiments whether Fe_2O_3 or Fe_3O_4 nucleate first, followed by processes leading to the inhomogenous precipitate. However, a natural process would be Fe_3O_4 nucleation and growth, which rapidly depletes the initially low concentration of Fe^{+2}. As the concentration of Fe^{+3} in the matrix increases, the growth of the Fe_2O_3 skin is to be expected.

Finally, the concentration of Fe_2O_3 in the metastable glass used in these experiments is close to the miscibility limit as determined from conventional wide angle x-ray diffraction and SAXS. The extensive precipitation observed even during the quench shows unexpectedly rapid phase separation; indeed glass in glass phase separation may be a precurser to the processes observed here. Further study is in progress.

ACKNOWLEDGEMENTS

This problem area has been variously investigated with J. Aiken, M. Fahmy, D. Moon, M.J. Park and M. Tomazawa, whose efforts are gratefully acknowledged. The interest and financial support of the Office of Naval Research under Grant NR 032-519 is appreciated as well as the facilities provided at the NASA interdisciplinary Research Center under Contract No. NGL 33-018-033.

REFERENCES

1. M. Tashiro, S. Sakka and T. Kakkubo, J. Ceramic Assoc. Japan 72, 92 (1964).
2. S. Sakka, T. Kakkubo and T. Kariyama, Discussion on Artificial Minerals Oct. (1964).
3. M. O'Horo and R. Steinitz, Mat. Res. Bull. 3, 117 (1968).
4. R.R. Shaw and J.H. Heasley, J. Amer. Ceramic Soc. 50, 297 (1967).

5. B.T. Shirk and W.R. Buessem, J. Amer. Ceramic Soc. 53, 192 (1970).

6. M. Tanigawa and H. Tanako, Osaka Kogyo Gigutsu Shikenjo 15, 285 (1964).

7. P. Shultz in Int. Commission on Glass-Amer., Meeting Toronto, Sept. 3-6 (1969).

8. M. Fahmy, M. Tomozawa and R.K. MacCrone, Amer. Ceramic Soc. Symposium on "Nucleation and Crystallization Revisited", Chicago (1971), in press.

9. M. Fahmy, M.J. Park, M. Tomozawa, and R.K. MacCrone, J. Phys. Chem. Glass 13, No. 2, 21 (1972).

10. A.A. van der Giessen, J. Phys. Chem. Solids 28, 343 (1953).

11. L. Neel, Ann. Physique 4, 249 (1949).

12. L. Neel, Rev. Mod. Phys. 25, [1], 58 (1953).

DISCUSSION

S.C. Moss: The glassy phase separation that seems to be the background in that pattern where there were larger clusters, presumably that is a phase separated region, or do I have it wrong? I am referring to the electron micrograph.

R.K. MacCrone: What I think you have is a glassy matrix of barium boron glass containing, say, about 5% iron, 10% iron, and in it small particles of Fe_2O_3 containing smaller particles of Fe_3O_4, and that they are two sized. There is a bimodel distribution.

S.C. Moss: One of them was looked quite uniform as if it had been arrived at by some sort of phase separation.

R.K. MacCrone: The slide is not as good as one would like but the original photograph shows that these are actually small spheres.

S.C. Moss: And crystalline?

R.K. MacCrone: I believe they are crystalline - Yes. We can see evidence for crystalline Fe_2O_3, we cannot see evidence for crystalline Fe_3O_4.

A.W. Simpson: Did you consider the possibility that you were forming Barium Ferrite, that is $BaO \cdot 6Fe_2O_3$?

R.K. MacCrone: Yes.

A.W. Simpson: This could have accounted for your double correspondance density. Is there any evidence, for this?

R.K. MacCrone: Well, the evidence (Ref. 6) is that as you decrease the iron concentration in these glasses you go through the sequence of Barium Ferrite, Fe_3O_4 and then Fe_2O_3 preferentially precipitating. Since we are at the range of very dilute iron, we believe that if there is any barium ferrite the concentration of it is very low.

S.C. Moss: The analogy with the alloy results is facinating. The idea that somehow there really are magnetic particles because of the magnetic ordering in, say CuMn, and that this is responsible in somewhat the same way perhaps for this hysteresis and displacement, etc. Would that be a reasonable conclusion?

R.K. MacCrone: Yes. Yes it is because the origin of the displacement in these systems is an exchange anisotropy between Fe_3O_4 and Fe_2O_3 compared to Co and CoO in some of these systems.

S.C. Moss: Also possibly wall effects - boundary effects?

R.K. MacCrone: Less likely I think.

MAGNETIC PROPERTIES OF AN IRON-RICH GLASS

G. R. Mather, Jr.

Corporate Research Laboratories
Owens-Illinois Technical Center
Toledo, Ohio 43601

ABSTRACT

Measurements of the magnetic moment of a glass containing 40 wt% Fe_2O_3 indicate that the superparamagnetic behavior
observed above 100°K can be explained by the presence of
antiferromagnetically ordered microdomains with a Néel temperature of about 700°K. Below 100°K the behavior is not
simply superparamagnetic, and a number of the thermal history and other effects observed are similar to those that
characterize mictomagnetic spin glasses.

INTRODUCTION

Recently we reported the results of Mössbauer effect
and electron spin resonance experiments which indicate that
a glass having the composition 40 wt% Fe_2O_3 : 12.5 wt% Li_2O :
47.5 wt% SiO_2 is superparamagnetic.[1] This was believed to
result from the presence of antiferromagnetically ordered
microdomains which arise because of the large amount of
Fe_2O_3 in the system and which possess small permanent moments because of imperfect compensation among the magnetic
ions. Inasmuch as x-ray and electron diffraction analysis
of the glass has revealed no evidence of a crystalline
magnetic phase, these microdomains are believed to be structurally disordered. Using a vibrating sample magnetometer,
we have now measured the magnetic moment of the glass at
temperatures ranging from 5°K to 900°K and in fields up to

87

10 kG. While the results over most of this temperature
range support our earlier interpretation, the magnetic behavior below 100°K is not simply superparamagnetic: thermal
quenching, thermal history, and other effects are observed
which are similar to those exhibited by magnetic alloy spin
glasses that Beck has recently termed mictomagnetic.[2]

RESULTS

Figure 1 shows typical magnetic moment-field isotherms
from the glass at several temperatures above 100°K. The
shapes of these curves are consistent with a distribution of
Langevin functions associated with domain moments that range
from about 1500 μ_B to 50 μ_B. There is reason to believe
that the moments arise largely from iron in the Fe^{3+} state,[1]
and if it is assumed that a domain containing n Fe^{3+} ions
has a moment $\mu = g\mu_B S\sqrt{n}$, with g=2 and S=5/2, n is found to
have values between about 10^5 and 10^2. If in addition it is
assumed that the domains have the same composition and density
as $\alpha-Fe_2O_3$, their dimensions range from approximately 150 Å
to 20 Å. These dimensions are comparable to those of the
phase separated regions detected by small angle x-ray scattering,[1] a fact which may mean that the magnetic microdomains
can be identified with the phase separated regions. The estimated domain sizes are reduced somewhat if, as the data
suggest, the number of uncompensated ions per domain is
greater than $\sqrt{n}$, but not so much as to rule out such an identification. When the isotherms in Fig. 1 are plotted against

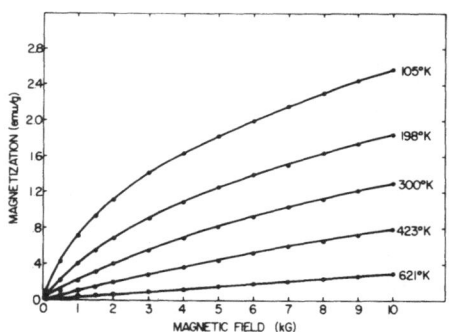
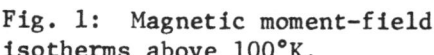
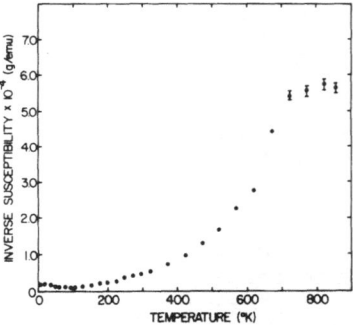

Fig. 1: Magnetic moment-field
isotherms above 100°K.

Fig. 2: Temperature
variation of the inverse
initial susceptibility.

H/T, they do not superimpose as for a simple superparamagnet but rather lie progressively lower with increasing temperature. This behavior could be due at least in part to domain moments which decrease as the temperature increases.

A temperature-dependent domain moment is also suggested by Fig. 2, in which the inverse initial susceptibility of the glass is plotted against temperature. Below 700°K there is no region over which the susceptibility clearly exhibits Curie-Weiss behavior, and in fact at 75°K the susceptibility passes through a maximum. The sharp break in the curve at 700 ± 25°K is believed to be indicative of the antiferromagnetic ordering of the magnetic ions in the domains. Above 700°K a ferrite phase begins to crystallize in the glass in amounts that increase slowly with time, and because of this the data at higher temperatures have the uncertainties indicated by the error bars. The effect of the crystallized phase on the moment of the sample at 10 kG is shown in Fig. 3. The ordering temperature of the ferrite is between 825°K and 850°K, and x-ray analysis reveals that it has the spinel structure. Work on materials with compositions similar to that of the present glass suggests that the phase is in fact lithium ferrite.

Figure 4 shows isotherms obtained after cooling the un-crystallized sample in a field of 10 kG, and it is apparent that the behavior below 100°K is not simply superparamagnetic. For fields less than 3 kG the isotherms are essentially independent of temperature between 105°K and 40°K, while for

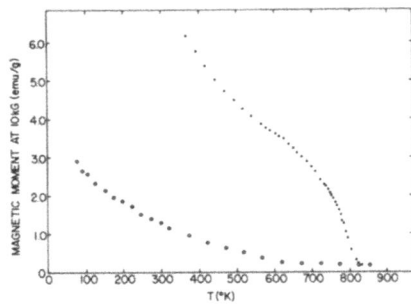

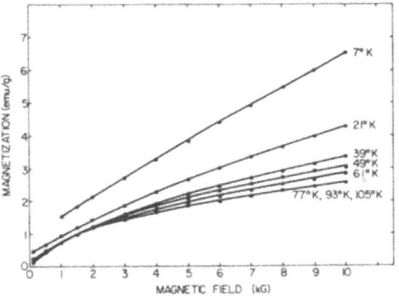

Fig. 3: Magnetic moment at 10 kG before (lower curve) and after (upper curve) crystallization.

Fig. 4: Low temperature magnetic moment-field isotherms.

fields greater than 3 kG the observed temperature dependence
is more complicated than would be expected for a simple
Curie-Weiss contribution to the magnetic moment. Such observations cannot be understood in terms of simple blocking
effects because the thermoremanence is much too small. Moreover, the isotherms near 100°K satisfy the Arrott criterion[3]
for a paramagnetic system, with no evidence of a tendency
toward magnetic ordering below 100°K. To the extent that
the Arrott criterion is a valid test in the glass, therefore,
there is no indication that the results can be explained by
a state with long-range magnetic order below 100°K.

Below 75°K hysteresis and thermal history effects are
observed, and the isotherms in Fig. 4 indicate the appearance of a thermoremanent magnetization which increases
rapidly with decreasing temperature, reaching a value of
about 1 emu/gm at 5°K. Such a remanence represents a polarization of less than 1% of the iron in the glass, and
accounts for only a small fraction of the total high field
moment, which Fig. 4 shows to be far from saturation at 10
kG. Fields of 5 kG are sufficient to saturate the remanence
at a given temperature when the glass is field cooled from
temperatures above 100°K, but fields in excess of 10 kG are
required to cause saturation at constant temperature. The
hysteresis loops obtained after the remanence has been
saturated by field cooling the sample are centered about the
zero field axis to within ±100 G, the approximate residual
field of the magnet, and are nearly antisymmetric about this
axis. A slowly increasing reverse field causes substantial
portions of the remanence to reverse in sudden steps at two
or more fields whose values are typically a few hundred
gauss. These fields are generally not the same for the positive and negative branches of the loop, and are not reproducible from experiment to experiment. Much larger fields are
required to reverse the remainder of the remanence, and in
fact reverse fields of nearly 10 kG are needed to establish
a remanence that is equal but opposite to that of the other
branch.

Other thermal history effects, similar to those exhibited by mictomagnetic alloys,[2] are also observed. When the
glass is cooled to 5°K in zero field and then warmed in a
field less than 3 kG, the magnetic moment is found to pass
through a maximum at about 75°K. When the sample is subsequently recooled in the same field, the moment reproduces
the values obtained during warming down to 75°K, below which

it remains nearly constant and therefore larger than the original values. In addition, the isotherm obtained after cooling the glass in zero field lies outside and generally below the hysteresis loops which result after the glass has been cooled in nonzero fields.

DISCUSSION

Although a consistent qualitative interpretation of most of the remanence, hysteresis, and other thermal history effects could be given in terms of the blocking of super-paramagnetic domains, the relative temperature independence, or thermal quenching, of the isotherms between 100°K and 40°K cannot be explained by the usual ideas of superpara-magnetism. Field-induced blocking of the domains and com-plicated multidomain structures in regions either with or without long-range magnetic order also appear to be incon-sistent with the data. Because the behavior of the glass between 100°K and 40°K cannot be understood in terms of simple ideas, it is quite possible that the effects observed below 75°K are not due to the simple blocking of superpara-magnetic domains. It should be noted, however, that the thermal quenching of the moment near 100°K, the maximum in the initial susceptibility at 75°K, and the thermal history effects at lower temperatures are all phenomena that are common in mictomagnetic materials.[2] Moreover, the present glass is similar to mictomagnets inasmuch as it appears to be a spin glass, or system in which there is no structural order among the magnetic ions. Although they do not exhibit long-range magnetic order at low temperatures, mictomagnets frequently have initial susceptibilities that obey a Curie-Weiss relationship at higher temperatures and it is possible that the initial superparamagnetic susceptibility of the glass can be explained by domains whose permanent moments are temperature dependent and whose superparamagnetic be-havior is characterized by a Curie-Weiss law. Assuming that the moment of a domain containing n Fe^{3+} ions is $\mu = g\mu_B S\sqrt{n} \; b(T)$, where $0 \leq b(T) \leq 1$ expresses the tempera-ture variation of the moment, it is easy to show that the initial superparamagnetic susceptibility is given by $\chi = Ng^2\mu_B^2 S^2 b^2(T)/3k(T-\theta)$. Here, N is the number of Fe^{3+} ions per gram and θ the paramagnetic Curie temperature; the other quantities have their usual meanings. Because of the assumption that the number of uncompensated ions per domain is $\sqrt{n}$, this expression is correct for an arbitrary distribu-

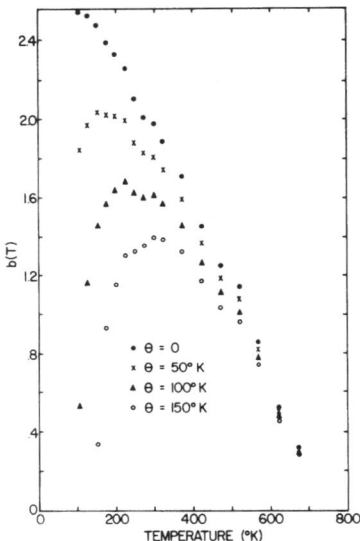

Fig. 5: Inferred temperature dependence b(T) of the micro-
domain moment.

tion of domain moments. Figure 5 shows curves for b(T) obtained by assuming different values of θ and making use of
the initial susceptibility data. The plot for θ=0 has the
qualitative shape expected of a permanent moment, but in
contradiction of the assumptions leading to the susceptibility expression it is clear that b(T) exceeds unity below
about 550°K. A possible explanation is the presence of
domains in which the number of uncompensated magnetic ions
is greater than $\sqrt{n}$. Indeed, initial attempts to fit the
isotherms with a distribution of Langevin functions corresponding to different domain sizes suggest that the number of
uncompensated ions is closer to $n^{2/3}$, for which b(T) is less
than one at all temperatures. When nonzero values of θ are
assumed, the other curves in Fig. 5 are obtained. All of
these have low temperature maxima which are probably arti-
facts associated with deviations in the superparamagnetic
susceptibility from Curie-Weiss behavior in the neighborhood
of the susceptibility maximum. It is clear that this sort
of analysis is inconclusive regarding the origins of the
susceptibility temperature dependence, a complete understanding of which will require further work, including a
detailed analysis of the isotherms in terms of various domain

moment distributions. Possibilities not yet fully explored include Klein's mean random field (MRF) concept[4] and the domain size growth model of Kneller et al.[5] Although the predictions of the MRF theory in its present form are not in agreement with the susceptibility data, it is conceivable that a model based on MRF ideas could account for the results. The domain growth model is based on magnetic ion concentration gradients which lead to a distribution of ordering temperatures among the ions in the boundary regions of the domains, and therefore to an effective growth in domain size with decreasing temperature. To date, the model has been applied only to the case of ferromagnetic alignment among the ions, whereas the case of antiferromagnetic alignment is more complicated because the contributions to the moment by ions above their Néel temperatures are not negligible.

ACKNOWLEDGEMENT

I would like to thank Mr. E. F. Hinebaugh for his valuable assistance in preparing the samples and collecting the data.

REFERENCES

1. G. R. Mather, Jr., H. O. Hooper, P. F. Kenealy, and R. R. Bukrey, AIP Conf. Proc. $\underline{5}$, 821 (1971).
2. S. Chakravorty, P. Panigrahy, and P. A. Beck, J. Appl. Phys. $\underline{42}$, 1698 (1971); P. A. Beck, Met. Trans. $\underline{2}$, 2015 (1971).
3. A. Arrott, Phys. Rev. $\underline{108}$, 1394 (1957).
4. M. W. Klein, Phys. Rev. $\underline{188}$, 933 (1969); L. Shen and M. W. Klein, AIP Conf. Proc. $\underline{5}$, 1155 (1971).
5. E. Kneller, M. Wolff, and E. Egger, J. Appl. Phys. $\underline{37}$, 1838 (1966).

DISCUSSION

I.S. Jacobs: You have a lot of interesting data there.
You may have a number of magnetic components contributing,
perhaps, and I think you are already open to that kind of
suggestion. It struck me that the data in the lowest temperature range - that is to say, at about 100°K and below -
shows a susceptibility that is nearly independent of temperature and that indeed would be characteristic of a multidomain ferro- or ferrimagnetic component. You also appear
to be seeing a contribution which is proportional to 1/T
in a crude kind of sense: that is, a contribution that increases with decreasing temperature and results in larger
and larger magnetizations at the higher fields. You may in
addition be seeing some single domain behavior which further enhances the magnetization and causes a remanence at
the lowest temperatures. These are contributions to think
of along with the other things you have mentioned.

G.R. Mather: These are contributions that I have considered, and in view of the complicated low-temperature behavior
they certainly have to be regarded as possibilities. The
temperature dependent contributions that we see at the lowest temperatures could arise from the Curie-law susceptibilities of ions whose number is changing for some reason,
while multidomain structures would, of course, be expected
to result in contributions that are relatively temperature
independent but nonlinear in field. However, when we
analyze the temperature variation of the data with multidomain structures in mind, we find no clear, consistent
evidence for the presence of such structures, whereas the
general low-temperature behavior does seem to be consistent
with a number of the mictomagnetic effects that Professor
Beck discussed earlier. I might also note that we observe
no Mössbauer hyperfine splittings above 40°K, so that if
multidomain structures are present above this temperature
they account for only a tiny fraction of the total iron.

ELECTRON SPIN RESONANCE STUDIES OF FERRIMAGNETIC PHASES PRECIPITATED IN SIMULATED LUNAR GLASSES HEAT TREATED IN THE PRESENCE OF OXYGEN[*]

D. L. Griscom and C. L. Marquardt

Naval Research Laboratory

Washington, D. C. 20390

Since the return of the first lunar samples by the Apollo 11 astronauts, lunar material has been investigated by virtually every well-known chemical, physical, and mineralogical technique.[1] It is known, for instance, that lunar soils comprise up to 50% glass and that these lunar glasses have compositions similar to those of the "moon rocks", i. e., 40-50% SiO_2, 30-40% $(CaO+MgO+FeO)$, 10-20% Al_2O_3, 0-8% TiO_2 and < 1% other individual constituents by weight. It is also known that lunar material is highly reduced with respect to terrestrial rocks; hydrated minerals are absent, or present in trace amounts only. There is ample evidence[1] that ~ 99% of the iron dissolved in lunar glasses is present as Fe^{2+}, while up to ~ 0.5 wt % metallic iron may be included as microscopic or submicroscopic suspensions. It is hardly suprising that, when lunar soils were found to exhibit an intense "characteristic" ferromagnetic resonance signal[1], it was ascribed to small, spherical particles of metallic iron.[2] Indeed, it was shown[2] by computer simulation techniques that the "characteristic" resonance could be characterized by a positive anisotropy constant, $2K_1/M_s$ ~ 500 Oe, as for iron metal. By contrast, most familiar iron ferrites have negative anisotropy constants. But, just when the case appeared to be closed, it was demonstrated[3]

[*]Work supported under NASA Order No. T-4735A.

that a resonance resembling the "characteristic" resonance could be obtained in simulated lunar glasses by
annealing at $\sim 650^{o}C$ (somewhere near the glass transition
temperature T_g) in the presence of a partial pressure of
oxygen $\gtrsim 0.1$ Torr. One purpose of the present paper is to
present addition detailed evidence which bears on the conclusion[3] that the "characteristic" resonance, paradoxically
enough, may be an oxidation effect. Hopefully, though, the
use of electron spin resonance (ESR) techniques to monitor
the growth and metamorphism of ferrimagnetic precipitates
in silicate glass matrices will be of wider interest to workers in "amorphous magnetism".

The simulated lunar glass used in this investigation is
described in detail elsewhere[3], along with the experimental
techniques employed. It was a typical lunar composition,
including 11.6 wt % FeO and 2.4 wt % TiO_2, except that all
other transition elements normally present in trace amounts
were deliberately excluded. An ESR signal at g $\approx$ 4.3 in
the glass as quenched from the melt betrayed[4] the presence
of some isolated Fe^{3+}. Comparison of the ESR intensity
with that of a calibrated control sample[3] placed the ferric
ion concentration at ~ 0.1 wt %. This was corroborated by
Mössbauer measurements.[5] Since most _actual_ lunar glasses
do not show the g $\approx$ 4.3 signal but do exhibit several other
resonances, it was decided to subject samples of the simulated glass to a variety of irradiations and heat treatments
(such as might have occurred on the moon) in the hopes of
producing lunar-like resonances. Remelting in an oxidizing
hydrogen-oxygen flame followed by rapid quenching altered
the existing ESR signal only slightly, as did heating a dry-
box-ground powder sample in a pre-dried, evacuated,
sealed quartz tube ("in vacuo") at $650^{o}C$ for >100 hours.
Remelting in a reducing flame gave rise to an intense ESR
signal quite unlike the "characteristic" resonance; this
"Type-II" resonance has been ascribed to small particles
of metallic iron.[3] However, heating the powdered sample
to $650^{o}C$ in air at 1 atm. or a pressure as low as ~ 0.5 Torr
resulted in the growth of an enormous resonance at g $\approx$ 2.
The latter bore a strong resemblance to the lunar "characteristic" resonance in such important aspects as linewidth,

line shape, intensity, g value, and the frequency and temperature dependences of these quantities. Similar heat treatments carried out on an otherwise-identical control glass wherein the FeO was suppressed in favor of CaO+MgO resulted in no such new resonance, nor did ^{60}Co γ-irradiations up to 4×10^7R performed on the simulated glass. Since the only potentially magnetic elements in the simulated glass were iron (mainly as Fe^{2+}) and titanium (the latter of which was also present in the control sample) it was concluded[3] that the new $g \approx 2$ resonance results from the conversion of Fe^{2+} to Fe^{3+} in the solid state under the influence of an oxidizing atmosphere. Indeed, Mössbauer studies[5] of a sample heated to 650°C for 67 hours in air at ~ 0.5 Torr demonstrated the appearance of a quadrupole doublet at room temperature attributable to $\sim 20\%$ of the total iron in (octahedral) paramagnetic Fe^{3+} states. However, the X-band ESR

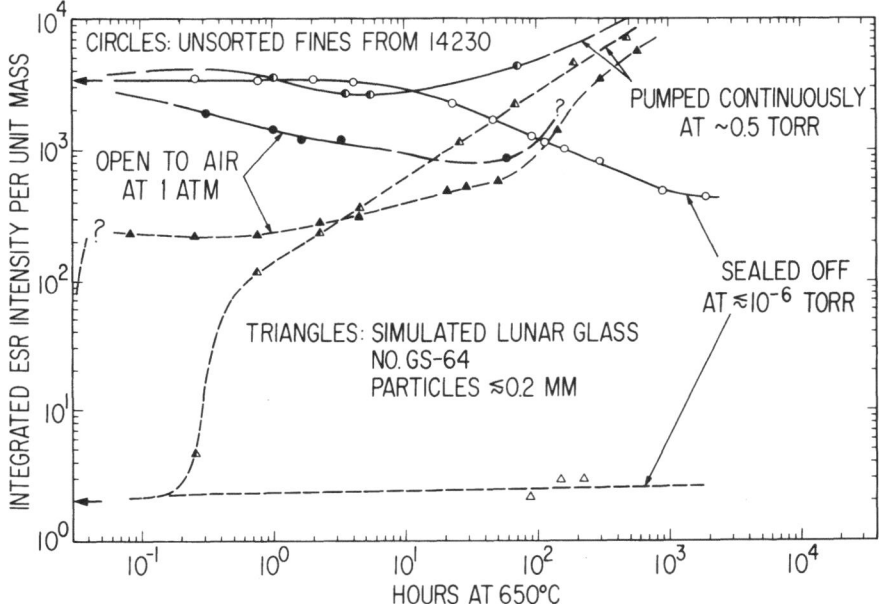

Fig. 1. Effects of isothermal anneals upon ESR intensities of simulated and actual lunar fines. Units of the ordinate are arbitrary; a value of $\sim 2 \times 10^2$ would be equivalent to 100% of the iron ions in <u>para</u>magnetic states; still larger values imply ferromagnetic resonance.

signal of this sample was ~100 times more intense than
expected for paramagnetic resonance. It is to be inferred,
then, that most of the resonance intensity here arises from
phases sufficiently condensed (or with sufficiently long range
exchange coupling) as to have net ferromagnetic moments.[2,3]
The precipitation in silicate matrices of ferrite phases (often
in superparamagnetic particle sizes) is indeed well-known[6].
The novel aspect here is the crucial influence of atmosphere
during the sub-solidus anneals.

In order to better elucidate the precipitated ferrite
phases and to judge their potential relationships with the
"characteristic" resonance, a number of simulated and actual lunar materials were subjected to carefully controlled
isothermal anneals at $650^{\circ}C$. An ESR spectrum was obtained
after each treatment. The second integrals of the experimental first-derivative spectra were obtained numerically and
these ESR absorption intensities were normalized to a "unit
spectrometer gain" and "unit sample mass" to obtain "specific
intensities" which can be compared directly. The results are
illustrated in Fig. 1. It can be noted that the very weak
resonance in the as-quenched simulated glass does not grow
when the sample is heated in vacuo, but when a continuous
supply of air is available at only ~0.5 Torr the ESR intensity
grows steadily as the 2/3 power of time after the first hour.
This is quite suggestive that resonance is occurring in the
surface layers of precipitated particles whose volumes are
increasing linearly with time. Such would be apropos of
Fe_2O_3 which is antiferromagnetic in bulk but is expected to
have a sizeable ferromagnetic moment in sufficiently small
particle sizes due to uncompensated surface spins.[7] An air
pressure of 1 atm. appears initially to accelerate and then
to inhibit the growth of ESR intensity.

Figure 2 (top row of spectra) displays the evolution of
X-band line shapes for continued heating at the low air
pressure. This figure also illustrates the further growth
and metamorphism of the line shapes when the air supply is
cut off and heating is continued in vacuo. The astonishing
feature is the 1-to-2-order-of-magnitude increase in intensity during the latter treatment. The ultimate intensities

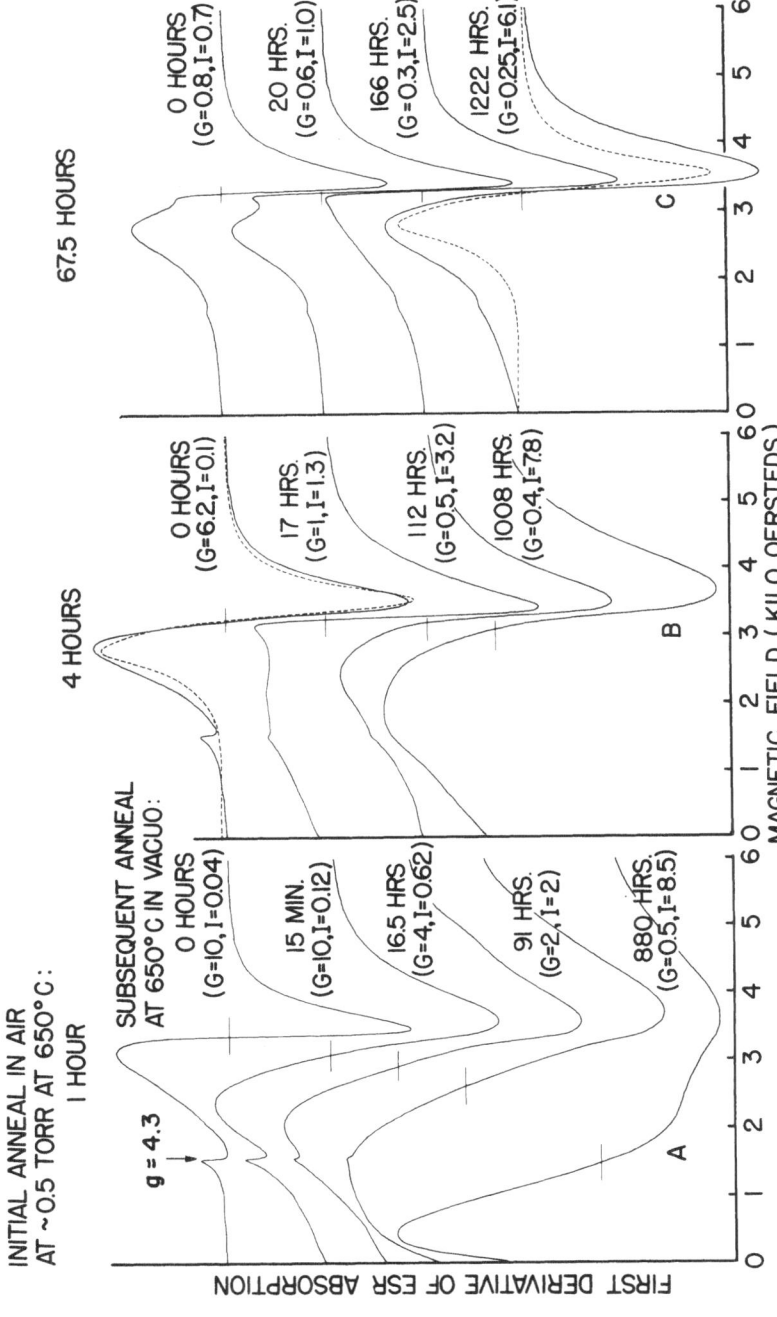

Fig. 2. ESR spectra obtained at 9 GHz and 300°K for powders of simulated lunar glass subjected to a sequence of heat treatments. Relative spectrometer gains and integrated ESR intensities are indicated by G and I, respectively. Dashed curves = Apollo 12 fines (G=0.3, I=1).

$(I \sim 7)$ correspond to ~ 0.7 wt % iron ions coupled ferro-magnetically. Since the actual magnetic phases are probably ferrimagnetic with some spins antiparallel, it is inferred that well over 10% of the iron ions initially present have become involved in magnetically ordered phases. To get an idea of what these phases are, additional experiments were performed on the sample exhibiting resonance C. Static susceptibility measurements indicated both paramagnetic and ferromagnetic components; the latter was characterized by a room-temperature saturation magnetization of ~ 0.8 e.m.u./g. An x-ray powder pattern displayed faint but unmistakable lines corresponding to magnetite or some other spinel-like phase. The ESR, susceptibility, Mössbauer, and x-ray results are all in accord with the presence of 1-3 wt % pure magnetite or a proportionately larger amount of spinel with diamagnetic cation substitutions.

A careful examination of the ESR line shapes leads to other interesting conclusions. First, all spectra illustrated in Fig. 2, with the possible (but not certain)[8] exception of "A", are consistent with positive magnetocrystalline aniso-tropy constants $2K_1/M_s$. (The sense of the spectral asymmetry determines the sign of the anisotropy constant; its magnitude is about equal to 3/5 of the overall linewidth, provided that anisotropy is the principal broadening mechanism.)[2] To confirm the anisotropy mechanism, spectra of portions of samples A, B, and C were obtained at 35 GHz (Fig. 3). In Fig. 3, the asymmetries of B and C again indicate positive, but slightly larger, anisotropy constants — just as for the "characteristic" resonance. This kind of frequency dependence is expected[9] for ferrites containing both Fe^{2+} and Fe^{3+}. "Pure" magnetite has a negative anisotropy constant for $T > 140°K$. However, in $x \, Fe_2TiO_4 \cdot (1-x) \, Fe_3O_4$ positive anisotropy constants have been observed up to $300°K$ for $x = 0.68$.[10] The attribution of the "characteristic" resonance to such titanium-rich phases would be consistent with an apparent[3] increase in linewidth with increasing TiO_2 contents of fines returned from different lunar sites.

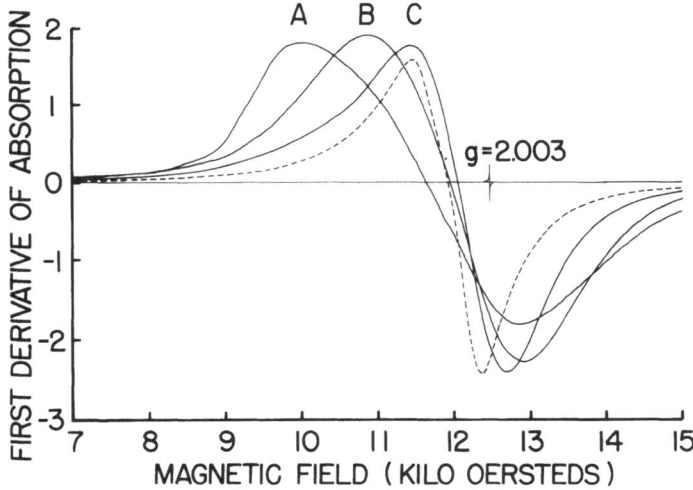

Fig. 3. ESR spectra obtained at 35 GHz and 300°K for samples A, B, and C of Fig. 2; dashed curve = "characteristic" resonance of a < 1-mm fraction of Apollo 12 fines (NASA No. 12001,15; same sample as shown in Fig. 2).

It is interesting that no "magnetite-like" phases seemed to precipitate in simulated lunar glasses heated only in vacuo, even though some Fe^{3+} was present in the as-quenched glass. It is speculated that Fe^{2+}, presumably incorporated as a network modifier,[11] can be oxidized in the solid state to Fe^{3+} which due to its interstitial nature is capable of diffusing toward nucleation centers more readily than (network forming?) Fe^{3+} originally present. Alternatively, it may be that Fe^{2+} and Fe^{3+} are phase separated in the melt. Either way, the influence of atmosphere on the sub-solidus unmixing products may be of technological interest. Beyond that, the present results support the conclusion[3] that the "characteristic" ferromagnetic resonance of returned lunar soils results from a range of titanium-substituted ferrites characterized by variable stoichiometries (dependent on local soil chemistry) but similar crystal structure, indicating that lunar surface materials from different Apollo landing sites have shared similar thermal and "weathering" histories.

REFERENCES

1. The best entry to the appropriate literature is through the Proceedings of the Annual Lunar Science Conference, in Geochim. Cosmochim, Acta, Suppls. 1-3, Vols. 1-3, (1970, 71, 72).
2. F.-D. Tsay, S.I. Chan and S.L. Manatt, Geochim. Cosmochim. Acta 35, 865 (1971).
3. D.L. Griscom and C.L. Marquardt, Geochim. Cosmochim. Acta, Suppl. 3, Vol. 3 (in press).
4. T. Castner, G.S. Newell, W.C. Holton, and C.P. Slichter, J. Chem. Phys. 32, 668 (1960).
5. D.W. Forester, private communication (1971).
6. R.R. Shaw and J.H. Heasley, J. Am. Ceram. Soc. 50, 297 (1967); D.W. Collins and L.N. Mulay, ibid., 53, 74 (1970).
7. L. Néel, J. Phys. Soc. Japan, 17, (Suppl. B-I), 676 (1962).
8. The appropriate considerations for deciding this case are given by E. Schlömann, J. Phys. Chem. Solids 6, 257 (1958).
9. R.M. Bozorth, B.B. Cetlin, J.K. Galt, F.R. Merritt, and W.A. Yager, Phys. Rev. 99, 1898 (1955).
10. Y. Syono and Y. Ishikawa, J. Phys. Soc. Japan 19, 1752 (1964).
11. H. Rawson, "Inorganic Glass-Forming Systems", Academic Press, New York (1967).

NOVEL AMORPHOUS MAGNETIC MATERIALS: MAGNETIC AND MÖSSBAUER
STUDIES ON IRON DISPERSIONS IN ZEOLITES*

L.N. Mulay† and D.W. Collins††

Materials Research Laboratory and
Solid State Science Program
The Pennsylvania State University
University Park, Pennsylvania 16802

INTRODUCTION AND EXPERIMENTAL

Our interest in this area stems from the extensive
studies we reported on the dispersions of iron species (ions,
atoms, etc.) in various vitreous silicates[1,2]. One of our
objectives in studying superparamagnetic systems has been
to devise novel synthetic methods for controlling the particle size of the magnetic species by "stuffing" them inside
some cage structures.

Some work has been reported on the dispersion of nickel[3,4] (metallic or ions) and on the dispersion of iron[5-8]
in zeolite matrices. However, there are difficulties[9] in
introducing iron species (atoms or ions) which appear to
break down the zeolite structure.

---

*Early phases of this work were supported by the AEC [Con-
tract AT(30-1)-2581]. Details of this work will be published elsewhere.

†All enquiries should be addressed to this author.

††Present address: Bell Telephone Laboratories, Murray Hill,
New Jersey 07974.

In this paper we report a synopsis of studies undertaken since 1966 on introducing iron via organometals such
as $Fe(CO)_5$, ferrocene, $[(C_5H_5)_2Fe]$ and covalent compounds
like $FeCl_3$. The Linde Molecular Sieve 13 X (hereafter ab-
breviated M.S.) was chosen because it offers a large internal (700-800 m^2/g) and a small external (1-3 m^2/g) area
with a relatively uniform pore diam $\sim$ 10 Å, which is large
enough for these compounds to slip through.

Exploratory experiments were carried out using $Fe(CO)_5$,
the only liquid iron carbonyl known. Addition to the molecular sieve described before was rather crude at first, but
the initial results were promising enough to continue research in this area. Subsequently, in an attempt to control the experimental conditions more closely, an experiment
was devised such that an "aerosol" of iron pentacarbonyl
was sprayed onto the M.S., which was constantly agitated.
This was found to be quite successful when the magnetic
measurements were made, as discussed below. A spray atomi-
zer was used to produce the aerosol, and a vibrating plate,
on which a crucible containing the M.S. powder was placed,
was used to provide the agitation. Other methods of adding
the iron pentacarbonyl included attempts to adsorb vapors
of $Fe(CO)_5$ onto the sieve and attempts to adsorb a solution
of $Fe(CO)_5$ in ether. The addition of ferrocene via its
solution in benzene has been successful. Anhydrous $FeCl_3$
dissolved in anhydrous ether, was also aprayed onto the M.S.
$FeCl_3$ is also known to be covalent and was expected to be
stable under vacuum. This experiment was also successful.
Finally, samples of each addition product were heat-treated
in air at 500°C and at 900°C for 5 minutes and for 30 min.
in order to (a) convert the addition product to Fe_2O_3 and
(b) to cause a growth of their particles.

Analysis of the molecular sieve - $Fe(CO)_5$ addition product indicated 11.8 weight percent Fe (expressed as Fe_2O_3);
in the M.S. - $FeCl_3$ addition product, 5.6 weight percent Fe
(expressed as Fe_2O_3) was present. The magnetic measurements
were made employing the well-known Faraday method, described
by Mulay[10]. A standard spectrometer was used to obtain the
Mössbauer spectra for Fe^{57} relative to sodium nitroprusside
(National Bureau of Standards) as a standard.

X-ray diffraction established that M.S. showed indeed
the faujasite pattern observed previously (ASTM Powder Pattern # 12-246) and that the M.S. structure was unaltered by

the addition of $Fe(CO)_5$ or by $FeCl_3$. Aging the M.S.-$Fe(CO)_5$
addition product at 500°C immediately gave particles of α-
Fe_2O_3 large enough to be detected by X-ray diffraction; an
increase in the growth of the particles was observed with an
increase in heat treatment time. At 900°C, the M.S. was
transformed into a different structure, and again, a growth
of α-Fe_2O_3 was observed with time of heat treatment. In
the case of the M.S. - $FeCl_3$ addition product, the presence
of α-Fe_2O_3 was noted at 900°C only. No other iron species
were found to be present in either case.

An electron microscopic study was made in an attempt
to check whether there was any appreciable accumulation of
magnetic material on the exterior surfaces of the molecular
sieve. However, an RCA electron microscope, with magnifi-
cation to 15,800 was not able to disclose either the presence or absence of any nonzeolitic material on the exterior surfaces of the M.S.

MAGNETIC PROPERTIES AND DISCUSSION

We calculated the per gram saturation magnetization
(σ_s) by plotting the per gram magnetization (σ) versus the
reciprocal magnetic field (1/H) and extrapolating to 1/H=0.
The magnetizations are expressed in units of gauss cm^3/g.
Following the procedure given by van der Giessen[11], the
magnetic moment (μ in Bohr Magnetons) was then obtained
from a best fit to a plot of σ/σ_s versus $\mu H/kT$. The particle size (r) in Å was calculated from the following equation:

$$r = \left[\frac{\mu}{\sigma_s \rho} \times 10^{24} \right]^{1/3}$$

where σ is the density of bulk α-Fe_2O_3. The average Weiss
constant (θ) in °K was determined by plotting the reciprocal
of the magnetic susceptibility (χ) as a function of temperature. These values are listed in Table I.("h.t." refers to
the heat treatment conditions).

The calculated values of σ (uncorrected for diamagnetic
contributions)of the samples as a function of the H/T ratio
are shown in Fig. 1. For the sake of clarity, the data for
the "500°C - 5 minute" heat treatment has been omitted; it
follows very closely the behavior of the "500°C - 30 minute"

Table I. Magnetic Parameters for the M.S. - Fe(CO)$_5$ System

Sample	θ	σ_S	μ	r
A. Addition Product	-122	0.45	140	85
B. h.t. 500°C - 5 min	- 31	0.40	190	95
C. h.t. 500°C - 30 min	-189	0.43	900	160
D. h.t. 900°C - 5 min	-140	1.62	800	100
E. h.t. 900°C - 30 min	-174	0.62	900	140

sample. Clearly, the approximate superposition of the data for this relatively "dilute" system conforms to the operational definition of superparamagnetism and suggests the presence of this phenomenon. The minor hysteresis that appears can be attributed to the ferrimagnetic behavior of the fine particle antiferromagnets as proposed by Néel[12] and discussed below.

The observed Mössbauer spectra as a function of the heat treatment parameters are shown in Fig. 1 and a spectrum for α-Fe$_2$O$_3$ is also included for comparison.

Plots of saturation magnetization (σ_S) as a function of the calculated particle size (r) [not shown here] displayed an increase in σ_S upto $\sim$ 100 Å and a subsequent decrease. This behavior suggests a growth of fine antiferromagnetic particles. Below a critical size, greater spin uncompensation is assumed to increase with increasing particle size; Néel[12] has suggested that the number of uncompensated spins (p) in a fine particle antiferromagnet increases as $\sqrt{N}$, where N is the total number of spins. In effect, the system appears to behave as a ferrimagnet up to a critical particle size. At a critical size, represented by the maximum in the above curve, antiferromagnetic single domains are beginning to form; the magnetic data for this point, which represents the highest degree

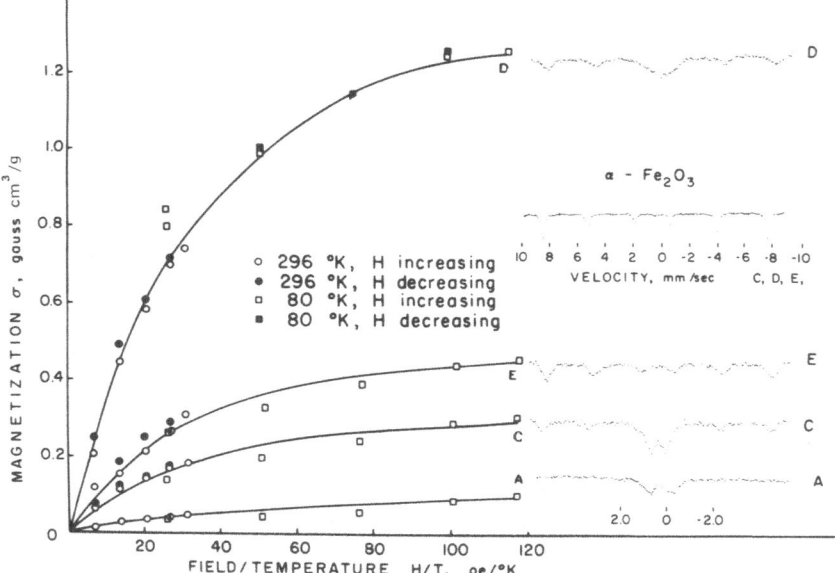

Figure 1. Magnetization as a function of H/T and the
Mössbauer Spectra (∿25°C) for the heat-treated samples in
M.S.-Fe(CO)$_5$ system [see Table I].

of ferrimagnetic character that the system achieves, exhibits minor hysteresis behavior even at 77°K. Above the
critical size, there is greater spin compensation occurring
in the direction of a bulk antiferromagnet. This is supported by the Mössbauer spectra in which hyperfine splitting, indicative of the presence of long range magnetic
order, first appears at the critical size. Thereafter, the
long range magnetic ordering increases at the expense of
the superparamagnetic particles. Similar particle size-de-
pendent Mössbauer spectra have been reported by several
workers and have been reviewed by us.[13] Hence, for our
M.S. - Fe$_2$O$_3$ system, it would appear that the critical particle size at room temperature is ∿100 Å, the radius at
which hyperfine splitting in the Mössbauer spectrum first
appears. Although we cannot discuss here all of our quantitative results on the Mössbauer spectra for the M.S. -
Fe(CO)$_5$ and the M.S. - FeCl$_3$ systems, it should be noted
that we have used Néel's well known relaxation equation[12]

$[1/\tau = f_0 \exp(-KV/kT)]$, Van der Woude and Dekker's[14] approach and an anisotropy constant of $k = 3 \times 10^4$ erg/cm^3 given by Naik and Desai[15] to obtain a critical radius of $\sim$110 Å for our superparamagnetic system. This result is likely to be fortuitous; nevertheless, it indicates a possibility that should be investigated in greater detail. Interestingly, Creer[16] has also suggested from static magnetic studies that 100 Å is the critical size for α-Fe$_2$O$_3$.

The concept of the growth of fine antiferromagnetic particles is also supported by the increase in the Weiss constant (θ), which is a measure of the exchange interactions, in the direction of bulk α-Fe$_2$O$_3$.

Thus the dispersion of iron species in zeolites shows not only superparamagnetic behavior but also provides a new way for studying the magnetic ordering within superparamagnetic clusters.

REFERENCES

1. D.W. Collins and L.N. Mulay, J. Amer. Ceram. Soc. 54, 69 (1971); 54, 52 (1971); 53, 74 (1970).
2. D.W. Collins, L.N. Mulay and W.F. Fisher, Bull. Amer. Ceram. Soc. 44, 343 (1965); Jap. J. Appl. Phys. 6, 1342 (1967).
3. W. Romanowski, Z. anorg. allgem. Chem. 351, 180 (1967).
4. C.S. Brooks and G.L.M. Christopher, J. Catal. 10, 211 (1968).
5. L.S. Singer and D.N. Stamires, J. Chem. Phys. 42, 3299 (1965).
6. V.I. Gol'danskii, I.P. Suzdalev, A.S. Plachinda and L.G. Shtyrkor, Proc. Acad. Sci. USSR-Phys. Chem. Div. 169, 511 (1966).
7. R.W.J. Wedd, B.V. Liengme, J.C. Scott and J.R. Sams, Sol. State Comm. 7, 1091 (1969).
8. W.N. Delgass, R.L. Garten and M. Boudart, J. Phys. Chem. 73, 2970 (1969).
9. C.K. Hersh, Molecular Sieves, Reinhold Publ. Co., New York (1961).
10. L.N. Mulay, "Techniques for Magnetic Susceptibility," in Physical Methods of Chemistry (A. Weissberger and B.W. Rossiter, Eds.) John Wiley & Sons, New York (1972).

11. A.A. van der Giessen, J. Phys. Chem. Solids 28, 343 (1967).
12. L. Néel, Compt. Rend. Acad. Sci. 252, 4075 (1961); 253, 9 (1961); 253, 203 (1961); J. Phys. Soc. Japan 17, 676 (1962); Ann. Geophys. 5, 99 (1949); Low Temperature Physics, (C. DeWitt, B. Dreyfus, and P.G. de Gennes, Eds.) Gordon & Breach, New York (1962).
13. D.W. Collins, J.T. Dehn and L.N. Mulay, "Superparamagnetism and Mössbauer spectroscopy," in Mössbauer Methodology, Vol. III (I.J. Gruverman, Ed.) Plenum Press, New York (1967).
14. F. van der Woude and A.J. Dekker, Phys. Stat. Sol. 9, 775 (1965).
15. Y.G. Naik and J.N. Desai, Ind. J. Pure Appl. Phys. 3, 27 (1965).
16. K.M. Creer, Geophys. J. 5, 16 (1961).

DISCUSSION

E.J. Siegel: Do the natural Zeolites have these properties?

L.N. Mulay: I don't know. I haven't looked at the natural zeolites but it would be interesting to look at those. I mentioned that Dr. Senftle at the Howard University and the U.S. Geological Survey in Washington, D.C., has looked at some of the minerals in which he sees evidence of superparamagnetism. I didn't do that work myself.

P. Horn: Is this technique safe? . . . How are you sure you don't break the whole thing?

L.N. Mulay: It all depends on how good a cook you are. The point is that if you don't go to the breaking point of the zeolite matrix which is about 900°C you are safe and again the other variable you have at your command is the concentration of the iron that you put in to start with. If you were to put in a tremendous amount of iron pentacarbonyl or that kind of a organometal, we have evidence that the whole structure might break apart but because of all the variables you can produce a system which is superparamagnetic.

NOVEL DISPERSIONS OF IRON IN AMORPHOUS GLASSLIKE CARBONS:[*] EXPLORATORY MAGNETIC STUDIES

Alan W. Thompson, P.L. Walker, Jr. and L.N. Mulay[+]

Materials Research Laboratory and the Solid State Science Program, The Pennsylvania State University, University Park, Pa. 16802

INTRODUCTION AND EXPERIMENTAL

One of us (LNM) reported extensive studies on the super-paramagnetic behavior of dispersions of iron species (atoms, ions) in the silicate[1-3] and zeolite type matrices[4]. In a search for novel methods for dispersing iron, this author noticed the possibility of dispersing Fe species in a diamagnetic carbon host lattice. As such this exploratory work was undertaken. A number of samples of amorphous glasslike carbon containing iron were supplied to us by Professor R. Kammareck[++] and Dr. Nakamizo. A schematic of their synthesis is shown in the following diagram[++].

---

[*]Work supported by the Advanced Research Projects Agency [ARPA-DAH15 17 C 0290].

[+]All inquiries should be addressed to this author.

[++]Details of the synthesis will be presented separately at the 11th Biennial Carbon Conference (Gatlinberg, TE) and published in its proceedings. We are very grateful to our distinguished polymer chemists for supplying these samples.

Schematic of Synthesis

(FA) copolymerize with (FDA)* or (VF)*

[at 60°C under acid catalysis]

↓

(Product-I)

[cured in vacuum at 250°C, 48 hrs]

↓

(Product-II)

[pyrolysis in inert atmosphere up to 970°C

rate of temperature increase from 5°C/hr to 60°C/hr]

↓

(AMORPHOUS GLASSLIKE CARBON

WITH DISPERSION OF Fe)

FA = furfuryl alcohol

(polymerizes to poly(furfuryl alcohol)(PFA))

FDA = { ferrocene dicarboxylic acid

VF = [vinyl ferrocene

*Note: The concentrations of FDA varied from 1 to 3% and that of VF from 1 to 10% by weight.

TABLE I: SAMPLES STUDIED AND TYPE OF MEASUREMENTS

Sample	Preparation Temperature			
	500°C	625°C	700°C	970°C
PFA	M	M	M	M
PFA + 1% FDA	M	M	M	M
PFA + 3% FDA	-	M,S	M	M,S
PFA + 1% VF	M	M	-	M
PFA + 10% VF	M,S	M,S	M,S	M,S

The particular samples studied are given in Table I. In this table, M indicates magnetization measurements, and S indicates susceptibility measurements.

The vibrating sample magnetometer technique was selected because of its ability to furnish rather quickly the magnetization as a function of the field and temperature, especially for samples with large magnetic moments. A magnetometer of this type, manufactured by the Princeton Applied Research Laboratory, Princeton, N.J., was employed for all magentization measurements reported here. Several high temperature susceptibility measurements (from room temperature to 750°C) were carried out in the Chemistry Department at the University of Pittsburgh* using an automated Faraday balance described by Butera, et al.[5] The sensitivity of this particular balance was fairly low. This limited the samples to be measured to those with high iron content and to those available in large quantity. The large amount of sample rendered the absolute measurement of the susceptibility to be somewhat ambiguous. Thus, only relative susceptibilities are reported here. However, the Curie points of the materials were accurately established.

---

*We are grateful to Drs. W.E. Wallace and R.S. Craig for providing these facilities.

MAGNETIC PROPERTIES AND DISCUSSION

Magnetization versus field measurements were made on all samples. The samples that contained iron displayed weak ferromagnetism. For all samples save one, saturation occurred at 5-6 K gauss. The 500°C VF sample saturated at 2-3 K gauss. The saturation magnetization per gram (σ) of Fe for various samples containing iron was measured. Results are shown in Figure 1a.

The FDA samples show similar behavior with an approach at the high temperature end to the saturation magnetization of Fe_3C (cementite), which was found in these carbons by a combination of electron microscopy and x-ray diffraction. It is interesting to note that the "dilute" (1% FDA) sample shows relatively higher saturation magnetization than the "concentrated" (3% FDA) sample over a wide range of their preparation temperatures. Thus it may be surmised that the 1% FDA sample has relatively small particles (probably superparamagnetic) of iron species, which are more difficult to saturate. In both cases, the final product obtained by preparing the samples at a higher temperature ($\sim$1000°C) turns out to be Fe_3C, which is formed by the conglomeration of such particles,

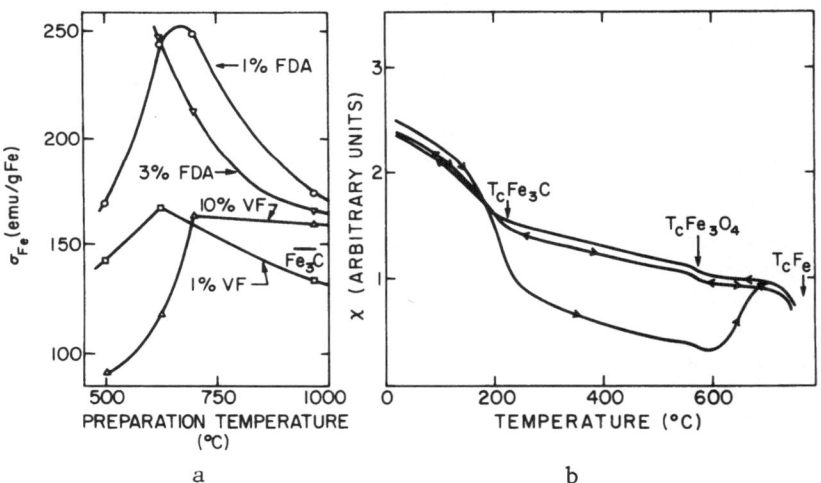

a   b

Figure 1. (a) Saturation magnetization measured at 19K gauss and 25°C for various glassy carbon samples as a function of their preparation temperature. (b) Magnetic susceptibility in arbitrary units as a function of temperature for the glass-like carbon prepared with 10% VF (970°C).

large or small, existing at preceding temperatures of preparation.

While the 1% VF and 10% VF samples show a similar trend over the low temperature region of preparation (500 to 700°C) this trend is reversed over the higher range (700 to 1000°C. This is suggestive of an unusual mode of conglomeration for the VF system, which is quite different than the FDA system.

In any case, it is noteworthy that the 10% VF system also approaches the formation of Fe_3C. The 1% VF samples appear to show the same trend as the FDA sample; however, the magnetization at 970°C is below that of Fe_3C. This may be due to incomplete conglomeration of fine particles to form bulk Fe_3C.

Electron micrographs of the 500°C, 10% VF sample show small "cubic" areas of high density material. Micrographs of the samples treated at higher temperatures show the material in these areas to have migrated out. It is apparent that the trend in magnetization reflects this migration. However, this species as of now has not been identified.

Despite the large error involved in measuring diamagnetic magnetization on the magnetometer, a definite trend to larger diamagnetic magnetization with increasing preparation temperature was observed. This increase in the diamagnetic magnetization is probably due to a decrease in paramagnetic centers (i.e. free radicals) with heat treatment.

SUSCEPTIBILITY MEASUREMENTS

Susceptibility versus temperature measurements were made on several of the samples, temperatures not exceeding the preparation temperatures. Except for the 500°C - 10% VF and 625°C - 10% VF samples, the samples all displayed similar behavior. The samples were heated and cooled several times as susceptibility measurements were being made. Figure 1b shows the observed behavior. Besides the Curie point for Fe_3C at 222°C, the curve also shows a Curie point at about 565°C, that for magnetite (Fe_3O_4). As the samples were heated above 600°C, there was observed an increase in the susceptibility with a turnover above 700°C indicating the Curie point of metallic iron at 770°C. Cooling back to room temperature, a marked decrease in the amount of Fe_3C is observed. Since Fe_3C is an unstable compound, it must be concluded that the iron in Fe_3C is being decomposed to metallic iron when heated above 600°C. This is interesting, since these samples, in their preparation, have previously been heated, in some cases, to above this temperature. In some way, the preparation method forces the formation of Fe_3C. It has been thought that these glasslike carbons have closed pores, thus being impervious to gases. This may not be so, since these materials have been stored in air for six months or better before susceptibility measurements were made. During this time, air may have entered the pores or some gas produced during the preparation may have escaped.

The susceptibility of the 500°C-10% VF and the 625°C-10% VF samples (Figure 2) both showed similar behavior, somewhat different from the other samples. The Curie points observed are those for Fe_3C and Fe_3O_4. Unlike the previous samples, the Fe_3C decomposes to Fe_3O_4 instead of iron. Here again, the heat treatments during preparation have forced Fe_3C formation. However, the small cubic areas of high density material revealed by electron microscopy cannot be determined from the data.

Further work along the lines indicated here is being continued in the hope of producing good superparamagnetic systems.

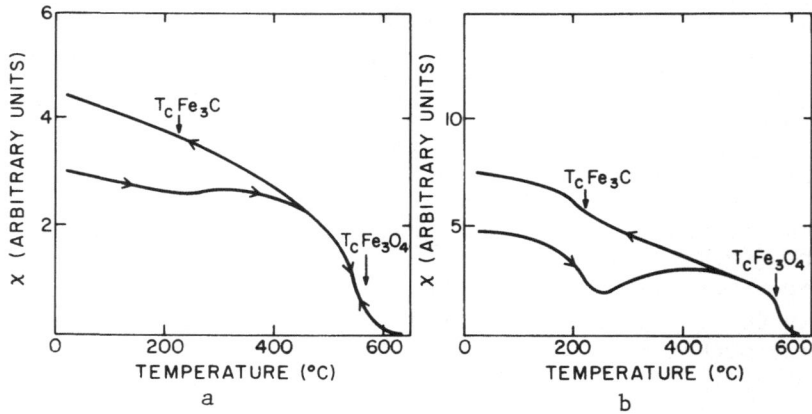

Figure 2. Magnetic susceptibility in arbitrary units as a function of temperature for the glasslike carbon prepared with 10% VF (a-500°C, b-625°C).

REFERENCES

1. D.W. Collins and L.N. Mulay, J. Amer. Ceram. Soc. 54,
    69 (1971); 54, 52 (1971); 53, 74 (1970).
2. D.W. Collins, L.N. Mulay and W.F. Fisher, Bull. Amer.
    Soc. 44, 343 (1965); Jap. J. Appl. Phys. 6, 1342 (1967).
3. D.W. Collins, J.T. Dehn and L.N. Mulay, "Superparamag-
    netism and Mössbauer Spectroscopy," in Mössbauer Metho-
    dology, Vol. III, Ed. I.J. Gruverman, Plenum Press,
    New York, 1967.
4. Another paper on the dispersion of iron in zeolite ap-
    pears in this Proceedings volume.
5. R.A. Butera, R.S. Craig and L.R. Cherry, Rev. Sci.
    Instr. 32, 708 (1961).

DISCUSSION

C.W. Rector: In the glassy carbons would you estimate the
sample bulk density of the spherules as compared to the
thread-like structures?

L.N. Mulay: I don't remember these figures, but I have a
report with me and I would be happy to look up the values
and give them to you.*

E.J. Siegel: Have you ever made glassy carbon with a high
nitrogen content?

L.N. Mulay: No. The other variable which I neglected to
mention in the synthesis of glass-like carbons is that some
work is being done by Professor Dachille, at Penn State University, under high presures which again produces all kinds of
of porosity.

---

*Spherules, (1.68–1.79 g cm^{-3}); thread-like structure,
 (1.47–1.5 g cm^{-3}). These values are extracted from the
 ARPA Semi-Annual Report   (Contract DAH-15-71-C-0290)
 "Glassy Carbon, Alloys", July 27, 1972, (P.L. Walker, Jr.),
 Penn State University, University Park, Pa. 16802.

MAGNETIC SUSCEPTIBILITY OF CHALCOGENIDE GLASSES

J. Tauc

Division of Engineering, Brown University

Providence, Rhode Island and

Bell Laboratories

Murray Hill, New Jersey 07974

F. J. Di Salvo, G. E. Peterson and D. L. Wood

Bell Laboratories

Murray Hill, New Jersey 07974

We shall report on the studies of the differences of the magnetic susceptibility X of pure semiconductors in the amorphous and crystalline states, and on the effect of doping by impurities. In Fig. 1 a typical behavior of $X(T)$ is shown.[1] In the crystalline state X can be a complicated function of T but in the amorphous state $X(T)$ can usually be divided into an almost temperature independent diamagnetic part X_{dia} and a much more temperature dependent paramagnetic part X_c. X_c is observable at low temperatures and can be described by the Curie formula C/T except at very low temperatures. We shall discuss both parts separately, and then present some new optical and ESR results in amorphous semiconductors. Our own experimental work was done on simple chalcogenide glasses As_2S_3 and As_2Se_3 (pure and doped) but we shall include in our discussion also results obtained on other materials.

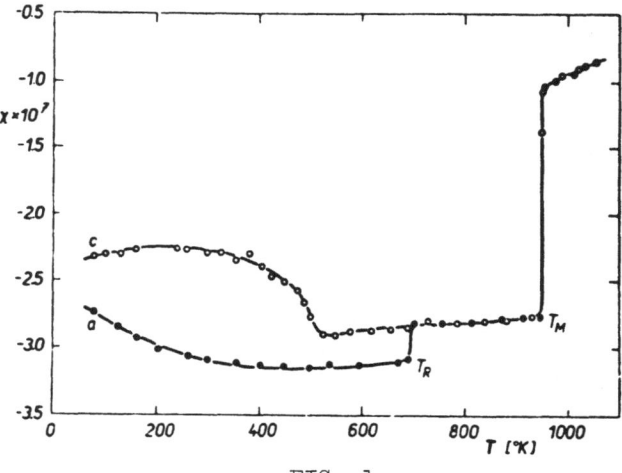

FIG. 1

Magnetic susceptibility of crystalline (c), amorphous (a) and liquid (above T_m) CdGeAs$_2$. After Cervinka et al.[1].

DIAMAGNETIC PART

It is generally observed (but not without exceptions) that a material in the amorphous state is more diamagnetic than in the crystalline state. A quantative comparison is difficult because the $\chi(T)$ of a crystal is often not a simple function of T and because of other complications.[2] (An example is shown in Fig. 1.) In simple chalcogenide glasses the difference of $\chi_{dia}^{crystal}$ and χ_{dia}^{glass} is a few percent (Table 1) but appears to be larger for some other glasses. A very large difference has been reported[3] for Ge: $\chi_a/\chi_c = 2.9 \pm 0.3$.

Two reasons for the additional diamagnetism in glasses have been suggested[4]: Diamagnetism of localized states and changes in the Van Vleck term produced by disorder.

In the presently used model of the electronic states of an amorphous solid (Mott-CFO model shown in Fig. 2) the states in the mobility gap are localized over increasingly smaller volumes as we move from the mobility edges deeper into the gap. The localized states produce a diamagnetic

TABLE I

Magnetic Susceptibility
(in units of 10^{-6} emu/gm)

Material	Polycrystalline	Glassy	Ref.
S	-.485	-.485	6
Se	-.272	-.291	
$CdAs_2$	-.258	-.269	7
As_2S_3	-.345	-.355	2
As_2Se_3	-.287	-.292	

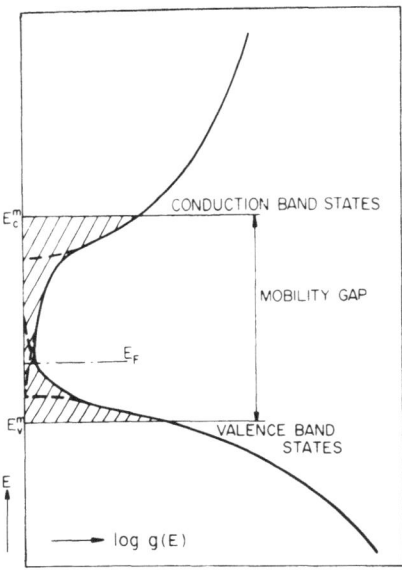

FIG. 2
Density of states g(E) in amorphous semiconductors.

Langevin contribution to the susceptibility. The states
close to the mobility edge of the valence band contribute
most because they are localized over the largest volumes
and their density $g(E_V^m)$ is largest.[4]

The Van Vleck term which produces a positive contribution to χ is likely to be reduced by disorder. This

trend is easy to see[4] with a simple description of the electronic states in amorphous solids, the random phase model.[5] The influence of disorder is expected to be larger in semiconductors than in insulators but the enormous difference of χ in amorphous and crystalline Ge is not explicable with this model.

LOW TEMPERATURE SPIN TAILS

In the Mott-CFO model of amorphous semiconductors (Fig. 2) the valence states are implicitly assumed to be doubly occupied (as the valence band states in crystals), even if they are localized. However, it was pointed out[8,9] that the Coulombic interaction between electrons localized over a small volume raises the energy of the second electron, eventually above the Fermi level. One therefore expects that states sufficiently deep in the gap will be singly occupied by electrons, and contribute a Curie-term to the magnetic susceptibility.

We were interested in whether the loss of the long-range order produces singly occupied states in the gap, and chose for our studies good glass formers As_2S_3, As_2Se_3 and some related chalcogenide glasses. We observed[10] Curie-like terms in $\chi(T)$ corresponding to free spin concentrations of $10^{17}-10^{18}$ cm^{-3} which we associated with the singly occupied localized states in the gap produced by disorder. We were aware that we had some impurities in the sample (in particular a few ppm of Fe), however, we believed we could separate their contribution from the contribution of disorder produced states by the following experiment: We sealed crystalline As_2S_3 in a quartz ampoule, measured $\chi(T)$, melted and quenched the sample in situ to transform it into the amorphous state, and measured $\chi(T)$ again. We found a substantial increase of the low temperature tail, which we ascribed to states produced by disorder.

However, more accurate studies[2] have shown that this explanation cannot be correct. We observed deviations from the Curie-like behavior of the tail (Fig. 3) which could not be understood if we assume the expected parameters for the disorder produced states, $g \approx 2$, $S = 1/2$. The most likely explanation[2] of the deviation appears to be crystal field splitting which is possible only for $S > 1/2$. Samples of As_2S_3 intentionally doped with Fe showed the same

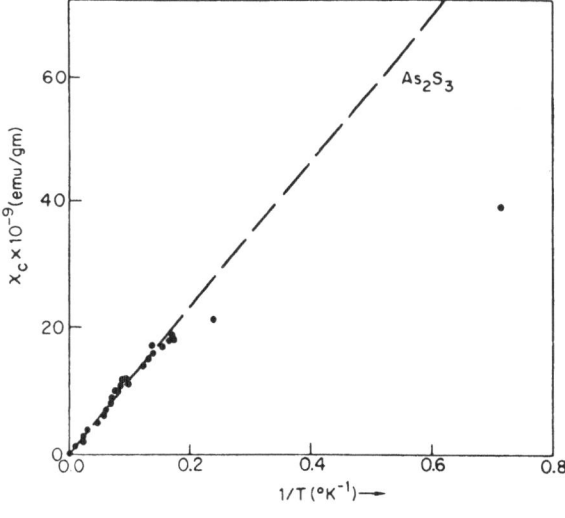

FIG. 3

The paramagnetic part of magnetic susceptibility X_c for
a nominally pure As_2S_3 vs. inverse temperature. After
Di Salvo et al.[2].

saturation effects. We are confident now that the tails
observed were due to Fe. If the sample was prepared from
ultrapure materials in an extremely pure quartz ampoule
which was enclosed in another quartz ampoule, only very
small tails were found in As_2S_3 and As_2Se_3 (Fig. 4). In
fact, these results show that the concentration of singly
occupied states in these materials must be below the sensitivity of the equipment, 3×10^{16} cm^{-3}. Measurements on
a-As_2Se_3 down to 0.015 K are in agreement with this conclusion.[11] The same result was found for S and Se.[6]

We must therefore conclude that good glass formers
like S, Se, As_2S_3 and As_2Se_3 have a very low concentration
of singly occupied states in the gap (less than 3×10^{16}
cm^{-3}). This may mean that the concentration of states in
the gap in these glasses is small. It has recently become
clear that the older theories of the electronic states in

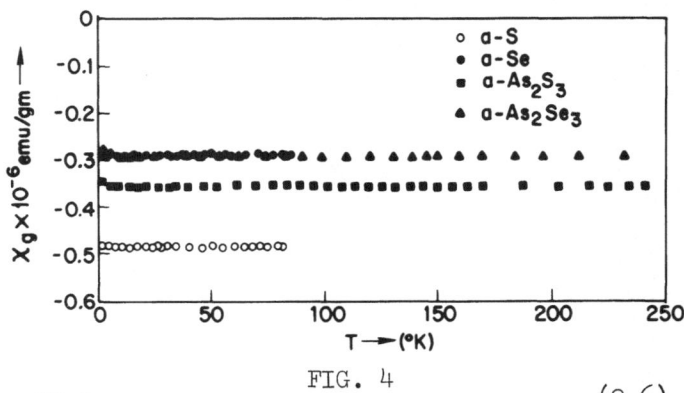

FIG. 4

$X(T)$ of very pure chalcogenide glasses.[2,6]

the gap (e.g., Mott-CFO theory) did not take into account the relaxation of the lattice which, as it is believed now but not actually proved, tends to remove the states from the gap.[12,13] A gap with a small concentration of states is in agreement with the high transparency of glasses below the absorption edge; however, the interpretation of the optical data is difficult because we do not know quantitatively how the transition probabilities change when we go from the crystal to the glass.

Therefore an alternative possibility cannot be discarded at present, namely that the concentration of states in the gap is higher than our estimate from the magnetic data because the states are doubly occupied. As Anderson[13] pointed out it is possible that the lattice interaction may be strong enough to overcome the Coulombic repulsion and make the double occupancy of states possible. Agarwal[14] suggested testing the existence of doubly occupied states by measuring the ESR response of illuminated samples. His measurements did not show evidence of doubly occupied states.

The spin tails in $X(T)$ of glasses are introduced by defects and impurities. Matyas[1,15] ascribed the large Curie terms in $CdGeAs_2$ glasses to broken bonds of Ge atoms. Non-annealed amorphous Ge films show large ESR signals which were associated with dangling bonds on the internal surfaces of voids.[16]

Good glass formers, however, appear to be free of broken bonds. Bagley et al.[6] found it impossible to produce Curie terms in S or Se by rapid quenching from the melt.

Impurities in Chalcogenide Glasses

The experiment in which a much larger tail was observed in amorphous As_2S_3 than in crystalline As_2S_3, although the concentration of Fe was the same, must be interpreted so that Fe is in a different state in the crystal than in the glass. As we could not prepare crystalline As_2S_3, we did more detailed experiments[2] with As_2Se_3 doped with Fe. It is seen in Fig. 5 that Fe in As_2Se_3 is in a paramagnetic state but in polycrystalline As_2Se_3 in an antiferromagnetic state with Neel temperature near 20 K. In crystalline material, Fe separates out of the bulk to form an iron rich phase, probably at zone boundaries, even in small concentration (10 ppm or more).

In our previous experiments[16] we observed that samples of As_2S_3 doped with Cu, Au or Ag did not have the Curie term.[17] However, when we introduced simultaneously Fe and Cu (or Au) into a-As_2S_3 or As_2Se_3 we did not observe any large influence of Cu on the magnetic behavior. We believe that these observations suggest that Fe penetrates

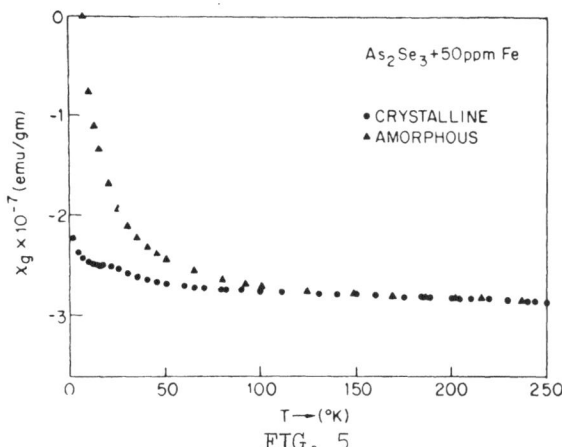

FIG. 5

$\chi(T)$ of amorphous and polycrystalline As_2Se_3 doped with 50 ppm per weight Fe. After Di Salvo et al.[2].

into the sample during the preparation of the sample from
the quartz ampoule. If the material contains Cu, Au or
Ag it does not stick to the wall of the ampoule as it does
otherwise, and probably the penetration of Fe into the
sample is reduced.

ESR of Iron Ions in Amorphous As_2S_3

Electron spin resonance is potentially an attractive
way of studying the concentration and valence state of
transition metal impurities in amorphous materials. Fe^{3+},
with a d^5 configuration is frequently detected at room
temperature with a g value in the neighborhood of 4. Fe^{2+},
on the other hand, because of rapid relaxation, will only
be observed at cryogenic temperatures. Unfortunately,
early attempts by other workers at Bell Laboratories to
detect iron ESR in 120 ppm doped As_2S_3 glass at 10 GHz
were not successful.

It has recently been found that low frequency (510
MHz) ESR may at times be beneficial in locating and resolving
resonances in amorphous and nonstoichiometric solids.[18,19]
It appears that there is some reduction in line width or an
alteration of line shape that aids in detection. We have
now observed a number of cases where reducing the frequency
has been helpful.

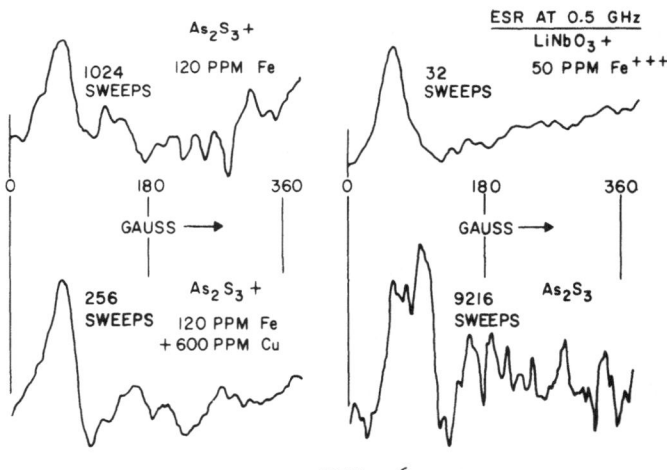

FIG. 6
ESR of Fe^{3+} (see text).

At 510 MHz the 120 ppm iron doped As_2S_3 sample readily revealed ESR at room temperature ascribable to Fe^{3+} (Fig. 6). The addition of Cu to the sample increased the intensity of the Fe^{+3} line by a factor of 4. This is clear from Fig. 6 where we see that the same signal to noise ratio is obtained with 1/4 the number of sweeps of the CAT (Computer of Average Transients). Signals were also detected in the undoped sample as well, but as can be seen the iron level is quite low.

To estimate the amount of Fe^{+3} in the As_2S_3 sample some sort of standard is necessary. Those usually employed, such as DPPH or $CuSO_4$ have line shapes very different than iron in this matrix and thus seem inappropriate. Iron doped lithium niobate powder has a shape quite similar to the glass and can be almost 100% converted to the trivalent state by oxygen treatment at 1000°C. Using this standard we estimate that only about 1% of the total iron concentration in As_2S_3 is in the trivalent state. The rest of the iron atoms must be in a paramagnetic state, otherwise the paramagnetic contribution of iron to the magnetic susceptibility of the doped As_2S_3 glasses would not be explicable. This shows that most of the iron in vitreous As_2S_3 is in the divalent state.

Optical Absorption of Fe Ions in Amorphous As_2S_3

We have observed that the additions of Fe to amorphous As_2S_3 produces optical absorption tails extending from the Urbach edge toward lower energies, and our results are shown in **Fig. 7.** The lowest curve is for a relatively pure sample containing about 5 ppm Fe, while the upper curves are for intentionally doped samples with 26 and 120 ppm Fe (by weight). These tails are not due to light scattering from a separate phase, but represent true optical absorption. The frequency dependence of the absorption in the tail is exponential, as **Fig. 7** shows, and the dependence on Fe concentration is approximately linear.[20]

The ESR and magnetic susceptibility results support the assumption that the Fe ions are present principally in the divalent state. Neither the broad frequency dependence nor the large oscillator strength for the additional absorption is consistent with the possibility of crystal

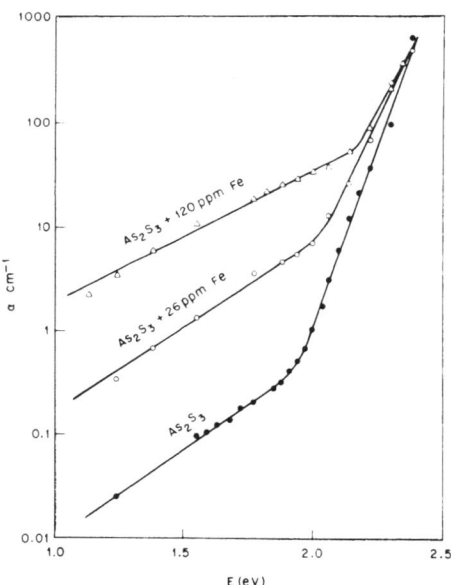

FIG. 7
Optical absorption of Fe-doped amorphous As_2Se_3 in the
vicinity of the absorption edge.[20]

field transitions within the Fe^{2+} ion as an explanation of
the effect. The behavior is, however, remeniscent of the
very strong optical absorption near the band edge for Fe^{2+}
in oxide crystals such as the garnets or orthoferrites[21,22].

In the present case we propose that there are Fe^{2+}
ion donor levels in the As_2S_3 band gap and that the optical
transitions originating on the Fe^{2+} ion promote an electron
to the conduction band with a large oscillator strength.
The frequency dependence will be essentially determined by
the density of states in the conduction band, and the concentration dependence would, of course, be linear as
observed.

We found essentially no effect on the optical absorption after adding 600 ppm of Cu to the material of the
upper curve in **Fig.** 7 containing 120 ppm Fe. This agrees
with the ESR results which show that the concentration of

the major Fe^{2+} species responsible for the optical
absorption is only slightly modified by the Cu addition.

CONCLUSIONS

We have reviewed the recent work on the magnetic susceptibility of amorphous semiconductors and presented new
results on ESR and optical absorption in As_2S_3. The glasses
are generally more diamagnetic than the corresponding crystals but the difference in good glass forming chalcogenides
is small. No evidence for disorder induced singly occupied
states with concentrations above approximately 3×10^{16} cm^{-3}
is found in these glasses. The previously reported paramagnetic contributions to the magnetic susceptibility were
traced to iron impurities. ESR measurements show that the
majority of the iron is in the Fe^{2+} state, only the small
fraction in the Fe^{3+} state being observable by this method.
The interpretation of the optical absorption induced by
iron in As_2S_3 is consistent with the results of the ESR
and susceptibility measurements.

REFERENCES

1. L. Cervinka, A. Hruby, M. Matyas, T. Simecek, J. Skacha,
   L. Stourac, J. Tauc and V. Vorlicek, J. Non-Crystalline
   Solids 4, 258 (1970).
2. F.J. Di Salvo, A. Menth, J.V. Waszczak and J. Tauc,
   Phys. Rev., to be published.
3. S.J. Hudgens, Bull. APS, 17, 232 (1972).
4. R.M. White, P.W. Anderson, Phil. Mag. 25, 737 (1972).
5. N.K. Hindley, J. Non-Crystalline Solids 5, 17, 31 (1970).
6. B.G. Bagley, F.J. Di Salvo and J.V. Waszczak, Solid
   State Communications 11, 89 (1972).
7. M. Matyas, Phys. Stat. Sol. 43, K 63 (1971).
8. M. Pollak, Discussions Faraday Soc. 50, 13 (1970).
9. T.A. Kaplan, S.D. Mahanti and W.M. Hartmann, Phys. Rev.
   Letters 27, 1796 (1971).
10. J. Tauc, A. Menth and D.L. Wood, Phys. Rev. Letters
    25, 749 (1970).
11. D.U. Gubser and P.C. Taylor, Physics Letters, 40A, 3
    (1972).
12. J.C. Phillips, Phys. Stat. Sol. (b) 44, K1 (1971).
13. P.W. Anderson, Nature Phys. Sci., 235, 163 (1972).
14. S.C. Agarwal, Phys. Rev., to be published.

15. M. Matyas, J. Non-Crystalline Solids, 8/9, 592 (1972).
16. M.H. Brodsky and R.S. Title, Phys. Rev. Letters 23, 581 (1969).
17. J. Tauc, A. Menth, J. Non-Crystalline Solids 8/9, 569 (1972).
18. G.E. Peterson and C.R. Kurkjian, Solid State Communications, to be published.
19. A.M. Glass, G.E. Peterson and T.J. Negran, Proc. ASTM: NBS Symposium on Damage in Laser Materials, 1972; NBS Special Publication, to be published.
20. In Ref. 10 the absorption tail in nominally pure As_2S_3 was tentatively associated with the disorder induced states in the gap. Our later work indicates that at least a part of this absorption may be due to the presence of iron.
21. D.L. Wood and J.P. Remeika, J. Appl. Phys. 37, 1232 (1966); 38, 1038 (1967).
22. D.L. Wood, J.P. Remeika and E.D. Kolb, J. Appl. Phys. 41, 5315 (1970).

DISCUSSION

S. Von Molnar: If I understand the argument correctly, by not seeing the paramagnetic tail in the pure materials you assert that the band tails are doubly occupied with the electrons paired antiparallely, right?

J. Tauc: There are two possibilities: either the concentration of states in the tail of the valence band is very small, or these states are doubly occupied.

S. Von Molnar: I just want to make a comment.   Have you thought at all about exciting one of those electrons in the presumed doubly occupied states?   This would leave one of them behind and you could test that by measuring susceptibility.   I was thinking about exciting the system optically and then to look at the Faraday effect or some other effect which would sense the susceptibility.   Have you thought about that?

J. Tauc: S.C. Agarwal (University of Chicago) had the same idea and did ESR measurements on illuminated As_2S_3 and other chalcogenide glasses. He did not find any evidence of singly occupied states produced by photon absorption. However, his method may not be sensitive enough, and we plan to use the low-frequency method mentioned in our talk for this experiment.

Optical measurements show that pure defect-free glasses are very transparent below the absorption edge, and if one analyzes the data in the same way as in crystals one obtains very small concentrations of states. However, theoretical arguments suggest that the optical transition probabilities in glasses may be significantly smaller than in crystals (J.D. Dow and J.J. Hopfield, J. Non-Crystalline Solids 8/10, 664, 1972) and therefore it has not been possible to draw any quantitative conclusions about the state densities in glasses from the optical data.

W.M. Hartmann: Even if the states are removed from the gap by some sort of relaxation processes as Anderson suggests they still might be singly occupied so long as they are open.

J. Tauc: Experiments show that in pure good glass formers the concentration of singly occupied states of any origin is very low.

S. Ovshinsky: Did you make any measurements or do you know of any measurements in telluride glasses?

J. Tauc: No, we didn't.

J. Wong: What was the form of the iron that was doped in As_2S_3, metallic Fe or an iron sulfide?

J. Tauc: Metal.

J. Wong: Then, seeing that Fe is oxidized, what was reduced in the As_2S_3 host?

J. Tauc: Vitreous As_2S_3 can exist in a non-stoichiometric composition in a very wide range. This makes the accommodation of impurity ions possible.

J. Wong: A comment on the optical absorption of Fe doped As_2S_3: Fe(II)-S and Fe(III)-S must have quite different charge transfer absorption.

J. Tauc: That's right. We believe that our optical results agree with the conclusion of the magnetic measurements that most iron is in the Fe^{2+} state.

I.S. Jacobs: Did you give the temperature at which the ESR data were taken?

J. Tauc: It was at room temperature.

I.S. Jacobs: Did you have any comment about the mechanism by which the line width was so dependent on frequency?

J. Tauc: George Peterson suggested a simple explanation (Ref. 18 and 19). The basic idea is that in glasses the g-factor has a distribution because of the fluctuations of the structure. The line width ΔH corresponding to a certain width of the g distribution is proportional to frequency.

LONG AND SHORT RANGE MAGNETIC INTERACTIONS AND ELECTRIC

SWITCHING IN HIGHLY DOPED, "AMORPHOUS" FERROMAGNETIC

SEMICONDUCTORS

P. Wachter

Laboratorium für Festkörperphysik, ETH-Zürich

8049 Hönggerberg, Switzerland

INTRODUCTION

Pure and crystalline EuO is a well known ferromagnetic semiconductor with T_C = 67 K. If Gd substitutionally replaces Eu, the Gd acts as a donor, and one obtains an increase in conductivity and of T_C. If the Gd concentration is of the order of percents, Gd clusters begin to form. Within the clusters the energy levels of the donor may be degenerate with the bottom of the conduction band, and outside a cluster localized donor states may still exist. Since the ionic radii of Gd^{3+} and Eu^{2+} are different, local distortions will further modify the energy levels. The energy of the bottom of the conduction band and possibly also the 4f-states will thus be warped as a function of the spatial coordinates. As a consequence, these materials will display an "amorphous" behavior. It is the purpose of this paper to cite various experiments concerning the optical, magnetic and conducting properties to prove that these materials can indeed be assumed amorphous, in spite of the fact that the X-ray pattern clearly shows crystalline structure.

MAGNETO-OPTICS AND SHORT RANGE MAGNETIC INTERACTION

The undoped Eu-chalcogenides and especially EuO are characterized by an energy level diagram sketched in Fig. 1. A valence band is formed from p-states of the anions and

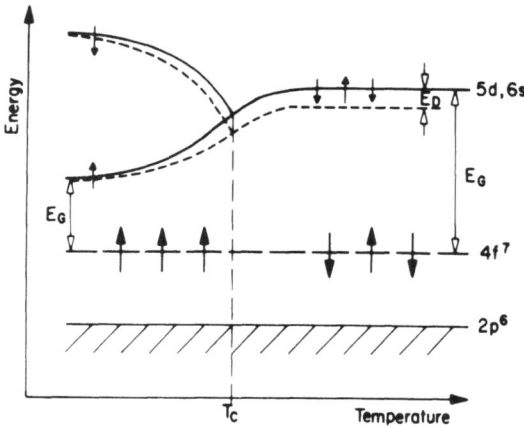

Figure 1:   Model of the Absorption Edge Shift for EuO.

conduction bands are formed by wave function with 5d- and
6s character. Between these bands the localized $4f^7$ ($^8S_{7/2}$)
states of the Eu^{2+} ions are found. The electric dipole
permitted, longest wavelength optical transition is thus
$4f^7$ - 5d, resulting in a sharp absorption edge. In the
free Eu^{2+} ion with S = 7/2, excited 5d states can have spin
parallel or antiparallel to S. According to Hund's rule
the parallel configuration is lower in energy. If in the
solid the wave functions of the excited 5d states overlap,
the bottom of the so formed conduction band will be an
average in energy between the spin up and spin down states.
The conduction band will be spin degenerate in the paramagnetic state. In the completely ordered ferromagnetic state
the spin degeneracy will be lifted and the band splits at
T_c as shown in Fig. 1. Detailed calculations[1] show that
the temperature dependence of the bottom of the conduction
band closely follows the spin correlation function,
$< S_i \cdot S_j >/S^2$, which is a measure of the short range magnetic interaction. It is evident from Fig. 1 that upon
cooling through T_c the absorption edge will display a strong
"red shift".[2,3] The experimental curve is shown in Fig. 2.

Introducing now Gd ions we find a donor activation
energy E_D of the $4f^7 5d^1$ state of Gd of some hundredths eV
which is temperature dependent[3] as indicated in Fig. 1.
(The $4f^7$ level of Gd is much lower in energy than the one
of Eu.) The optical absorption of the doped material consists of an absorption edge which, at paramagnetic temperatures, is displaced towards the red compared with undoped

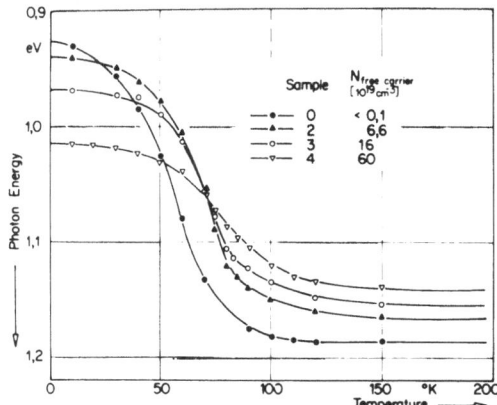

Figure 2:   Shift of the Absorption Edge of Gd Doped and
            Undoped EuO.

EuO, and of a free carrier absorption. (Detailed analysis
permits the evaluation of the carrier concentration.)[3] The
total "red shift" of the absorption edge upon cooling is
appreciably reduced in doped material (Fig. 2).

        The excited 5d states of the Eu ions will only be in-
fluenced by the magnetization if they are sufficiently de-
localized, that is, if the wave functions of neighbouring
Eu ions overlap and if a 4f-5d exchange mechanism exists.[4]
If in the limit the wave functions become completely localized, that is, if we are dealing with free ions, the optical
4f-5d transition will no longer be influenced by an alignment of neighbouring 4f spins. We thus interpret the reduction of the total "red shift" in the doped materials as
an indication of a reduced delocalization of wavefunctions.
This is caused by a loss in periodicity due to the high doping level. The large local distortions introduced by the
dopant, especially if there is a cluster formation, modifies
the energy levels of ground and excited states[5] and thus
results in an effective reduction of the energy gap for
$T \gg T_C$ as shown in Fig. 2. (The local distortion can be
much larger than the average reduction of the lattice constant.)

        We recall that the temperature dependence of the "red
shift" is described by the spin correlation function. This
function is proportional to the magnetic energy of the system. The temperature derivative of the magnetic energy is

proportional to the magnetic part of the specific heat $C_{v,m}$

$$< S_i \cdot S_j >/S^2 \sim U_m, \quad dU/dT \sim C_{v,m}$$

We thus can derive from the optical absorption the magnetic specific heat and normalize it with an independent measurement of $C_{v,m}$ on the undoped material. The sharp Λ-anomaly, typical for pure material, disappears and the peak becomes more and more rounded upon doping.[6] This can be interpreted as a distribution of exchange constants, which indicates that simple molecular field analysis is not suitable for these highly doped materials.

Comparing highly doped with undoped samples, from this chapter we draw the following conclusions:

1. A reduced delocalization of excited wave functions due to a loss of periodicity.
2. Reduction in the energy gap due to local lattice distortions.
3. Distribution of exchange constants.

LONG RANGE MAGNETIC INTERACTIONS

To obtain the long range magnetic interaction we have chosen the Faraday effect which, in these materials, is proportional to the magnetization. In Fig. 3 we compare the normalized magnetization of a undoped and doped sample under identical conditions with the same magnetic field applied. T_c has been determined in zero external field by the onset of depolarization of polarized light passing the sample. We realize that the doped sample at for e.g. T_c has only about half the magnetization of the pure sample. The doped material is harder to magnetize. The same result is obtained in a measurement of M versus H. In these compounds we do not expect magnetic anisotropy. Therefore we relate this curve to an antiferromagnetic contribution upon doping: The Gd spin is not parallel with the Eu spins[7] Recently it has been shown[8] on single crystals of the mixed system $Eu_{1-x}Gd_xS$ that the saturation magnetization versus x decreases linearly with x and extrapolates to zero for $x = 1$. The conclusion has been drawn that Gd enters these materials as antiferromagnetic pairs or clusters. This also seems to apply for Gd in doped EuO. We must realize that M_s at 4 K can only be reduced in the order of percents which is within the limits of error. However, at intermediate temperatures (Fig. 3) antiparallel Gd clusters can appreciably hinder the free alignment of the Eu spins.

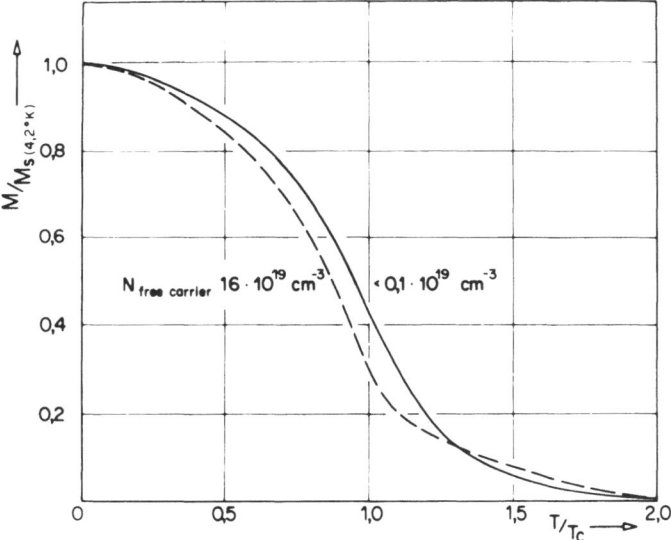

Figure 3: Magnetization of Gd Doped and Undoped EuO.

From this chapter we draw the conclusion:
4. Existance of antiparallel Gd clusters.

PHONONS AND COUPLED MODES

The optical reflectivity of pure and Gd doped EuO single crystals has recently been measured between 200 μ and
13 eV[9] and a Kramers-Kronig analysis of the dielectric constants has been performed. In Fig. 4 the real part of the
d.c. is shown only in the infrared spectral region. The
zeros of ε' are the resonances of the various modes. Indicated are longitudinal and transverse optical phonon modes,
plasmons and coupled plasmon-LO phonon modes. In this paper
let us concentrate on the only pure mode (TO) which can be
investigated as function of doping. We realize that for
"weak" doping, (the Gd concentration is the one effectively
measured in the sample by means of X-ray fluorescence) the
TO mode hardly changes, however, for higher doping levels
an appreciable shift of 10% to higher energies is found.
The TO mode becomes harder. In a simple model we can understand this as a reduction of the ion chain length connected
with the loss of periodicity, the formation of Gd clusters
and the relaxation of the k-selection rule.

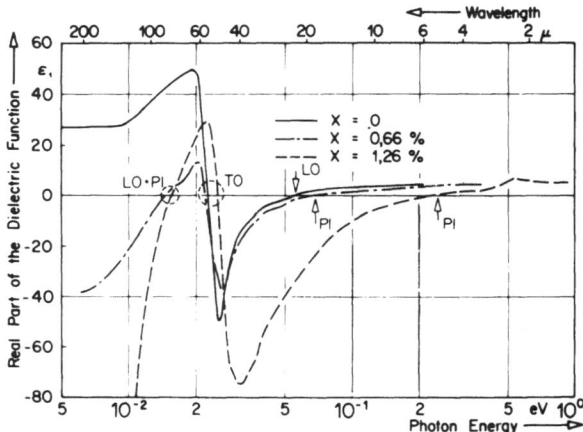

Figure 4:   Real Part of the Dielectric Function of
$Eu_{1-x}Gd_xO$

From this chapter we draw the conclusion:
5.   Loss of periodicity, more localization of modes
and relaxation of selection rules.

AMORPHOUS MODEL

If one accepts points 1-5 as arguments in favor of the
assumption that highly doped EuO can be considered to be
amorphous, we want to propose the following model, visualized in Fig. 5.   In function of space we plot the bottom of
the conduction band and the localized 4f states.   Within
and near a Gd cluster (not shown in the Fig.) the donors
form an impurity band which is degenerate with the conduction band and the electrons quasi form a sea in the now appearing dip of the conduction band.   Outside a cluster localized Gd donors may still exist.   Different electron seas
are connected with constriction paths.   Depending on the resistance of these paths and the number of still localized,
unionized donors, samples can show semiconducting or metallic behavior.   With the exception of the localized 4f states
Fig. 5 represents quite a common model for an amorphous semiconductor.   It has been suggested that amorphous materials
exhibit electrical switching phenomena, known as the Ovshinsky effect.[10]

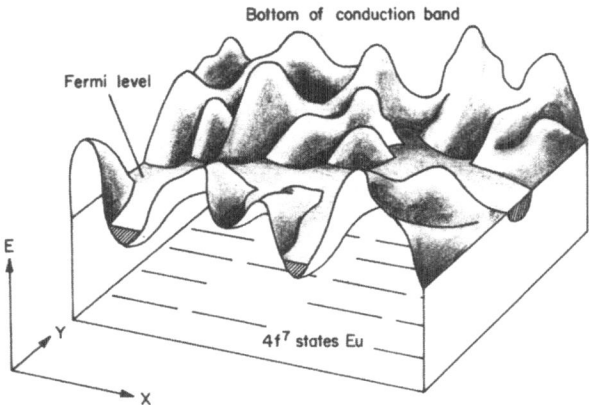

Figure 5: Bottom of Conduction Band of "Amorphous" EuO

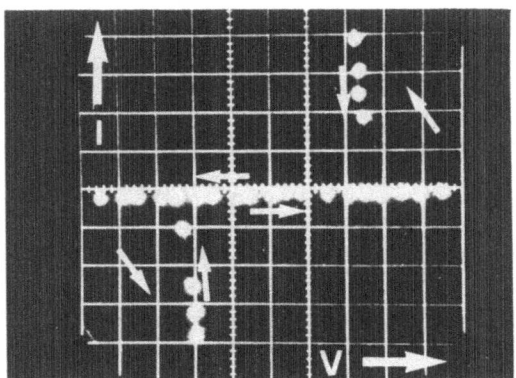

Figure 6: Current-Voltage characteristics of EuO+0.5% Gd

ELECTRICAL SWITCHING

Silver contacts applied to highly doped single crystals of EuO indeed displayed bidirectional reversible threshold switching when operated with[11] 50 Hz. Sometimes the same samples could also be operated in the bistable memory mode. It is remarkable that neither pure nor weakly doped EuO (up to 10^{18} cm^{-3}) shows any switching effect at room temperature. Apparently band warping and Gd cluster formation is necessary to observe any switching. In order to exclude thermal hysteresis in switching, we stepwise increased the voltage to achieve thermal equilibrium (Fig. 6). We clearly observe the threshold voltage necessary to flip the switch into the high conducting state and the holding current before the flip back. While these experiments are performed at room temperature it is evident that magnetic switching will be observed near T_c. Here energy gap and donor activation energies can be appreciably reduced by a magnetic field. Magnetic switching is then the application of an electric field not sufficient to flip the switch. Now turning on a magnetic field will induce the transition. A reduction of either the magnetic or the electric field will cause the flip back.

REFERENCES

1. F. Rys, J.S. Helman and W. Baltensperger, Phys. kondens. Materie 6, 105 (1967).
2. P. Wachter, CRC Crit. Rev. Solid State Sciences 3, 189 (1972).
3. E. Kaldis, J. Schoenes and P. Wachter, AIP Conf. Proc. 5, 269 (1972).
4. J.B. Goodenough, Magnetism and the Chemical Bond, New York (1963).
5. G. Busch, M. Campagna and H.Ch. Siegmann, J. Appl. Phys. 42, 1779 (1971).
6. V.L. Moruzzi, D.T. Teaney and B. van der Hoeven, Solid State Comm. 6, 461 (1968).
7. E. Bayer and W. Zinn, Z. Angew. Physik 32, 83 (1971).
8. T.R. McGuire and F. Holtzberg, J. Appl. Phys. 43, (1972), in press.
9. G. Güntherodt and P. Wachter, Phys. Letters (1972), in press.
10. S.R. Ovshinsky, Phys. Rev. Letters 21, 1450 (1968).
11. P. Wachter, Phys. Letters (1972), in press.

DISCUSSION

S. von Molnar: Would you like to describe the actual physical appearance of the switching device?

P. Wachter: It consists of single crystals about 2x3 mm with silver painted electrodes, about 2mm apart. The silver electrodes are mechanically clamped in the sample holder.

MAGNETIC ORDERING IN NaCl-LATTICES WITH STRUCTURAL DISORDER

G. Busch, M. Campagna, and H.C. Siegmann

Laboratorium für Festkörperphysik der

E.T.H.-Zürich, Schweiz

ABSTRACT*

EuS, EuSe, EuTe and GdP, GdAs and GdSb crystallize in the NaCl-structure and have a large and highly localized spin moment due to the half filled 4f-shell of the Eu- or Gd-ion. Thin films of these substances have been evaporated in UHV onto substrates at 4.2°K. The measurement of the yield of photoelectrons indicated that a metastable state was obtained which transformed into the polycrystalline one at temperatures between 50-150°K. The photoelectron threshold changed by up to 0.8 eV for the most ionic compound EuS. The measurement of the spin polarization of the photoelectrons as a function of both the photon energy and the external magnetic field strength yielded information on the energy of the electron states and their magnetic interactions: EuS and EuSe retain their ferromagnetic order in the metastable state. The antiferromagnets show a complex behavior, with even a complete destruction of antiferromagnetism in metastable GdP. In the insulators with ionic bonding new electron states appear in an energy range that is forbidden in the crystalline state and part of the new states carry a spin moment. The observed changes of photoelectric and magnetic properties on introducing disorder correlate with the degree of covalency of the bonds.

---

*A full report will appear in the International Journal of Magnetism.

THE LOW TEMPERATURE MAGNETIC PROPERTIES OF GLASSY AND CRYSTALLINE Pd$_{.775}$Cu$_{.06}$Si$_{.165}$

B. G. Bagley and F. J. DiSalvo

Bell Laboratories

Murray Hill, New Jersey   07974

Since Klement et al[1] prepared the first metallic glass* there has been much interest in the properties of these new materials; magnetic properties receiving particular attention.[2-10] Glass formation in the binary palladium-silicon system was first observed by Duwez et al,[11] and Chen and Turnbull[12] reported that small additions of noble metals to the binary Pd-Si alloys greatly enhanced glass formation and stability. The composition Pd$_{.775}$Cu$_{.06}$Si$_{.165}$ is a particularly stable glass,[12] and recently the low temperature specific heats of glassy and polycrystalline materials of this composition have been reported.[13,14] In this note we present the results of our investigations on the low temperature magnetic properties of glassy and polycrystalline Pd$_{.775}$Cu$_{.06}$Si$_{.165}$. The purity of our starting materials, sample preparation, and crystallization procedure duplicate that used by Golding et al[13] for their specific heat measurements, and differ in all three aspects from that used by Chen and Haemmerle.[14]

A melt was prepared from elements whose purities were 5-9's or better. The reaction (to form the melt) and subsequent homogenization were done under vacuum in a quartz container at a temperature of 1373°K. The glass was then

---

*We use the definition of a glass as an amorphous solid obtained by cooling the liquid.

obtained by rapidly quenching this melt (contained in a
quartz capillary under vacuum) into ice water. The resulting sample was a rod approximately 1 mm in diameter and .5
gm in mass. As prepared, the glass exhibited an amorphous
x-ray diffraction pattern; and a calorimetric examination
of a small piece of the sample indicated a glass transition
at 643°K and crystallization at 683°K (20°K/min calorimetric
scanning rate) which is consistent with previous work.[12]

The metal glass sample (∿.5 gm) was placed in an
ultrapure quartz container (4 mm inside diameter), alter-
nately evacuated and back filled with helium several times,
and finally sealed with an internal helium pressure of 100
torr to provide good thermal contact. The Faraday method
was used to measure the temperature dependence of the
magnetic susceptibility down to 1.6°K. The magnetic field
for these temperature scans was constant at 12.8 kG. In
addition, at room temperature the sample susceptibility was
measured as a function of field between 12.8 and 2.5 kG
(8 data points) and found to be field independent. This
field independence indicates that the sample was not contaminated by ferromagnetic material.[15]

After the completion of magnetic measurements on the
glass, it was crystallized by heat treating in situ for 16
minutes at 673°K; transformation kinetic studies having
indicated that the material is in its first metastable
crystalline state after such a treatment.[16] Susceptibility
vs temperature measurements were then made on the polycrystalline sample. Again, the susceptibility was field
independent at room temperature insuring that the sample had
not become ferromagnetically contaminated during treatment.

Finally, the sample container was opened, the polycrystal removed, and the contribution of the empty container
measured. The sample susceptibilities are the measured
values minus the container contribution. The container was
so constructed that its contribution was less than 50% of
the total measured value. From the repeatability of our
measurements and the use of several known calibration standards (including single crystal Ge, liquid Hg, polycrystal
In, chrome alum, and pyridine) we conclude that our accuracy
is 3%.

An x-ray diffraction pattern of our polycrystalline
sample (metastable) is the same as that reported by Röschel

and Raub[17] for the crystalline phase Pd_5Si, and no other
phases were detected. For our heat treatment, metallographic
examination has indicated a crystallite size of 1-10
microns,[18] and a Debye-Scherrer x-ray diffraction photograph of our polycrystalline sample showed it to be isotropic with respect to crystal orientations. This result
insures that the true average susceptibility was measured
in the polycrystalline state, even though the structure is
apparently not cubic, and non-cubic crystals do not have
isotropic susceptibilities. X-ray emission spectroscopic
chemical analysis on the sample indicated the presence of
3 ± 2 ppm of Fe and 5 ± 4 ppm of Co.

Our results are shown in Fig. 1, where the susceptibility per gram (χ_g) is plotted versus temperature for the
glassy and polycrystalline phases. At 295°K we measure
$\chi(\text{glass}) = -.045\times10^{-6}$ emu/gm and $\chi(\text{polycrystal}) = -0.126\times10^{-6}$
emu/gm. We observe that above 100°K the susceptibility of
both the crystalline and glassy materials increases linearly

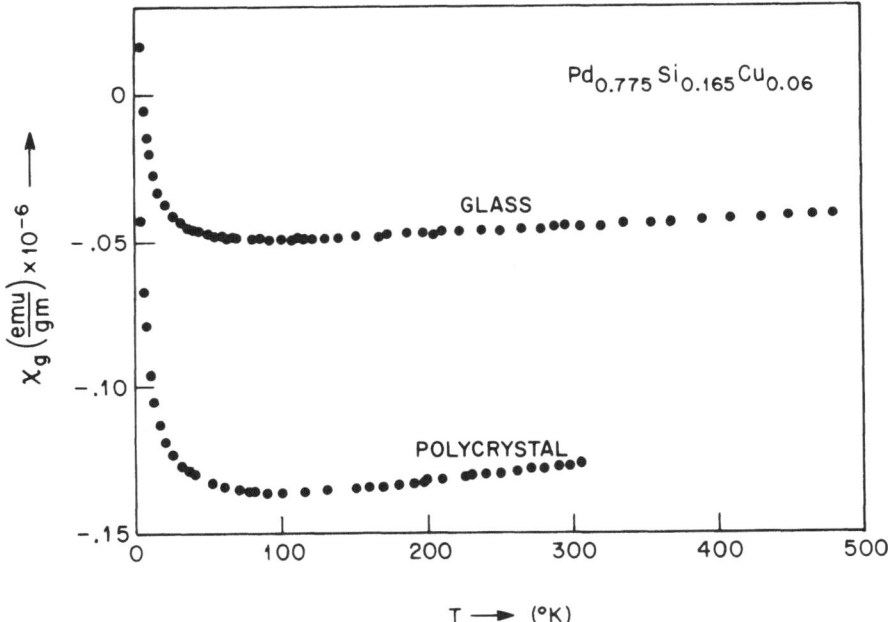

Fig. 1 The temperature dependence of the magnetic susceptibilities (χ_g) of polycrystalline and glassy
$Pd_{.775}Cu_{.06}Si_{.165}$. The polycrystal is obtained by
heat treating the glass for 16 minutes at 673°K.

with increasing temperature, becoming less diamagnetic.
The slope of the susceptibility vs temperature for the
glass is 2.1×10^{-11} emu/gm-°K, and for the crystalline material is 4.7×10^{-11} emu/gm-°K. Below 100°K, a paramagnetic
contribution is observed which increases with decreasing
temperature. This "Curie-tail" can be accounted for by
approximately 7 ppm (atomic ratio) of iron (assuming S =
5/2, g=2), which is consistent with the purity of our starting materials and the chemical analysis of our sample.

Contrary to the usual result for semiconducting glasses
for which χ(glass) $\lesssim \chi$(polycrystal),[19] we observe that
χ(glass) $> \chi$(polycrystal) for this metallic glass. It is
difficult at best to interpret this change in susceptibility
upon crystallization, because many of the terms in a
generalized susceptibility expression are expected to differ
from glass to crystal. However, the measured susceptibility
may be crudely divided into two contributions; one from the
valence bands and the other from the conduction band. If
we assume (consistent with the results on the semiconducting
glasses)[19] that the valence band contribution does not
change significantly upon crystallization, then it would
appear that the observed shift reflects a change in either
the Van Vleck or Landau-Peierls parts of the conduction
electron susceptibility. A large change in the Pauli spin
susceptibility due to a change in the density of states at
the Fermi level is unlikely, because the electronic specific
heat coefficient is approximately the same for both phases.[13]

In contrast to the weak diamagnetism of these Pd rich
alloys, the susceptibility of elemental Pd is large and
paramagnetic (χ_g (273°K) = $+5.15 \times 10^{-6}$ emu/gm). This indicates that much of the d character of the wave functions at
the Fermi surface of Pd is lost; the states at the Fermi
surface of $Pd_{.775}Cu_{.06}Si_{.165}$ are likely of s and/or p character. This conclusion is in agreement with the magnitude
of the measured electronic specific heat coefficient.[13,14]
The temperature dependence of the glass and polycrystal
susceptibilities indicates that there is some structure in
the density of states within approximately kT of the Fermi
surface. Although such structure is common in d band
materials, it has also been identified as the cause of the
temperature dependent susceptibilities in a few s/p band
metals (e.g. the $AuGa_2$ family).[20]

We thank Eva M. Vogel and J.V. Wasczczak for their assistance.

REFERENCES

1. W. Klement, Jr., R.H. Willens, and Pol Duwez, Nature 187, 869 (1960).
2. C.C. Tsuei and Pol Duwez, J. Appl. Phys. 37, 435 (1966).
3. Pol Duwez and S.C.H. Lin, J. Appl. Phys. 38, 4096 (1967).
4. C.C. Tsuei, G. Longworth, and S.C.H. Lin, Phys. Rev. 170, 603 (1968).
5. C.C. Tsuei and R. Hasegawa, Solid State Comm. 7, 1581 (1969).
6. R. Hasegawa, J. Appl. Phys. 41, 4096 (1970).
7. R. Hasegawa and C.C. Tsuei, Phys. Rev. B2, 1631 (1970).
8. R. Hasegawa, Phys. Rev. B 3, 1631 (1971).
9. A.K. Sinha, J. Appl. Phys. 42, 5184 (1971).
10. T.E. Sharon and C.C. Tsuei, Phys. Rev. B5, 1047 (1972).
11. Pol Duwez, R.H. Willens, and R.C. Crewdson, J. Appl. Phys. 36, 2267 (1965).
12. H.S. Chen and D. Turnbull, Acta Met. 17, 1021 (1969).
13. B. Golding, B.G. Bagley, and F.S.L. Hsu, Phys. Rev. Letters 29, 68 (1972).
14. H.S. Chen and W.H. Haemmerle, J. Non-Crystalline Solids, in press.
15. P.W. Selwood, "Magnetochemistry", 2nd ed. Interscience, New York (1956).
16. B.G. Bagley, to be published.
17. E. Röschel and C.J. Raub, Z. Metallkunde 62, 840 (1971).
18. B. Miller, private communication.
19. J. Tauc and F.J. DiSalvo, these Proceedings.
20. W.W. Warren, Jr., R.W. Shaw, Jr., A. Menth, F.J. DiSalvo, A.R. Storm, and J.H. Wernick, to be published.

DISCUSSION

S.G. Bishop: To shift for a moment from the magnetic properties to the transport properties of these materials you refer to as metallic glasses. In amorphous semiconductors you encounter low mobilities; of the order of 1 (or less) cm^2/volt-sec. What happens to the transport properties of these metals when they become disordered? Do you get the same evidence of extremely low mobility?

B.G. Bagley: At present, there is little data available
on the transport properties of metallic glasses, and even
less on the changes which occur upon crystallization.
Duwez et al. (Ref. 11) observed the room temperature resistivity of a $Pd_{83}Si_{17}$ glass to be 60 $\mu\Omega$-cm and only weakly
temperature dependent between 2° and 300°K. Upon crystallization, the room temperature resistivity decreases by a
factor of about 3 and becomes more temperature dependent.
R.H. Willens (private communication) observed that a Pd_{80}
Si_{20} glass has a Hall coefficient of -9.3×10^{-11} m^3/cou-
lomb; a magnetoresistance of < 1 part in 10^6 at 10 kG; and
a thermoelectric power (at 273°K) of -1.8 μV/°K. W.A. Reed
and H.S. Chen (private communication) observed that a
$Pd_{0.775}Cu_{0.06}Si_{0.165}$ glass has a room temperature resistivity of 95 $\mu\Omega$-cm; a Hall coefficient of approximately 10^{-10}
m^3/coulomb; and a magnetoresistance of less than 1 part in
10^4 at 20 kG. In these two studies, the properties of the
crystalline material were not measured so a comparison cannot be made. The change in transport upon disordering a
metal is not as marked as in the semiconductors, where on
crystallizing a glass a factor of 10^8 decrease in resistivity is not uncommon.

AMORPHOUS AND CRYSTALLINE SELENIUM - A MÖSSBAUER EFFECT STUDY

P. Boolchand

Physics Department, University of Cincinnati

Cincinnati, Ohio 45221

The Mössbauer effect (ME) on Te^{125} impurity imbedded in Se, has been used to investigate the amorphous and crystalline phases of the host. Addition of small amounts of Te in Se is known to form copolymeric chains (c-chains) and mixed rings at the expense of Se_n chains and Se_8 rings.[1] ME spectra of $Se_{0.98}Te_{0.02}$ in both amorphous and crystalline phases, reveal the presence of large quadrupole splitting (QS) of 12.72(40) mm/s and 10.33(8) mm/s in mixed rings and c-chains. The QS in hexagonal $Se_{0.98}Te_{0.02}$ was attributed to that of c-chains.

The large QS in mixed rings is intriguing and may be compared to that in c-chains. In both cases the dominant contribution to the EFG comes from the unbalanced p electron density which is determined by the symmetry and covalent bond lengths Te forms with its two near neighbor Se.[2]

We propose that the large QS in mixed rings results due to a shorter covalent Te-Se bond in the ring geometry. Se_8 rings are considerably smaller in size than c-chains, and one may expect them to be stiffer than Se chains in accomodating a larger Te atom for Se. It is therefore likely that the Te-Se bond length will be smaller in mixed rings than in c-chains. This interpretation is supported by isomer shift measurements that show a larger shift in

149

mixed rings (+ 1.0(2) mm/s) than in c-chains (+ 0.4(2) mm/s).
Clearly the effect of a smaller covalent Te-Se bond leads to
a larger time average of both the unbalanced 5p electron
density and the s electron density at Te. One may also
expect the recoil free fraction in mixed rings to be substantially higher, a fact which is supported by the ME
results.

REFERENCES

1. A.V. Tobolsky and G.D.T. Owen, J. Polymer Sci. 59,
    329 (1962).
2. P. Boolchand and P. Suranyi, Phys. Rev. B (to be
    published).

DISORDER AND STRONG CORRELATIONS IN NARROW ENERGY BANDS

F. Brouers, P. Lederer and M. Héritier

Laboratoire de Physique des Solides, Université

Paris-Sud, Centre d'Orsay - 91, Orsay (France)

INTRODUCTION

We have studied the density of states and the mobility of a hole in a disordered magnetic insulator, in order to investigate the joint effect of atomic disorder and spin disorder on the properties of magnetic insulators.

Our reason for undertaking this study is that a number of recent, and less recent, experiments point out the necessity of taking into account simultaneously atomic disorder and electron electron interactions in binary alloys of narrow band materials as well as amorphous semiconductors. For example, in the mixed system $V_{1-x} Nb_x O_2$, the high temperature metallic phase of VO_2, which is Pauli paramagnetic, becomes insulating and magnetic, with a good Curie susceptibility, when the concentration of Nb exceeds 0.15. In this narrow d-band material, which contains one electron per atom, this insulating phase cannot be understood[1] unless both Coulomb interactions and atomic disorder are taken into account. A similar situation has been discussed for the case of VO_x[2]. In the case of amorphous semiconducting alloys, such as SiP, one cannot understand the metal-insulator transition, and the corresponding magnetic properties, without taking into account both effects[3].

The reason is simply that for such systems, the width of the band, the width of the distribution of atomic levels, and the Coulomb interactions have comparable magnitudes. This causes great difficulties in the theoretical description of the influence of electron-electron interactions on the localization of electrons in amorphous systems, or the influence of disorder on the Mott transition[4]. This is the reason why we have chosen to study here the case of a binary alloy with almost one electron per atom and infinite Coulomb interactions, which we consider as a starting point for the more interesting case of finite interactions. The experimentally relevant systems would be mixed transition metal oxides of Ni, Co or Fe, which can be good Mott insulators, rather than of V or Cr which exhibit Mott transitions with temperature or pressure, and amorphous semiconductors such as SiP, below the critical concentration.

MODEL

The Hamiltonian for our problem is

$$(1) \quad H = \sum_{i,\sigma} t_{ii} n_{i\sigma} + \sum_{i,j,\sigma} t_{ij} c^+_{i\sigma} c_{j\sigma} + U \sum_i n_{i\uparrow} n_{i\downarrow}$$

where t_{ii} can take two values E_A or E_B according to whether site i is occupied by an A atom or by a B atom. t_{ij} is taken to be a constant if i and j are nearest neighbours, zero otherwise. U is the intra atomic Coulomb integral. Taking U infinite amounts to neglecting configurations with double or zero occupancy of atomic sites. With one hole present, the Hamiltonian which describes this magnetic insulator becomes

$$(2) \quad H' = P \left(\sum_{i,\sigma} t_{ii} n_{i\sigma} + \sum_{i,j,\sigma} t_{ij} c^+_{i\sigma} c_{j\sigma} \right) P$$

where P projects on the subspace of wave functions

$$(3) \quad \psi_{i\alpha_i} = (-1)^i c^+_{1,\sigma_1} c^+_{2,\sigma_2} \cdots c^+_{i-1,\sigma_{i-1}} c^+_{i+1,\sigma_{i+1}} \cdots c^+_{N\sigma_N} |0>$$

where $|0>$ is the vacuum and α_i denotes the spin configuration σ_1, σ_2, $\ldots \sigma_N$. The convention of sign $(-1)^i$ for $\psi_{i\alpha_i}$ allows to discuss in a parallel way the case $N_e > N$ and the case $N_e < N$ by considering holes in the former case instead of electrons in the latter case, and by changing the sign of t. For s.c. and b.c. c. lattices, one can divide the lattice into two sublattices, such that all the nearest neighbours of a lattice point on one sublattice belong to the other sublattice[5]. Then t_{ij} changes sign if we introduce the phase factor $(-1)^i$ to the atomic wave functions at the lattice points on one sublattice.

GROUND STATE FOR THE DISORDERED ALMOST HALF-FILLED NARROW s-BAND

We have proved the following theorem :
For s.c, b.c.c. lattices and f.c.c., h.c.p. lattices with $N_e = N - 1$ and $U = \infty$, the ferromagnetic state with the maximum total spin is the ground state of the system.

The demonstration of this theorem follows closely the proof given by Nagaoka for the narrow almost half filled s-band in the pure case[5]. It will be given in **detail** elsewhere[6]. We shall only sketch it briefly here.

First one finds the lowest energy state when $N\uparrow = N_e$ and $N\downarrow = 0$, using the localization theorems for a disordered binary alloy[7]. Then one shows that there is no state with energy lower than this state and then that there is no state with the same energy and $S < S_{max}$. Whence it follows that the ground state has maximum spin and is the completely ferromagnetic state.

The physical meaning of Nagaoka's result in the pure case has been discussed recently by Izuyama[8]. In our case as well, the main interest of this theorem is that it exists but it has little importance in physical systems since it breaks down for finite U when $\frac{N - N_e}{N} < \frac{t}{U}$, and it has no consequence on the free energy at finite temperatures.

SINGLE PARTICLE EXCITATIONS IN DISORDERED MAGNETIC INSULATORS

A) Retraceable Path Approximation.
We have computed the density of states of the extra hole, as Brinkman and Rice[9] did for the pure case. We have examined three configurations i- ferromagnetic (F) ;

ii - antiferromagnetic (AF) and iii - random (R).

The F case reduces to a simple tight binding binary alloy band. For AF and R configurations, one can argue that the dominant contribution to the density of states comes from retraceable paths, most closed loops being suppressed because they change the spin configuration[9].
Let us write the Green's function

(4) $\quad G_{ii}^{\alpha_i} (\omega)^{-1} = \omega + E_i - \sum_i^{\alpha_i} (\omega)$

The simplest approximation to the self energy $\sum^{(1)}(\omega)$ is a walk to the nearest neighbour and an immediate return. Let x be the concentration of atoms A, (1 - x) that of B, and z the coordination number. Then

(5) $\quad \sum^{(1)}(\omega) = zt^2 \left(\dfrac{x}{\omega + E_A} + \dfrac{1 - x}{\omega + E_B} \right)$

The modification of $\sum^{(1)}(\omega)$ if we include a walk from nearest neighbour to next nearest neighbour is

(6) $\quad \sum^{(2)}(\omega) = zt^2 \left[\dfrac{x}{\omega + E_A - (z - 1)t^2 \left(\dfrac{x}{\omega + E_A} + \dfrac{1 - x}{\omega + E_B} \right)} \right.$

$\left. + \dfrac{1 - x}{\omega + E_B - (z - 1)t^2 \left(\dfrac{x}{\omega + E_A} + \dfrac{1 - x}{\omega + E_B} \right)} \right]$

Finally by repeated application, we generate a continued fraction, which we can solve easily, obtaining a 3rd degree equation for $\sum(\omega)$:

(7) $\sum(\omega) - zt^2 \left[\dfrac{x}{\omega + E_A - \dfrac{z - 1}{z} \sum(\omega)} + \dfrac{1 - x}{\omega + E_B - \dfrac{z - 1}{z} \sum(\omega)} \right] = 0$

The density of states is computed from

$\overline{G}(\omega) = \dfrac{x}{\omega + E_A - \sum(\omega)} + \dfrac{1 - x}{\omega + E_B - \sum(\omega)}$

(7) has been solved numerically for various values of x and

$(E_A - E_B)/t$. The results agree remarkably well with the self consistent C. P. A. approximation described below.

B) C. P. A. approximation.

This approximation amounts to treating the medium as a translationnally invariant averaged medium as far as the lattice potential is concerned. In the expression for $\sum(\omega)$, one replaces all atomic energies E_i by a self consistent potential $\alpha(\omega)$. Thus the self energy becomes

$\sum(\omega) = \sum°(\omega + \alpha(\omega))$ where $\sum°(\omega)$ is the self energy for $E_A = E_B$. Then one writes the average Green's function

$$\bar{G}_{ii}(\omega) = x\, G_{ii}^A + (1 - x)C_{ii}^B = 1/\left[\omega + \alpha(\omega) - \sum°\left(\omega+\alpha(\omega)\right)\right]$$

(8)

$$= x/\left[\omega + E_A - \sum°\left(\omega+\alpha(\omega)\right)\right] + (1-x)/\left[\omega + E_B - \sum°\left(\omega-\alpha(\omega)\right)\right]$$

Then for F configuration $\sum°(\omega)$ is the **unperturbed** tight binding band self energy, while for the AF and R case we take the Retraceable Path Approximation of Brinkman and Rice's for the pure system.
The self consistent approximation has been solved numerically for various concentrations and atomic potentials. Detailed results will be given elsewhere ; we only show one typical result from the C. P. A. calculation, for both F and R configuration (see Fig. 1). One interesting result is that the band narrowing found in the pure case when going from F to R or AF configuration has the same order of magnitude for the alloy for the total distribution of energy states, but does not hold for each sub band : the gap between minority and majority sub bands depends very much on spin configuration, and can be smaller for R and AF case than for the F case. This opens the possibility for large magnetic field effects on the single particle density of states in various magnetic insulators[10].

MOBILITY OF A HOLE IN THE DISORDERED MAGNETIC SEMICONDUCTOR

Within the approximation that the dominant paths for the motion of the hole are those with no closed loops, in the R and AF configurations, Brinkman and Rice[9] have calculated the motion of a hole in the pure case. They find that the hole undergoes Brownian motion because of the strong scattering due to non ferromagnetic arrangements of the spins.

In our case, atomic disorder does not change significantly the picture. Indeed we find that the mobility of a hole is reduced only by a factor 0.9 to 0.25 at most compared to its value when $E_A = E_B$ in the R and AF case. For concentrations of the order of 0.30 and $E_A - E_B \sim zt$, the change in the mobility due to atomic disorder is of about 25 % at low temperatures, and 15 % at high temperature. The mobility of the hole, defined as $\mu = \sigma \Omega /e$ is calculated from the following expression

$$\mu = \frac{2\beta\, ea^2\, t^2 \displaystyle\int_{-zt}^{+zt} \frac{d\omega}{\pi}\, e^{-\beta\omega}\, \bigl(\mathrm{Im}\, G(\omega - i\delta)\bigr)^2}{\displaystyle\int_{-zt}^{+zt} \frac{d\omega}{\pi}\, e^{-\beta\omega}\, \mathrm{Im}\, G(\omega - i\delta)}$$

Within this approximation, the mobility is proportional to the average over the thermally occupied states of the hopping probability $zt^2\, \mathrm{Im}\, G(\omega - i\delta)$.

One typical result is shown on fig. 2. Two different behaviours occur for high temperature and low temperatures. At high temperature, the mobility decreases weakly with increasing concentration of impurities. At low T, the mobility may decrease sharply if there is a small concentration of low energy impurities, and it increases again with increasing impurity concentration, which reflects the increase in the density of states at the bottom of the band.

More detailed results on the mobility will be given elsewhere[10].

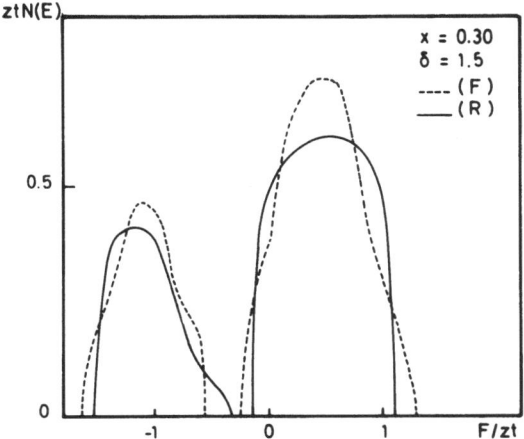

Fig. 1. Density of states for F and R configuration for a concentrated alloy with $\delta = (E_A - E_B)/zt$.

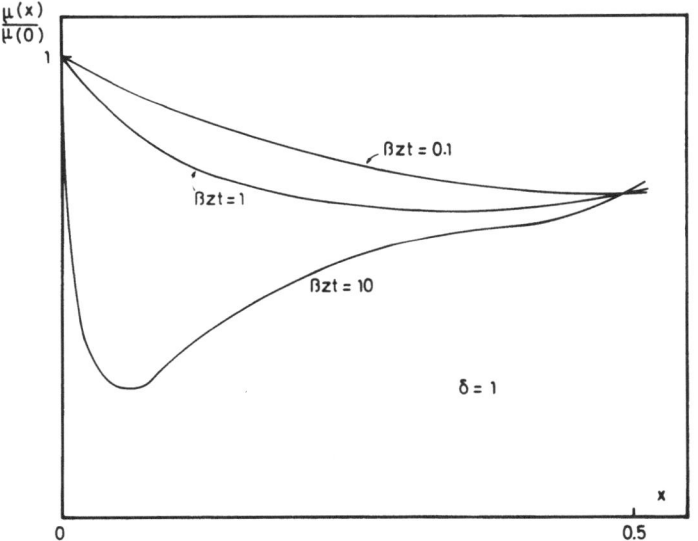

Fig. 2. Mobility vs. concentration for three different temperatures. $\beta = 1/k_B T$.

REFERENCES

(1) P. Lederer et al., J. Phys. Chem. Sol. 33, 1969 (1972).
(2) N. F. Mott, Phil. Mag. 24, 935 (1971).
(3) N. F. Mott (private communication).
(4) M. Héritier, P. Lederer, to be published.
(5) Y. Nagaoka, Phys. Rev. 147, 392 (1966).
(6) F. Brouers, M. Héritier, P. Lederer, to be published.
(7) D. J. Thouless, Journ. Phys. C. 3, 1559 (1970).
(8) T. Izuyama, Phys. Rev. B 5 , 190 (1972).
    In our case, there is no inversion symmetry of the
    space coordinates, but the extension of our result to
    the thermodynamic limit would be as difficult as in
    Nagaoka's case.
(9) W. F. Brinkman and T. M. Rice, Phys. Rev. B 2, 1324
    (1970).
(10) P. Lederer, F. Brouers and M. Héritier, to be publi-
    shed in Sol. St. Comm.

DISCUSSION

D.M. Esterling: I have two questions. One is: Brinkman
and Rice had calculated two particle correlation functions
and approximated by $G(1) G(1)$. I wondered whether you did
the same thing. The second is: Would you care to comment
on your technique of averaging over all possible random
spin configurations and comparing the result of that with
the ferromagnetic state rather than comparing the paramagnetic state which has an equal number of up and down
spins, and which is quite different from the random state,
with the ferromagnetic state.

P. Lederer: For the first question essentially the Brinkman-Rice approximation is even more valid when you have
atomic disorder, so we use the same. You can show that
higher order corrections are even smaller. For your second
question, I didn't quite follow.

D.M. Esterling: The difference between a random configuration and a paramagnetic configuration, I think, may be
quite important. A paramagnetic configuration is a situation where you have equal number of up and down spins. A
random configuration, if I understand the Brinkman-Rice de-

finition of that, is essentially averaging over all possible
numbers of up and down spins.

P. Lederer: The probability of having a spin up or a spin
down is 1/2, so that I think that number of up-spins and
down-spins is the same. Of course, you don't have dynamics
there, and so I agree, it's a crude approximation.

D.M. Esterling: No, I don't think they use an equal number
of up and down spins in the random averaging. I may be
mistaken but I don't think so.

P. Lederer: I think so.

E.J. Siegel: Have you ever tried for finite U, in other
words, as I asked before, have you ever tried to put in the
Pauli Principle for parallel spins. It seems as if you are
neglecting a lot of pairs of parallel spins, and that may
give a large contribution.

P. Lederer: The reason why you can make this theory is the
Pauli Principle. Essentially you cannot occupy one site
with two like spins since this is an s-band, and two antiparallel spins cannot sit on the same site because the interaction is infinite between opposite spins. Just to answer your question in a partial way we have investigated the
stability of the ground state and obviously there is just
one simple small change with the Nagaoka theory, which is,
that if the difference between the energy levels of this
binary alloy is smaller than U then the result of Nagaoka
essentially remains that the stability of this ferromagnetic
ground state is given by the condition that the number of
holes over the total number of sites must be larger than
T/U. This is about all we did for the finite U case. So,
in fact, I should say that this theorem is not very useful,
because in all practical cases in real life you will not be
in that limit, but this is a theorists theorem.

OCCURRENCE OF LOCAL MAGNETIC MOMENTS IN DISORDERED MATERIALS

M. Cyrot

Laboratoire de Physique des Solides, Université

Paris-Sud, Centre d'Orsay, 91405-Orsay, France

There is a growing interest in magnetism in dilute magnetic systems and in amorphous materials due to new experimental results. The possibility of amorphous magnetic materials was first point out by Gubanov[1] over a decade ago and the existence of such materials has been confirmed[2][3]. Dilute magnetic systems were studied both experimentally by alloying a ferromagnet with a non magnetic material and theoretically by using a Heisenberg model where the spins occupy only part of the lattice sites[4]. In the following, we are interested not in the appearance of an ordered magnetic phase, assuming that moments exist on the sites, but on the appearance of these local magnetic moments in these disordered structures. Thus, we start from a band scheme, and introducing correlations between electrons, study the appearance of magnetism.

As a model, we take a material whose band is described in a tight binding scheme. We take into account correlations between electrons only when they are on the same site as in the Hubbard model. Disorder is introduced by making inaccessible to electrons random sites in the lattice. Thus our hamiltonian is :

$$H = \sum_{ij\sigma}{}' T_{ij\sigma} c_{i\sigma}^{+} c_{j\sigma} + \sum_i U n_{i\uparrow} n_{i\downarrow} . \qquad (1)$$

The atomic level on a concentration c of the sites is taken

infinite in order to prevent a jump of the electrons on these sites. Our model can be applied to the appearance of magnetism in dilute systems or to the effect of alloying on magnetic materials. Some of our results may also be qualitatively valid for amorphous materials because their atomic density is less than in the crystalline state.

In a series of papers,[5] we discussed the appearance of magnetism in the Hubbard model for a crystalline material. For this purpose, we separated in the last term of Eq. 1 the charge and magnetization part:

$$U n_{i\uparrow} n_{i\downarrow} = -\frac{U}{4}(n_{i\uparrow} - n_{i\downarrow})^2 + \frac{U}{4}(n_{i\uparrow} + n_{i\downarrow})^2. \qquad (2)$$

The complicated functional integral formalism that we used[6] can be avoided for a qualitative description by making a Hartree-Fock approximation on the two terms of Eq. 2. The charge appears to be always one, in the case of one electron per atom, so we consider only the magnetic part in the following by rewiting Eq. 2 as

$$-\frac{U}{2}\mu_i(n_{i\uparrow} - n_{i\downarrow})$$

Having solved the linearized hamiltonian, Eq. 1, we determine self-consistently the magnetic moment on site i by

$$\mu_i = <n_{i\uparrow}> - <n_{i\downarrow}>. \qquad (3)$$

In the crystalline case, the linearized hamiltonian, Eq. 1, turns out to be the model hamiltonian used in alloy problems. The best approximation now available for this purpose, is the coherent potential approximation (C.P.A.).[7] By using it, we calculated the self-consistent μ_i as a function of $\frac{U}{W}$ (Fig. 1). W is the band width and in a simple cubic case is given by $2 Z T_{ij}$, where Z is the number of nearest neighbors. For the numerical computation, we used a Hubbard band for the density of states in absence of correlations.

$$n(E)_{\text{cryst.}} = \frac{8}{\pi W^2}\left(\left(\frac{W}{2}\right) - E^2\right)^{1/2}, \quad |E| < \frac{W}{2}$$

$$= 0 \qquad\qquad\qquad |E| > \frac{W}{2}.$$

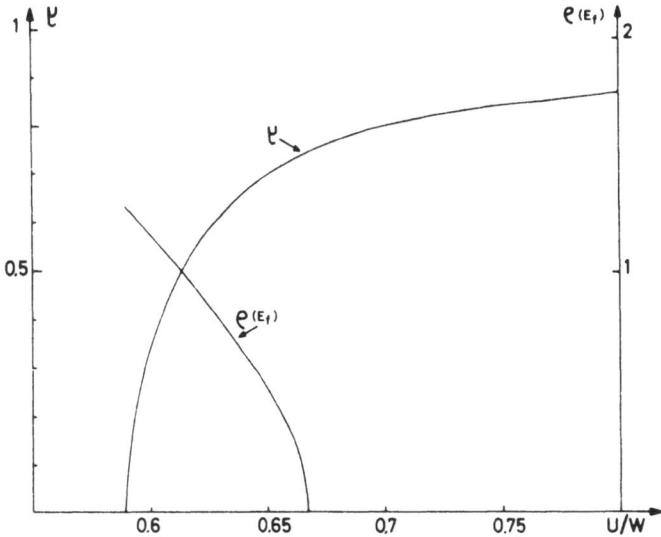

In our model hamiltonian for disordered systems, we have first to calculate the density of states for the system in absence of correlations. In the Hubbard model, we made inaccessible to electrons a concentration c of sites by making the atomic level infinite. For U = 0, we have a kind of alloy which can be treated using the C. P. A. For this particular "alloy" the self-energy is given by Soven's relation[7] where we take $\varepsilon_A = \infty$ and $\varepsilon_B = 0$. Thus

$$\sum(E) = -\frac{c}{F(E)}$$

with

$$F(E) = \int dE \frac{n(E')}{E - E' - \sum(E)} \quad .$$

This can be solved to give

$$n_{dis.}(E) = \frac{8}{\pi W^2} \left((\frac{W}{2})^2 (1 - c) - E\right)^{1/2} \text{ for } |E| < \frac{W}{2} \sqrt{1 - c}$$

$$= 0 \qquad\qquad\qquad\qquad \text{ for } |E| > \frac{W}{2} \sqrt{1 - c} \quad .$$

The interest of this result stems from the fact than in our
model of disorder, the shape of the band is not changed
compared to the crystalline case and the width is reduced

by a factor $\sqrt{1-c}$. Thus our results for the crystalline
case, are valid through a simple scaling of W. We emphasize
that this approximation, only described the reduction of
the average kinetic energy in the presence of strongly
repulsive energy, i.e., we neglect completely all the incoherent scattering effects of the impurities. This prevents
any application to the onset of Anderson localization.

The first consequence of this result is that there
exist a direct scaling between concentration and pressure
in the low concentration case. Low pressure or impurities
tend both to favour the appearance of magnetism. If we assume that U is not strongly dependent on pressure, the effect
of pressure on hamiltonian, Eq. 1, is just to change the bandwidth. In a tight binding approach

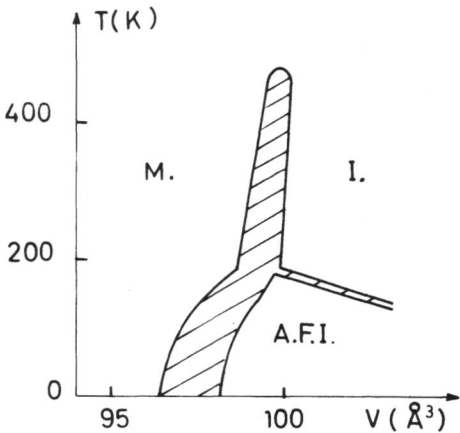

$$W = W_o \, e^{-q\left(\dfrac{V - V_o}{V_o}\right)}$$

q is proportional to the Slater coefficient. Thus

$$\frac{dW}{W} = \frac{q}{B} \, dp,$$

where B is the Bulk modulus. When introducing a concentration c of inaccessible sites:

$$\frac{dW}{W} = -\frac{c}{2}$$

Thus we have the equivalence between pressure and concentration

$$dp = -\frac{cB}{2q} \quad.$$

We apply this formula to the experiment of Mc Whan and Remeika[8] who showed that introduction of one per cent of Chromium in V_2O_3 is equivalent to a change of 3.6 kb. Since Cr is a stable trivalent ion, one might as a first attempt regard it simply as deleting a site from the conduction band of pure V_2O_3.[9] Using a value for B of 2.14 kb, we find q = 3. This is to be compared with the value of q = 2 that we obtain by fitting other experiments with theory.[10] These two values live within the region one can expect.[11] The critical ratio of U/T_{ij} for the appearance of moments is decreased in the disordered state because of the narrowing of the band. Appearance of moments is easier and their values are larger than in the crystalline case. On this basis, one should expect that a nearly magnetic crystalline material could become magnetic when in an amorphous state.

In a pure material we have shown[10] that if the bulk modulus per atom is small, there is an instability linked with the appearance of moments with decreasing pressure. This leads to a first order transition with change of volume. In this calculation we assumed that the only effect of pressure was through the change of bandwidth. Thus in the temperature versus volume plane, we have a region which corresponds to unstable phases (Fig. 2). When the transition is obtained by alloying instead of varying pressure, choosing concentrations corresponding to the unstable region, we have a miscibility gap. The system will become inhomogeneous and there will be two phases with two different concentrations. The miscibility gap is however very narrow because it is linked with the electronic instability. This seems to be the case in doped V_2O_3. Above a critical temperature which is given by

$$\frac{T_c}{U} = \frac{8}{3\sqrt{6}\,\pi^2}\,(1 - \frac{9\pi^2}{160}\,\frac{Y}{q^2})\ ,$$

where Y is the bulk modulus per atom, there is no instability. Thus it should be possible to cool from above the demixing temperature and obtain a continuous metastable transition with increasing concentration.

It is interesting to compute the interaction between moments. However for such a calculation, one must in principle take into account the incoherent scattering effects of the impurities. However, as soon as Hartree-Fock magnetic moments μ_i are stable, the interaction is strongly damped even in the pure case. The mean free path effect due to the impurities will not change qualitatively the features. We have found for $\frac{U_\mu}{W} \ll 1$ an interaction damped by the exponential factor

$$\exp\ (\ -\ 6\mu^2\ \frac{U^2}{W^2}\ \frac{r}{a}\ )\ ,$$

where a is the interatomic distance. A small decrease of W produces an increase of μ and the range of interaction decreases rapidly towards some interatomic distances. In this model, the interaction is always limited to nearest neighbors.

Up to now, we considered a simple scaling of the bandwidth. However, if we want to compute the ordering temperature for the magnetic moments, we have to take into account the structure. For instance it is well known that above a critical concentration of non magnetic sites, an ordered phase will not occur[4]. Thus the ordering temperature cannot be scaled with the crystalline one. We will not discuss this kind of problem.

REFERENCES

1. A.I. Gubanov, Sov. Phys. Solid State $\underline{2}$, 468 (1961).
2. C.C. Tsuei and Pol Duwez, J. Appl. Phys. $\underline{37}$, 435 (1966).
3. K. Tanura and H. Endo, Phys. Letters $\underline{29A}$, 52 (1969).
4. G.A. Murray, Proc. Phys. Soc. $\underline{89}$, 87 (1966).
5. M. Cyrot, Phys. Rev. Letters $\underline{28}$, 871 (1970); J. de Physique $\underline{33}$, 125 (1972); Phil. Mag. $\underline{25}$, 1031 (1972).
6. I.C. Kimball and J.R. Schrieffer, International Conference on Magnetism and Magnetic Materials, Chicago (1971).
7. P. Soven, Phys. Rev. $\underline{156}$, 809 (1967).
8. D.B. Mc Whan and J.P. Remeika, Phys. Rev. $\underline{B\ 2}$, 3734 (1971).
9. T.M. Rice and W.F. Brinkman, Phys. Rev. $\underline{B\ 5}$, 4350 (1972).
10. M. Cyrot and P. Lacour-Gayet, to be published.
11. J.C. Slater and G. Koster, Phys. Rev. $\underline{94}$, 1498 (1954).

ONSET OF MAGNETIC ORDERING IN RANDOM SUBSTITUTIONAL ALLOYS

B.R. Coles

Physics Department, Imperial College

London SW7 2BZ, England

In the early days of the theory of magnetism in transition metals considerable interest attached to the ways in which the magnetic order (usually ferromagnetism) of pure metals was modified by the addition of solute metals, and especially to the variations from solute to solute for a given host and from host to host for a given solute. More recently the large amount of work on the problem of dilute magnetic alloys has led to a renewed interest in the problem of how magnetic order appears when the concentration of transition metal or rare-earth metal solute is increased in solid solution in a host metal (or occasionally intermetallic compound) which does not itself undergo magnetic ordering at any temperature. We shall give here a brief review of the types of situation that can arise, drawing attention to any particularly interesting systems.

SOLUTE AND SOLVENT CHARACTER

Although the tremendous experimental and theoretical activity on the dilute alloy problem has not yet provided the promised clarification of magnetic order in metals, it has at least provided us with a basis for the classification of solid solution systems. Thus we can distinguish between "good moment" solutes, where susceptibilities resemble those of dilute magnetic salts; "no moment" solutes, where any change from the host susceptibility is temperature-independent and comparatively small; and "intermediate moment" systems where a temperature-dependent susceptibility is conferred by the solute, especially at high temperatures, but any fit to a Curie-Weiss expression seems to involve a θ value which is quite significant $(10° - 10^2$ K) but concentration independent.

We shall not dwell here on the present state of received doctrine in the Kondo problem; but we shall recognize that, whether approached from the s-d exchange model (using the Kasuya-Kondo Hamiltonian), or by the postulate that virtual bound states of the Friedel-Anderson type, while satisfying some criterion for being non-magnetic, are capable of undergoing local spin fluctuations, it is possible to specify approximately some temperature T^* above which local moments are detected in various properties (couplings of which to and via the conduction electrons are a minor perturbation). The form T^* will be used to avoid arguments over whether it should be regarded as T_K or T_{sf} and we shall ignore factors (of the order 3) which sometimes connect theoretical values of these temperatures with characteristic temperatures yielded by fitting data to

$$\chi = C/(T - T^*) \quad \text{or} \quad \rho = \rho_o(1 - (T/T^*)^2)$$

The local spin fluctuation concept has the advantage that T^* can be regarded as associated with an impurity yielding a scattering resonance in the band structure at the Fermi surface (a virtual bound state) or with an impurity differing from the host atoms mainly in the magnitude of the local exchange enhancement arising from on-site intra orbital Coulomb repulsion. The Kondo approach has the advantage that explicit expressions for a characteristic temperature emerge from theoretical work, and can be obtained by those brave enough from fits to data

We may then rank our "magnetic" solutes according to the value of T^* appropriate to the dilute (concentration-independent) limit. Solvents are arbitrarily classified into:

(a) simple materials with no partly filled d- or f- shells

(b) simple transition metals where there is no evidence for significant exchange enhancement, and

(c) "enhanced" transition metals where the susceptibility is significantly larger than that estimated from the electronic specific heat.

T^* (K)	10^{-3}	10^0	10^3
Solute Type / **Host Type**	Good moments	Kondo moments	No moments
Simple metal	AuGd* CuMn AuMn	ZnMn CuFe AuFe	AuCo AlMn AlFe CuNi
Simple transition metal		MoFe RhFe	MoCo NbFe TiFe VFe RhCo RhNi
Enhanced transition metal	PdFe	PtFe PtCo PdCo	PtNi PdNi

*
The concept of a characteristic temperature is probably not appropriate for good moments on 4f shells where a positive spin coupling to conduction electrons is normal.

For completeness we should include as a separate host type transition metals where there is evidence, experimental or theoretical, of a wave-vector dependent susceptibility $\chi(q)$ that peaks at some non-zero value of q; such materials will be discussed briefly later but are not included in the table given above. This table is only schematic but serves as a basis for the discussion of the effects of increasing solute concentration where metallurgical considerations permit this to take place without phase separation.

INTERACTIONS BETWEEN "GOOD" MOMENTS

It was recognized many years ago that exchange interactions between local moments and conduction electrons would lead to a long-range coupling between the local moments on different atoms, and such an effect will exist whether the local-conduction coupling is a "true" exchange or an "effective" exchange that arises from local state-conduction electron mixing. Such coupling is normally described as arising from the RKKY (Ruderman-Kittel-Kasuya-Yosida) interaction, and shows itself in many dilute alloy properties as well as in the pure rare-earth metals* to which Kasuya applied it and the nuclear magnetic relaxation phenomena for which it was introduced by Ruderman and Kittel.   Because of the sharp cut-off at the Fermi level of the conduction electron distribution the polarization induced by a local moment in the conduction band oscillates with wave-number $2k_F$, and no simple general conclusion can therefore be drawn concerning the character of any magnetic order that results from it in an alloy of finite concentration.   (It is important to note that, as emphasized by Anderson, a random distribution of local moments over the points of a lattice does not in general lead to a vanishing of the Curie-Weiss θ.)   There appear to be cases, e.g. $LaAl_2$ - $GdAl_2$ and GeTe - MnTe, where the first clearly visible magnetically ordered state that results is ferromagnetic; but many systems of interest, e.g. Au-Fe, Cu-Mn, Mo-Fe, seem to be characterized by the co-existence of couplings of both ferromagnetic and antiferromagnetic character and it is for such systems that the term spin glass has come into use.   For such good moment systems the temperature characteristic of such interactions will be greater than any possible Kondo temperature for all concentrations of solute above a small fraction of 1%, and for Cu -Mn effects due to these interactions seem to dominate in the millidegree, parts per million regime of temperature and concentration.

It should be emphasized that our understanding of the microscopic spin structure of the low temperature state is very unclear; in what follows we shall regard a spin glass as a random solid solution of moment-bearing atoms in a non-magnetic host, which after cooling to very low temperatures in zero applied field has the solute moments "frozen" in local molecular fields, these fields having a distribution of magnitudes and directions such that the net magnetization of any region containing a few tens of solute atoms is zero.   It follows naturally that the spin entropy has frozen out at low temperatures, that the susceptibility as the spins lock in will fall below the Curie -Weiss behaviour, probably giving rise to a maximum, and that cooling in a large applied field through the spin glass temperature range will result in a biassing of

---

*
How large a role it plays in pure transition metals where d-d overlaps are significant is still not clear.

the distribution of molecular fields such that a thermo-remanent magnetization and a displaced magnetization-field relationship will be found at low temperatures.   All these features were observed in the classic work of Kouvel[1] on the Cu-Mn and Ag-Mn systems and some of them have been found in the Au-Fe and Mo-Fe systems where T* is a few tenths of a degree in the dilute limit.   The first clear evidence that ordering effects extended to very low concentrations, where near-neighbour interactions were unlikely, was provided by the freezing-out of spin-disorder scattering[2] in the electrical resistivity of alloys of this type below a temperature of the order of $10c$ °K where c is the solute concentration in atomic per cent.[3]   Mössbauer effect measurements in suitable systems provided powerful supporting evidence.   There are some features, however, which have not yet received a totally satisfactory explanation and some of which may involve details of the metallurgical character such as tendencies to short range atomic ordering or to clustering.

One effect which is revealed in some detail by Dr. Cannella's paper at this conference is the sharp maximum in the initial susceptibility at a temperature in reasonable agreement with that indicated by Mössbauer measurements and higher field magnetization data.   The initial susceptibility data above these temperatures indicate that the entities ordering there are superparamagnetic clusters, but any naive spin-glass approach would even so lead one to expect a broad distribution of molecular fields and, it would seem, a rounded susceptibility maximum.

Excitations on Spin Glass Systems

A broad distribution of molecular fields has underlain the discussions of the low temperature specific heats of Cu-Mn and related alloys given by Blandin,[4] Overhauser,[5] Marshall,[6] Klein and Brout[7] and other authors.   The attention this topic has received is undoubtedly due to the intriguing feature of a large term in the specific heat, linear in temperature and independent of composition. (In the vicinity of the ordering temperature, which is of course a function of composition, this passes through a maximum and falls off slowly at higher temperatures.)   Early theories of this effect used an Ising model in which spin reversals against a local field, which had a finite probability of being zero, yielded the observed behaviour.

A Heisenberg treatment where phase space arguments reduce p(0) to zero would seem to spoil the agreement, but this has recently been shown not to be the case by Rivier (this conference) and Klein (L.T.13). It has also been emphasized that magnetic resonance data of Griffiths[8] for Cu-Mn imply the existence of reasonably long-lived spin-wave like excitations which must contribute to the specific heat. Dr. Souletie has remarked here that the Ising approach allows finite p(0) but denies us spin waves while the Heisenberg approach yields the reverse, and both may allow a proper analysis of the results if used self-consistently.

The Role of d-d Interactions in Spin Glasses

We have already remarked on the evidence in the initial susceptibility of Au-Fe for superparamagnetic clusters, and one might at first ascribe these to ferromagnetic coupling of near neighbour solute atoms by d-d interactions (as opposed to RKKY) which dominate at higher concentrations (above the percolation limit) to yield the observed long-range ferromagnetism. This seems in good accord with the failure of ferromagnetism to appear in more concentrated Cu-Mn and Au-Mn solid solutions, since Mn-Mn nearest neighbours are known often to yield antiparallel coupling; and it is interesting to note that the simple spin-glass features of these alloys in the dilute (RKKY) regime seem to persist at the higher concentrations where antiparallel d-d coupling must play a role in keeping the net magnetization at zero. (Only at concentrations above about 70% Mn does there seem evidence for straightforward antiferromagnetism modified by the presence of some Cu atoms on both sub-lattices.) It is therefore rather surprising to hear from Dr. Cannella that he finds well-defined superparamagnetic effects in Cu-Mn and Au-Mn at 5-10% solute. It is perhaps conceivable that atomic short-range order is able to keep Mn atoms apart from each other enough to yield ferromagnetic clusters of second nearest neighbours to play the role of those of nearest neighbours in Au-Fe.

INTERACTIONS IN KONDO SYSTEMS

It would be most interesting to compare the effects of increasing concentration in the Cu-Fe system (T^* dilute ~ 10°K) with those in Au-Fe (T^*dilute ~ 0.3°K) but, perhaps for that very reason,

the solid solubility in the former is too limited. The experiment
can be carried out however for Rh–Fe[9]which seems to have magnetic
similarity to Cu–Fe and (with metallurgical care) for Au–Co where
T* is at least an order of magnitude larger.[10] In both systems spin-
glass behaviour is found once interactions have driven T* below the
spin–glass ordering temperature, rather in the way that magnetic
order can be found in systems where crystal fields yield singlet
ground states once exchange energies exceed the ground state
isolation. What is not clear for the above alloys however is the
relative role of RKKY effects and d–d effects in lowering T*.
Concentration dependence in very dilute Cu–Fe, Rh–Fe and Au–V as
seen in resistivity behaviour seems to show that the former can be
effective but Tournier and the Grenoble group have given convincing
arguments for lowering of T* in Au–Co by the latter. It should be
pointed out that d–d interactions at larger solute concentrations in
Au–V are, paradoxically, clearly apparent in the demagnetization
(raising of T*) of the solute.[11] This contrast should perhaps have been
expected since d–d overlaps give a ferromagnetic band in pure Co
and a non–magnetic, even superconducting, band in pure V.
Presumably therefore a hypothetical continuous series of Au–Co solid
solutions would show a d–d driven transition from spin glass to ferromagnetism, but Rh–Fe at least up to 30% Fe appears to resemble
Au–Mn more than Au–Fe. It should be recalled that at 20% Fe
f.c.c. Au–Fe, Pd–Fe and Ni–Fe have ferromagnetic Fe–Fe coupling
while the two latter certainly have antiparallel Fe–Fe coupling in
the same crystal structure at high Fe contents, so average d–band
occupations may play a part.

THE MAGNETIZATION OF NON–MAGNETIC SOLUTES

For alloys on the extreme right–hand side of the table the onset
of magnetic order is of particular interest, since it would seem
appropriate to regard the solute in the dilute limit as just a site of
local exchange enhancement of the susceptibility. This approach[12,13]
has been used to discuss the susceptibility and transport properties in
that limit by Lederer and Mills and Kaiser and Doniach, and there
have recently been attempts to extend it to finite concentrations[14]
and the transition to ferromagnetism by consideration of the way in
which interacting local spin fluctuations bootstrap their way to
aligned moments. This is clearly an easier process when the host also
possesses some exchange enhancement (Pd–Ni) than in "simpler"

systems (Cu-Ni), but both these systems have attracted considerable attention recently. As usual measurements in finite magnetic field are of limited value unless theoretical guidance on meaningful extrapolations exist, and unfortunately we have as yet no calculations of the dependence of magnetization on field near the critical composition. Electrical resistivity measurements in zero field[15] and[16] specific heat results show an initial linear increase in effects due to spin fluctuations (non-interacting regime) followed by a steep rise to a maximum at c_{crit}. Although near neighbour interactions might be expected to yield superparamagnetic clusters there is as yet no evidence for a low temperature coupling of these in a spin-glass for any Ni-containing system. On the other hand in Rh-Co, which seems to belong in this no moment, local enhancement category, work in the author's laboratory (to be reported at LT 13) shows a critical composition for the appearance of a spin-glass at 20% Co while long-range ferromagnetism is delayed until about 38% Co. One suspects that this contrast is due to the part played by intra-atomic inter-orbital coupling.

GOOD MOMENTS IN ENHANCED HOSTS

We have referred above to the behaviour of "non-magnetic" solutes in enhanced hosts; another category of interest is given by the alloys in which the host is significantly exchange enhanced and the solute has a good moment, and of these Pd-Fe is the archetype. This system has been studied in many places and by many techniques, including neutron diffuse scattering that yields the spatial distribution of the magnetization, and a fairly clear picture is available.[17] The local moment on an Fe atom polarizes the matrix in its vicinity (perhaps as many as 100 Pd atoms) so that the resultant giant moments couple to one another to yield long-range ferromagnetism at very low iron concentrations: by 0.5 atomic% the Curie temperature has already reached 15°K. There is some evidence[18] that at very low concentrations, such giant moments couple only indirectly through RKKY interactions, yielding a spin glass but the details of the transition have not been explored. At higher concentrations the contribution of the palladium d-band polarization to the moment per iron atom falls but the d-d coupling of iron nearest neighbours remains ferromagnetic to more than 50% Fe. In contrast, Pd-Mn while showing giant moment ferromagnetism (with much lower Curie temperatures) at low solute concentrations has this ferromagnetism destroyed at more than about 5% Mn by the antiparallel coupling of Mn nearest neighbours, and spin-glass like behaviour then reappears.[19] The Curie temperature in the ferromagnetic regime reaches a maximum at about 8° K and 2½% Mn.   Pt-based alloys where the host enhancement is smaller have been less intensively studied, but Loram et al.[20] have produced resistivity indications that exchange effects have to overcome a low T* Kondo effect before long-range giant moment ferromagnetism can appear.   In a similar fashion a non-magnetic ground state in the dilute limit (with T* greater than that for Pt-Fe) leads to a significant delay in the onset of ferromagnetism in Pt-Co with increasing solute concentration,[21] although giant moments once stabilized couple more strongly than in Pt-Fe so that the Curie temperature-composition curves cross.

SPIN - DENSITY - WAVE   HOSTS

To complete the picture it should be pointed out that a very interesting type of magnetic order can appear when the conduction electrons are almost capable of supporting a helical spin-density-wave. This would appear to be the case for Yttrium[22] since the addition of as little as 10% Tb can yield a helix with a turn angle of 50° and a T_N of 50 K.   Thus Y which has no 4f electrons must have a conduction band structure very like that of the rare-earth metals, so that the presence of a small number of randomly distributed 4f moments are able to establish a situation of long range magnetic order by adopting an orientation dictated locally by the conduction electron spin-density-wave.

The spin-density-wave of chromium, on the other hand, seems insensitive to local moments on iron atoms and they to it, for the Curie-Weiss contribution they make to the susceptibility is little modified by passage through the Neel temperature.   At higher concentrations of iron the matrix antiferromagnetism disappears, and spin-glass behaviour is found at low temperatures, yielding to long-range ferromagnetism as in Au-Fe above about 15% Fe.

REFERENCES

1. J.S. Kouvel, J. Phys. Chem. Solids 24, 795 (1963).

2. B.R. Coles, Advances in Physics 7, 40 (1958).

3. D.K.C. MacDonald, W.B. Pearson and I.B. Templeton, Proc. Roy. Soc. (A) 266, 161 (1962).

4. A. Blandin and J. Friedel, J. Phys. Radium 20, 160 (1959).

5. A.W. Overhauser, J. Phys. Chem. Solids 13, 71 (1960).

6. W. Marshall, Phys. Rev. 118, 1520 (1960).

7. M.W. Klein and R. Brout, Phys. Rev. 132, 2412 (1963).

8. D. Griffiths, Proc. Phys. Soc. 90, 707 (1967).

9. A.P. Murani and B.R. Coles, J. Phys. C 3, S159 (1970) (Met. Phys. Suppl.).

10. B. Lecoanet and R. Tournier, Proc. L T XII (Acad. Press, Japan, 1971), p. 735.

11. J. Van Dam and P.C.M. Gubbens, Phys. Letters 34A, 185 (1971).

12. P. Lederer and D.L. Mills, Phys. Rev. Letters 20, 1036 (1968).

13. A.B. Kaiser and S. Doniach, Int. J. Magnetism 1, 11 (1970).

14. R. Harris and M.J. Zuckermann, Phys. Rev. B5, 101 (1972); K. Levin, R. Bass, and K.H. Bennemann, Phys. Rev., to appear (1972).

15. A. Tari and B.R. Coles, J. Phys. F. 1, L69 (1971).

16. G. Chouteau, R. Fourneaux, K. Gabrecht and R. Tournier, Phys. Rev. Letters 20, 193 (1968).

17. G.G. Low, Advances in Physics 18, 371 (1969).

18. G. Chouteau, R. Fourneaux and R. Tournier, Proc. LT XII (Acad. Press, Japan, 1971), p. 769.

19. H.C. Jamieson, R.H. Taylor, A. Tari, B.R. Coles, to be published.

20. J.W. Loram, R.J. White and A.D.C. Grassie, Phys. Rev. B5, 3659 (1972).

21. R.M. Bozorth, D.D. Davis, and J.H. Wernick, J. Phys. Soc. Japan 17, Suppl. B1, 112 (1962).

22. W.C. Koehler in Chapter 3 of "Magnetic Properties of Rare Earth Metals" Ed. R.J. Elliott, Academic Press 1972.

DISCUSSION

H. Alloul: I would like to comment on the interpretation
in terms of some type of damped spin-wave excitation of the
effects seen in EPR in C̲u̲Mn.   I suggest that the change in
character seen between about Tg/2 where the signal disappears and Tg (the susceptibility maximum temperature) is
not due to the observation of some collective spin motion,
but to the effective observation of only some parts of the
sample.   This is in accord with the loss of signal amplitude without great increase in width which indicates a loss
of intensity.

B.R. Coles: I rather doubt that.   The intensity of the resonance is certainly diminishing as the temperature is lowered, but it moves in a systematic way to very low fields
without much broadening.   Thus it cannot be due to the residual contribution of those spins not yet involved in the
ordering (or g would still be 2.0), and it cannot be due
simply to individual spins in a field shifted by the local
molecular field since such fields are likely to have a very
broad distribution while the resonance width is still only
100 oersted or so.

S.M. Shapiro: Have you considered any further what these
collective excitations could be, that is, what type of response function would characterize them.   Would they be
over-damped magnons of any type, and if so of what q-depen-
dence.

B.R. Coles: Only in a vague way.   Their energies might be
reasonably well-defined while q is not.   It is interesting
to speculate whether they can in fact, propagate or whether
one could have the type of localization produced by lattice
disorder.   Local magnon modes are rather an attractive idea.

P. Lederer: I am not sure that one can draw simple conclusions (one critical concentration for both local moments and
ferromagnetism) from the coefficient of the T^2 term in the
P̲d-Ni resistivity since magnetized pairs of solute atoms,
at less than critical concentration, would not show up in
the resistivity.   Did you suggest that the critical concentration depends on the type of measurement made?

B.R. Coles: No. Our magnetization and resistivity data a-
gree on the critical composition in this system. I don't
think one can talk about good moments on pairs or clusters
in the paramagnetic regime because they would have to order
at low temperatures and there is no evidence for low temperature spin glass or ferromagnetic ordering for c < 2 at.%.

P. Lederer: Do you dispute the saturation magnetization
measurements on these?

B.R. Coles: The magnetization-field data do not clearly indicate pair or cluster moments stable to arbitrarily low
temperatures. Their interpretation, especially at very
large fields, is quite a difficult matter.

B.R. Cooper: Could you give a simple definition of your
picture of the magnetic "glass" state and what you regard
as the definitive measurement.

B.R. Coles: An ideal magnetic "glass" state would be achiev-
ed at a very low temperature where "good" moments are coupled only by RKKY interactions to yield a zero net magnetization but finite local fields pinning moment orientations at
each solute site. I think $LaAl_2-GdAl_2$ should be of this
type, but in Au-Fe and Cu-Mn there will be near neighbor
d-d effects also, and we do not yet know if these are necessary to yield the characteristic shift of the magnetization curve produced by field-cooling. In Rh-Co my criterion
for the magnetic glass is whether I get the features (shifted magnetization curve, time-dependent decay of frozen-in
moment, rounded susceptibility maximum in moderate fields)
that are observed in Cu-5% Mn or Au-5% Fe. But as I indicated, some of these may not be quintessential features of
the magnetic glass.

D.E. MacLaughlin: The possible "delaying" of the spin glass
transition by superconductivity of the host in La-based
alloys is very interesting. It's true that in the (La-Gd)
Al_2 system the spin-glass ordering temperature vanishes for
Gd concentrations low enough to allow superconductivity;
however, this is not a general result [e.g., La-Gd, (La-Gd)
In_3]. It would seem in general that the number of electrons
involved in superconductivity (about 10^{-3} electrons/atoms)
would not be sufficient to appreciably affect the RKKY interaction, although I think that this is an experimentally

unresolved point at present. S.H. Liu treated this problem
theoretically a number of years ago, and found that a BCS
superconductor only affects RKKY to order (k_BT_c/E_F) to some
power.

B.R. Coles: I was not suggesting that superconductivity in
the host delayed the spin glass onset like a Kondo effect on
the impurity, although there might be special cases where
parts of the Fermi surface which play a vital role in the
RKKY interaction are also of particular importance to the
host superconductivity and there might then be a direct com-
petition for the interests of a particular group of messen-
gers.

D.E. MacLaughlin: It is a little hard to see how that would
come about. Superconductivity involves a small number of
electrons in Cooper pairs at the Fermi level, the RKKY coupling takes advantage of the whole Fermi sea.

B.R. Coles: That is, as I said, true in general; but as I
have emphasized the ability of rare-earth moments to give
rise to magnetic order in Yttrium arises because of special
features of the Fermi surface.

R. Tournier: I want to say two things. The first is that
all the dilute alloys well below the Kondo temperature show
the phenomenon of magnetic glass ordering. The Wheatley
group has shown that we observe in Cu-Fe magnetic ordering
which is proportional to c^2, it is magnetic ordering between
pairs of Fe which exists in Cu. At very low concentrations
an ordering temperature proportional to c^2 is observed, and
we have shown that this magnetic ordering has exactly the
characteristic expected from the RKKY interaction, the susceptibility, and specific heat being approximately independent of concentration, so we have thus magnetic glass ordering between pairs. Also, for a group of three atoms of
Co in gold-cobalt we observe an ordering temperature which
is proportional to the cube of the concentration and the
susceptibility and the specific heat are linear in temperature and then you have thus a magnetic glass ordering between magnetic clusters. The second question is about the
problem of saturation magnetization well below the critical
concentration. It is certain that the magnetic field builds
a giant moment because you know that Kouvel observed giant
moments in the ferromagnetic regime of Cu-Ni near the
critical concentration. But I think that it is possible

that the external field can create a giant moment below the
critical concentration because we very nearly have a magnetic cloud in zero field. We can observe giant moments in a
field, and here I reply to the question of Lederer that it
is difficult in the magnetization measurements to separate
saturation magnetization which is due to magnetic clusters
giving a giant moment and the magnetization which is built
by the external field and which corresponds to the saturation due to a nearly magnetic cloud - and this nearly magnetic cloud seems to have superparamagnetic behavior at high
temperatures (as Kouvel has observed in the Rh-Ni system)
and some sort of non-magnetic behavior at very low temperatures with T^2 resistivity.

B.R. Coles: I agree completely with both those comments and
especially, I thank you for answering Dr. Lederer that in
fact, it is very difficult to separate these two components.
Of course if one does not have the presence of a field,
which is as you say stabilizing the moments, one does then
have well defined phase transition within the theory of
phase transitions. There is not a sharp Curie point in the
field so that the argument about whether there is then long
range ferromagnetism below the critical concentration given
by resistivity data becomes academic. I also think that
your point about the clusters and pairs in Co in gold etc.,
shows very nicely how the spin glass can set itself up once
the pair interactions have, as it were, locally destroyed
or reduced T_{Kondo}. It is merely the question of what component then enters into the spin glass and I think you are
right that the pairs and triplets and so on (when the isolated atom has a high T_{Kondo}) can play the same role as
isolated atoms in systems where T_{Kondo} is low.

S.C. Moss: If the interaction in a spin glass is just RKKY,
then the nature of the ordering is not long range order like
that in iron below the Curie point or something like Cr below its Neel temperature. Is it possible that the spin
glass temperature is something like a glass transition temperature for the spin system and is as much a dynamical
phenomenon as anything else? A temperature below which some
tendency towards ordering which increases with decreasing
temperature is frozen in.

B.R. Coles: Yes, well that is how one would define the
spin glass in the absence of close neighbor interactions,
I think.

S.C. Moss: It is like a glass transition temperature for
the magnetic system then.

B.R. Coles: I think that Dr. Anderson would not hold that
the analogy holds exactly but as far as the experimentalist
is concerned, yes.

N. Rivier: I would like to ask you about this specific heat
at low temperature. You believe that a $T^{3/2}$ law characteristic of localized magnons would be in addition to a linear
law characteristic of the kind of spins which are free to
rotate in a field H_0 different from zero. If this is the
case then would you be able to observe it, because I think
a T-law in combination with a $T^{3/2}$-law would be very difficult to observe.

B.R. Coles: No, I believe what happens is probably something like the Kaiser and Doniach very low temperature transition from a T^2 regime to a T regime in the local spin
fluctuations scattering. It is not $T^{3/2} + T$, I think you
get something that goes like T^n rapidly becoming T. However,
that bit of my talk was written before I talked to Dr.
Souletie on the bus this morning. He made, I think, an extremely interesting suggestion, that the quasi T term of the
Ising model (which does not allow anything except the total
turnover of the local spin against the local field, which
is clearly non-physical in a way) vanishes of course, in the
Heisenberg treatment of the spin glass because the phase
space shrinks to zero at H=0. But then you are allowed excitations which correspond to partial rotation, in other
words to a spin wave like excitation and he suggested that,
in fact, this equivalence is not totally accidental; that
you can either do an Ising treatment and then get your linear term out that way, or you do a Heisenberg treatment and
allow you spin-wave-like excitations and you get your linear
term out that way. This is a very interesting idea and I
am just waiting for somebody to work it out.

N. Rivier: Can I make a comment? I think I wasn't clear in my talk yesterday that this is indeed what happens. You can make a Heisenberg model and you get a specific heat which is linear in temperature as well, and for the reason that you quoted Souletie as suggesting.

T.A. Kaplan: In the PdMn system you drew a phase diagram with a paramagnetic phase and a magnetic glass phase separated by a line. What is the character of the transition there?

B.R. Coles: That is phenomenologically defined in terms of the temperature for a rounded susceptibility maximum like that in the Rh-Fe system. Fortunately, in this system one can also study the EPR of the manganese and it is the effects in EPR that principally makes me invoke a large amount of short range order in the region between the ferromagnetic phase and the spin glass. There is a lot of specific heat there, there is a lot of line broadening of the resonance there, there is a clear diminution of the electrical resistivity there and so on. But on the other hand, it is not until you cross the line, you spoke of, that you get susceptibility maxima and glass-like behavior.

STRUCTURAL ASPECTS OF MAGNETIC INTERACTIONS IN METALLIC SPIN GLASSES*

D. J. Sellmyer[†] , J. M. Franz[§] , and
L. K. Thomas[**]

Massachusetts Institute of Technology
Cambridge, Massachusetts   92139

INTRODUCTION

The dilute magnetic alloy or spin-glass problem is a very interesting but apparently quite difficult statistical mechanical problem. There are several alloy systems, e.g., Cu(Mn), Ag(Mn) and Au(Fe) which show common features such as susceptibility and resistivity maxima at low temperatures. Marshall, Klein and Brout, Friedel and others have developed theories, based on RKKY coupling of magnetic moments on the impurity atoms, that are able to explain some of the experimental facts in the above systems.[1] However, there are also other alloys which show anomalous behavior and it is the purpose of this paper to review our experimental studies of several such systems.

RESULTS AND DISCUSSION

Consider first the paridigm for all spin glasses: Cu(Mn). Figure 1 shows the behavior of c/χ_i where c and χ_i are the impurity concentration and susceptibility. Clearly, above $10°K$ curves for all concentrations coincide so the spins are behaving essentially independently in a

* Supported by NSF Grant GP-21312 and by ARPA.
† Present address: Physics Dept., University of Nebraska, Lincoln, NE   68508
§ Present address: IBM Corp., Burlington, Vt.
** Present address: Technische Universität, Berlin.

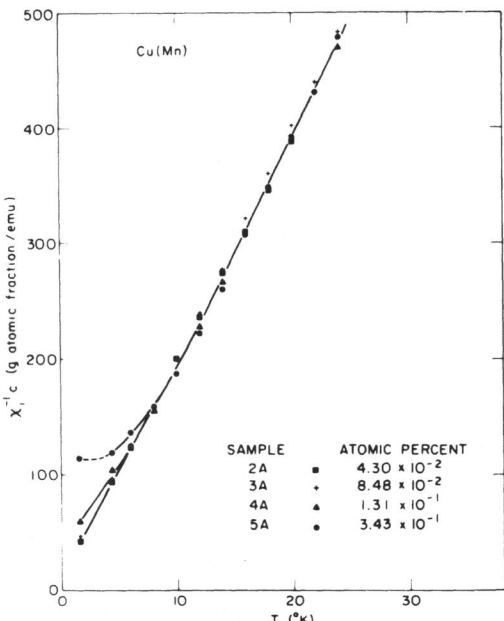

Fig. 1.   Inverse susceptibility for Cu(Mn).

Curie-Weiss manner. Below 10°K, however, there are evidently antiferromagnetic-type interactions operative: χ_i/c is seen to decrease as c increases. As explained in detail elsewhere,[2] we have compared our $\chi_i(c)$ and M(H) [magnetization] results to 100 kOe with the predictions of the recent mean-random-field theory of Klein[1] in the low T limit. This theory, which is an Ising model approximation coupled with the RKKY interaction, can explain in a semi-quantitative way the $\chi_i(c)$ and M(H) results, as well as phenomena such as susceptibility maxima at temperatures T_{max} proportional to c. Thus the spin-glass problem seems to be understood, in a first-order sense, for Cu(Mn) and similar systems.

The Cu(Fe) system, on the other hand, shows quite different behavior as seen in Fig. 2. In contrast to the Cu(Mn) results, Fig. 2 shows that in Cu(Fe): (a) the magnetic interactions are relatively strong in that they are not broken up to quite high temperatures (even to 300°K), (b) there exist ferromagnetic interactions - as c increases,

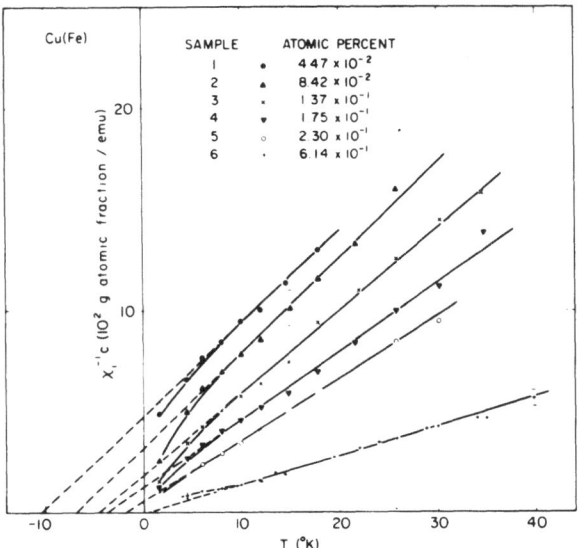

Fig. 2. Inverse susceptibility for Cu(Fe).

c/χ_i decreases, and (c) there exists evidence also for
antiferromagnetic interactions in the 6140 ppm sample –
the increase in c/χ_i above the Curie-Weiss extrapolation
below 10°K. Detailed studies of the concentration
dependence of χ_i and M(H) have been performed by Tholence
and Tournier (TT)[3] for c ≤ 300 ppm and by Franz and
Sellmyer (FS)[4] for 400 ≤ c ≤ 6000 ppm. It was shown that
M(H) can be separated into two parts due respectively to
isolated Fe atoms with spin $s_1 \simeq 3/2$ and to ferromagnetically coupled pairs of Fe atoms with $s_2 \simeq 3$. TT interpreted the pairs as being due to RKKY-coupled Fe atoms
separated by less than d ≃ 11 Å. FS, on the other hand,
showed that the density of pairs depends on the sample-
preparation method suggesting various d values, which
would be inconsistent with long-range RKKY coupling.
Further evidence for the important influence of structure
on magnetic properties was obtained by Svensson,[5] who
observed spin-glass transitions in Cu(Fe) alloys but only
if the samples were severely cold-worked. These transitions are indicated by resistivity and susceptibility
maxima at the temperatures shown in Fig. 3. All of these
results can be interpreted most reasonably as follows. The
ferromagnetic pairs consist of close or nearest-neighbor
pairs of Fe atoms rather than being RKKY coupled atoms to

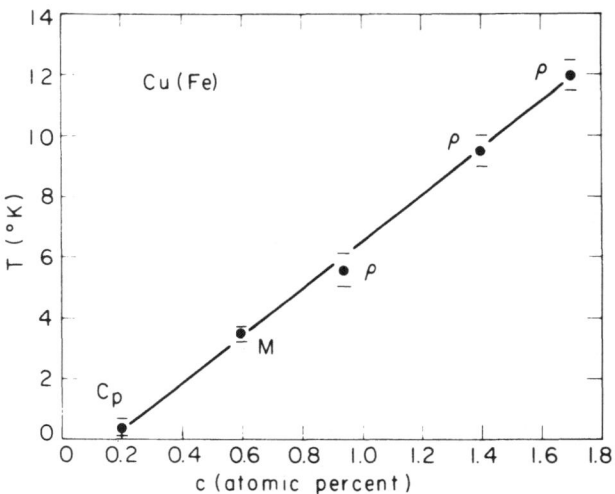

Fig. 3.   Apparent ordering temperatures
in Cu(Fe), (after Ref. 4).

distances of 11 Å. These pairs, which consist of chemical clusters whose density in these supersaturated alloys
depends sensitively on sample-preparation techniques, are
broken up by cold work so that the long-range RKKY-type
interactions responsible for spin-glass transitions, again
can be observed.[5] These as well as recent Mössbauer
results[6] suggest strongly that in Cu(Fe) the short-range
order of the magnetic impurities plays a crucial role in
determining the magnetic properties.

Dilute W(Fe) alloys show behavior which also differs
markedly from that of Cu(Mn). A plot of c/χ_i for W(Fe)
is qualitatively the same as Fig. 2 for Cu(Fe). $\chi_i(c,T)$
can be fit to an expression of the form $\chi_1 c + \chi_2 c^2$ where
the first term represents isolated Fe atoms and the second
term is due to ferromagnetic inter-impurity interactions;
the fit is shown in Fig. 4. The results, which are presented in detail elsewhere,[7] show that a Curie-Weiss
analysis of χ_1 leads to $p_{eff} = 3.9$ implying $s \simeq 3/2$ and
$\theta = 1 \pm 2°K$. Low temperature susceptibility maxima, e.g.,
at 8°K for c = 0.33 at. %, suggest magnetic transitions
of a spin-glass nature. Little is known yet about the
effect of short-range order in these alloys but, in any
case, W(Fe) represents another system in which simple
spin-glass behavior is not observed.

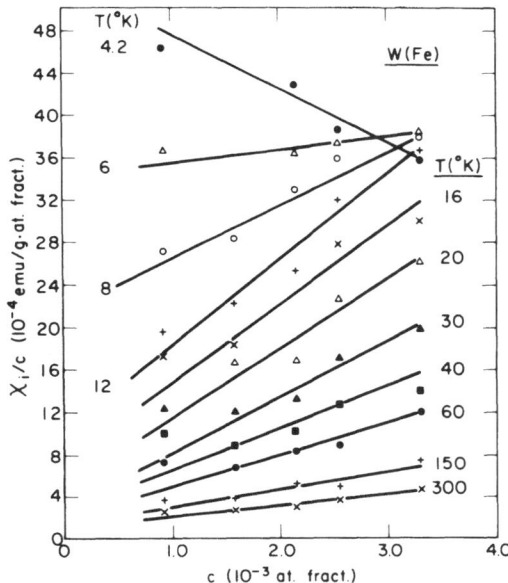

Fig. 4. χ_i/c vs. c for W(Fe) alloys.

Our final examples of magnetic transitions in disordered spin systems are dilute magnetic alloys based on compounds of the form TM where T represents Fe or Co and M is Al or Ga.[8,9] These compounds, which have the CsCl structure, are nonmagnetic at exact stoichiometry, but dilute solutions of excess T-metal atoms (here denoted as T_x) go into the lattice substitutionally for M atoms and form localized moments in FeAl(Fe_x), CoAl(Co_x), and CoGa(Co_x). The Kondo effect is observed in the latter two cases but not in FeAl(Fe_x). Also, in the Co compounds, as the Co_x concentration increases, ferromagnetism sets in with σ_{sat} and T_c increasing in a nonlinear way with c_x, the Co_x concentration. On the other hand, for FeAl(Fe_x) it appears that spin-glass transitions occur in that there exist susceptibility maxima at temperatures proportional to the Fe_x concentration.[10] Figure 5 shows schematically the coupling between T_x atoms in neighboring unit cells. The data suggest a ferromagnetic coupling in the Co compounds[8] whereas antiferromagnetic coupling is indicated for FeAl(Fe_x).[9,10] Thus in these cases we have alloys which are the same structurally but the nature of the magnetic states is quite different.

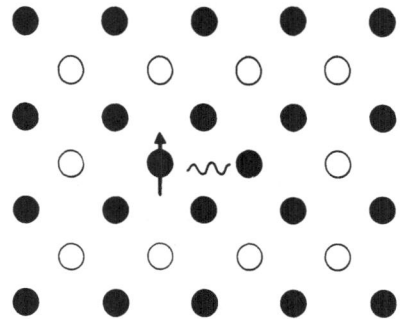

Fig. 5. Schematic representation of spin coupling between T_x atoms in $TM(T_x)$ alloys.

CONCLUSIONS

The above results suggest that it may be very difficult to understand metallic spin glasses generally unless the theory accounts specifically for details of the electronic structure of the host and the localized impurity states. Furthermore, it seems likely that it will be necessary to treat structural aspects such as near-neighbor configurations of two or more magnetic atoms, perhaps with spin-polarized cluster calculations, in order to truly understand the magnetic properties of these disordered systems.

REFERENCES

1. See: M.W. Klein, Phys. Rev. <u>173</u>, 552 (1968), and references therein.
2. J.M. Franz, and D.J. Sellmyer, to be published.
3. J.L. Tholence and R. Tournier, Phys. Rev. Letters <u>25</u>, 867 (1970).
4. J. M. Franz and D.J. Sellmyer, AIP Conf. Proc. No. 5 (1972), p. 1150.
5. K. Svensson, Proc. LT10, Moscow 1966, p. 267.

6. B. Window, J. Phys. C$\underline{3}$, S323 (1970), and to be publish-
    ed.
7. L.K. Thomas and D.J. Sellmyer, to be published.
8. D.J. Sellmyer, G.R. Caskey, and J.M. Franz, J. Phys.
    Chem. Solids $\underline{33}$, 561 (1972), and refs. therein.
9. G.R. Caskey, J.M. Franz, and D.J. Sellmyer, to be
    published, and refs. therein.
10. G.P. Huffman, J. Appl. Phys. $\underline{42}$, 1606 (1971).

DISCUSSION

R. Tournier: I am very surprised that the model we have
used to explain the Cu(Fe) properties can explain the behavior of your alloys at higher concentrations because the
model of pairs that we have used is only valuable at very
low concentrations, where the pair approximation is good,
when you have a very small number of pairs in the system.
I think it is pure chance for you to have obtained a c^2 law
at high concentrations because the number of pairs that we
have obtained is very large, about 100 c^2. At these concentrations the number of pairs is probably of the order of
the number of iron impurities that you have introduced in
the alloy. I am very surprised that you can analyze so
simply the impurities.

D.J. Sellmyer: The magnetization of our samples does <u>not</u>
contain simply c and c^2 terms; I very carefully said that
the quantity $\mu(c)$ that was shown was equal to M/c only in
the $c \to 0$ limit. Actually, there are (1-c) terms in the
denominator of $\mu(c)$. I did not have time in this talk to
give the details of how we separated the isolated and pair
contributions to M but it is done in our last Magnetism
Conference paper (Ref. 4). If you look at that paper you
will see how the data were analyzed. The important point
here is that the pair density and spin-glass transitions
in Cu(Fe) [Fig. 3] are quite sensitive to structural modifications (of the short-range order among Fe atoms) which
suggests that the magnetic pairs are, in fact, close-neigh-
bor pairs rather than being long-range RKKY coupled Fe atoms.

B.R. Coles: Your last point about the importance of the
electronic structure of host and matrix is rather interesting because your iron in tungsten data look rather like
iron in copper. Iron in Molydenum on the other hand looks
rather like iron in gold. The Kondo temperature in W(Fe)
appears to be kind of low and of course the intrinsic im-
probability is that there is a moment there at all because
of course on an naive view one wouldn't expect a narrow magnetic resonance for a d-element in a d-element and that
certainly is a feature that theory needs to take in hand.

S.C. Moss: I am not sure I understand if it is, in fact,
the case that one has magnetic ordering in the very dilute
alloys. Can one have ferromagnetism or antiferromagnetism
for very small concentrations?

D.J. Sellmyer: What do you mean by antiferromagnetism?

S.C. Moss: A critical temperature, long range order.

D.J. Sellmyer: No, nobody is saying that there exists long-
range magnetic order below the susceptibility maximum in a
spin glass. A simple interpretation of the susceptibility
maximum at T_{max} is that there is RKKY coupling between the
spins, i.e., both positive and negative spin coupling. As
the temperature is lowered through T_{max} the close neighbor
spins are "locked in" instead of flipping around thermally.
Since some of these locked-in spins are coupled antiferromagnetically, the susceptibility decreases as T decreases
below T_{max}.

I.S. Jacobs: I am a little bit old-fashioned about this.
I have been away from it for a long, long time but isn't
it important to point out for the audience that the metallurgical stability or the solubility of the transition ions
in the non-transition hosts is very marginal in some of the
systems and much more acceptable in some of the other systems? I think it is important in terms of putting the subject into a framework.

D.J. Sellmyer: Yes. If I didn't mention that I should have
mentioned that the Cu(Fe) alloys are certainly supersaturated. I have no evidence in the W(Fe) alloys which look
magnetically like Cu(Fe), whether there is chemical clustering; I mean, I haven't done experiments to try to pin that
down.

P. Horn: I seem to remember that in the past there were
some detailed calculations done on the Cu-Fe-Al system which
suggested a negative superexchange via the 3-d electrons of
the aluminum between iron atoms where there is an aluminum
atom in between. Do you know the status of this? Is it
directly observable in your results?

D.J. Sellmyer: I am not familiar with work on the system
you mention. However, Huffman has tried to work out an
RKKY-type model in the FeAl alloys I mentioned and he suggests that the excess iron atoms in neighboring unit cells
are, in fact, coupled antiferromagnetically which sort of
fits in with the fact that we see susceptibility maxima in
these systems. You can speak to him if you want to get the
latest word. I think he is still working on that problem.

W.B. Muir: I was just wondering in view of this metallurgical stability problem whether you had ever considered the
possibility that you not only had clusters in the Cu(Fe)
system but that you had downright precipitation of iron in
the grain boundaries, because we did some magnetoresistance
work some years ago now which I think pretty clearly shows
that even at the most dilute concentrations that you care
to go to, you have got to be very, very careful and as soon
as you start to get field dependences then you have to worry
about whether you have got a field dependence in the precipitates, which is nonsense, or real field dependence.

D.J. Sellmyer: Are you talking about γ-iron? [Muir: I
don't know what sort of iron.] We do not believe that precipitation plays a significant role in our results. It is
reasonable to interpret the magnetization in terms of only
two contributions: isolated Fe atoms and ferromagnetically
coupled pairs. The contribution in addition to the isolated
Fe atoms seems to be due to pairs and not precipitates because of its c^2 dependence and the fact that the spin associated with it is ≈3 which is about twice that for an isolated Fe atom in Cu. Furthermore, our interpretation is consistent with recent high-resolution Mössbauer studies of
Knauer at Sandia who is able to observe isolated Fe atoms
and a doublet whose relative area corresponds roughly to
our estimate of the pair concentration. His results, as
well of those of Window (Ref. 6), show another line which
becomes increasingly stronger as the samples are annealed
at low temperatures where precipitation is expected. If

one prepares the samples in a "reasonable" way, which I won't define here, one does not get a terrible problem with precipitation.

W.B. Muir: Yes, but Mössbauer sensitivity is going to be several orders of magnitude less for this effect than any susceptibility measurements. You will see these precipitates fiercely with the susceptibility while the Mössbauer will not see them at all.

D.J. Sellmyer: This is not the case for our concentration region because, as mentioned above, one sees approximately the same relative numbers of pairs and isolated Fe atoms in both experiments. This is explained more fully in Ref. 4.

B.R. Coles: There is an important distinction here because in fact, excess iron does come out as epitaxial intergrowth on the [111] planes not as bcc iron (α iron) in the grain boundaries.

D.J. Sellmyer: Knauer also has shown (private communication) that he obtained the __same__ Mössbauer spectrum from single crystal and polycrystalline Cu(Fe) specimens. Thus the evidence is that iron atoms at grain boundaries also are not playing an important role in either his or, by inference, our results.

LOW FIELD MAGNETIC SUSCEPTIBILITY OF DILUTE ALLOYS: ARE

DILUTE CuMn AND AuMn MAGNETICALLY ORDERED?*

V. Cannella

Wayne State University, Department of Physics

Detroit, Michigan 48202

ABSTRACT

The temperature dependence of the low field ($\sim$5 gauss) magnetic susceptibility, $\chi(T)$, of lower concentration (1, 5, 10 at.%) CuMn and AuMn alloys has been measured using a 155 Hz a-c technique. In contrast to the broad maxima found in earlier higher field measurements, sharp cusps were observed in $\chi(T)$ indicating the onset of antiferromagnetic ordering at well defined temperatures. The sharp transitions were found to be dependent upon the metallurgical state of the samples, and applied magnetic fields of $\sim$100 G rounded the transitions. Analyses of $1/\chi$ vs. T in the paramagnetic regions indicate substantial superparamagnetic clustering in the higher concentration samples.

INTRODUCTION

Early studies of the magnetic susceptibility of magnetic spin-glass alloy systems such as CuMn,[1,2] AuMn,[3] and AuFe[3] found broad rounded maxima in the temperature dependence of the susceptibility, $\chi(T)$, and these were believed to be characteristic of random alloy systems where oscillating RKKY interactions provided the coupling between the randomly located magnetic moments.[4] In an extensive, more recent study of the very low field susceptibility of quenched AuFe alloys[5] we found that, in contrast to the broad rounded maxima in $\chi(T)$ found by other authors, there existed

sharp peaks in $\chi(T)$ at temperatures in excellent agreement
with magnetic ordering temperatures found by Mössbauer
studies.[6] We found, moreover, that the application of magnetic fields of a few hundred gauss would destroy the sharp
peaks and produce broader maxima in agreement with earlier
measurements which were taken at higher fields. The existence of a Curie-Weiss law for $\chi(T)$ down to the ordering
temperature T_o, followed by a sharp transition with $\chi(T)$ approaching a non-zero intercept as T approaches zero, seemed
clearly to indicate some type of cooperative antiferromagnetic transition, although in a random alloy it is difficult
to conceive of any antiferromagnetic state other than a random spin-glass type of arrangement. To determine whether
this behavior was unique to the AuFe alloy system or common
to other spin-glasses we have extended these measurements
of the very low field susceptibility to other alloy systems.
We here present the results of our initial measurements on
the CuMn and AuMn alloy systems.

The experimental technique used was a low field (~ 5
gauss rms), low frequency (155 Hz) a-c mutual inductance
method. Our cryogenic arrangement allowed the sample to be
moved in and out of the sample coil, allowing the determination of the absolute value of $\chi(T)$. The electronic apparatus followed the design of Maxwell,[7] with modifications to
improve the resistive balance network. Temperatures were
measured using germanium and platinum resistance thermomet-
ers. Details of the method of taking measurements may be
found elsewhere.[8]

EXPERIMENTAL RESULTS AND ANALYSES

Typical experimental results for the low field $\chi(T)$
and $1/\chi$ vs. T are shown for CuMn with concentration C = 1
at.% and 10 at.% in Figures. 1 and 2, and for AuMn with
C = 5 at.% and 10 at.% in Figs. 3 and 4 (similar curves
were found for 5% and 9% CuMn, and for 2% AuMn). From
these figures it is clear that the sharp peaks in $\chi(T)$ are
found in CuMn and AuMn alloys for concentrations ~ 1 to 10
at.% Mn. Initial measurements on unannealed (cold-worked)
5 at.% and 10 at.% CuMn samples showed broad rounded maxima
rather than sharp peaks, and it was only after these samples
were annealed (24 hrs. at 900°C) and slowly cooled that the
sharp peaks occurred. Fig. 2 includes $\chi(T)$ for the cold
worked 10 at% CuMn sample. The data taken in various

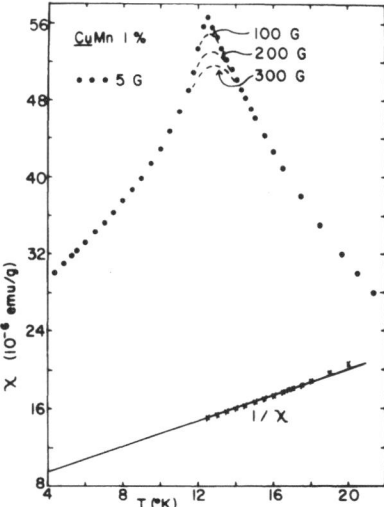

Figure 1. $\chi(T)$ and $1/\chi$ vs. T for annealed 1 at.% CuMn, including $\chi(T)$ data taken in applied fields of 100, 200, and 300 gauss. $1/\chi$ is in arbitrary units.

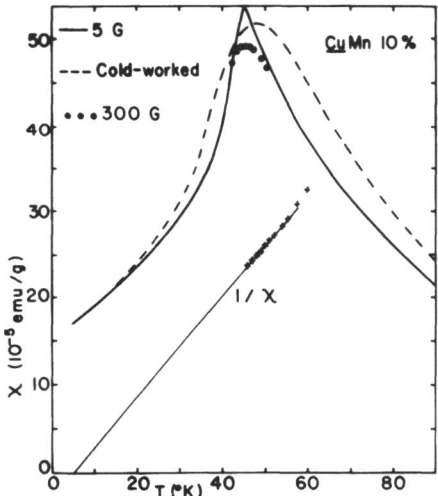

Figure 2. $\chi(T)$ and $1/\chi$ vs. T for annealed 10 at.% CuMn including data taken in an applied field of 300 gauss, and $\chi(T)$ for a cold-worked sample. Data were taken every 1/4°K near the peak and every 1/2 or 1° elsewhere. The scatter is less then the thickness of the lines. $1/\chi$ is in arbitrary units.

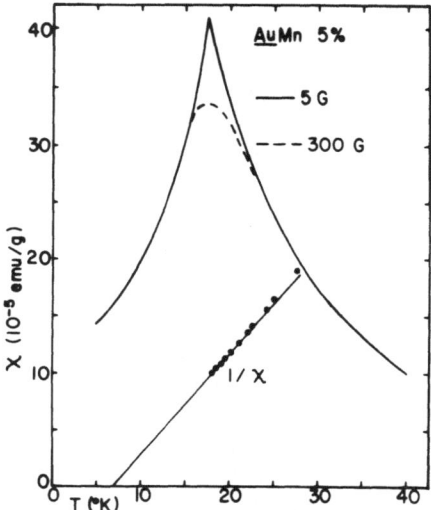

Figure 3.   χ(T) and 1/χ vs. T for annealed 5 at.% AuMn including data taken in an applied field of 300 gauss. The
data separation and scatter are as in Fig. 2. 1/χ is in
arbitrary units.

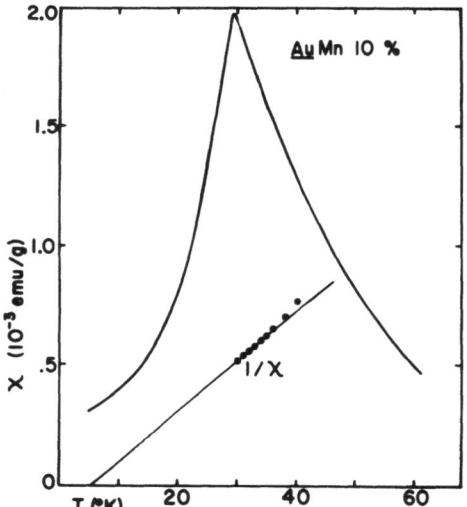

Figure 4.   χ(T) and 1/χ vs. T for 10 at.% AuMn. The data
separation and scatter are as in Fig. 2. 1/χ is in arbitrary units.

applied fields shown in Figs. 1, 2, and 3 show that the
application of relatively small applied magnetic fields
(300 gauss) will reduce and round the peaks in $\chi(T)$, similar
to the results found for $\underline{Au}Fe$.

From the intercept and slope of the plots of $1/\chi$ vs.
T we have found the paramagnetic Curie temperature, θ, and
the effective number of Bohr magnetons per magnetic atom,
p_{eff}, using classical molecular field analysis. The large
values found for p_{eff} in these alloys indicate considerable
superparamagnetic clustering, especially for the higher concentrations. Consequently, we analyzed $1/\chi$ vs. T between
T_0 and $2 T_0$ since we were interested in p_{eff} and θ characteristic of the interactions which occur near T_0, rather
than in the single magnetic atom moments and interactions
which determine the paramagnetic behavior at much higher
temperatures. Table I shows the values of T_0 (determined
by the maximum in $\chi(T)$), θ, and p_{eff} for the various concentrations of $\underline{Cu}Mn$ and $\underline{Au}Mn$, and includes the results of
our earlier work on quenched $\underline{Au}Fe$ alloys. The values of
p_{eff} for $\underline{Cu}Mn$ and $\underline{Au}Mn$ clearly show superparamagnetic clustering just as for $\underline{Au}Fe$, but a distinctly different behavior
occurs in each alloy system. The values of p_{eff} for quenched $\underline{Au}Fe$ alloys exhibit nearly single atom moments for 1 at.
%, but p_{eff} grows very rapidly with increasing Fe concentration indicating ferromagnetic interactions between Fe-Fe
first and perhaps second near neighbors.[9] The values of
p_{eff} for annealed $\underline{Cu}Mn$ samples level off as the concentration rises, due, probably, to antiferromagnetic interactions
between Mn-Mn near neighbors. It seems likely that the need
for annealing the higher concentration $\underline{Cu}Mn$ alloys in order
to produce sharp peaks is related to preferred local atomic
configurations of Mn in the Cu lattice. The very strong
superparamagnetism indicated by the high values of p_{eff}
found for annealed $\underline{Au}Mn$ alloys should be related to the
local atomic formation of known ferromagnetic superlattice
phases, for example, Au_4Mn.

The concentration dependences of the ordering temper-
ature T_0 for $\underline{Cu}Mn$, $\underline{Au}Mn$, and $\underline{Au}Fe$ are shown in Fig. 5. For
these alloys the concentration dependence of T_0 has the form
$T_0 = AC^m$, where A is 9.5, 10.6, and 5.0 °K/at.%, and m is
0.54, 0.63, and 0.75 for $\underline{Au}Fe$, $\underline{Cu}Mn$, and $\underline{Au}Mn$ respectively.
The concentration dependence of θ for $\underline{Cu}Mn$ and $\underline{Au}Mn$ does
not show any easily discernable pattern comparable to the
linear dependence of θ upon C for quenched $\underline{Au}Fe$ with

Table I. The values for T_o, θ, and p_{eff} for various concentrations of annealed <u>Cu</u>Mn, annealed <u>Au</u>Mn, and
quenched <u>Au</u>Fe (Ref. 5).

Alloy	C (at.%)	T_o(°K)	θ(°K)	p_{eff}
<u>Cu</u>Mn	1.3	12.3	1.5	5.5
<u>Cu</u>Mn	5.2	27.8	−6.5	10
<u>Cu</u>Mn	8.8	42.5	6	10
<u>Cu</u>Mn	10.3	45	9	11
<u>Au</u>Mn	1.14	5.7	1	7.3
<u>Au</u>Mn	2.0	8.3	−1.5	10.6
<u>Au</u>Mn	5.7	17.5	7	11.5
<u>Au</u>Mn	10.3	29.3	5	25
<u>Au</u>Fe	.91	8.5	−2	3.4
<u>Au</u>Fe	2.0	13.9	−1	3.6
<u>Au</u>Fe	4.9	22.2	1.5	4.8
<u>Au</u>Fe	7.6	27.9	4	7.1
<u>Au</u>Fe	12	36	38	11.4
<u>Au</u>Fe	13	38	40	15.5

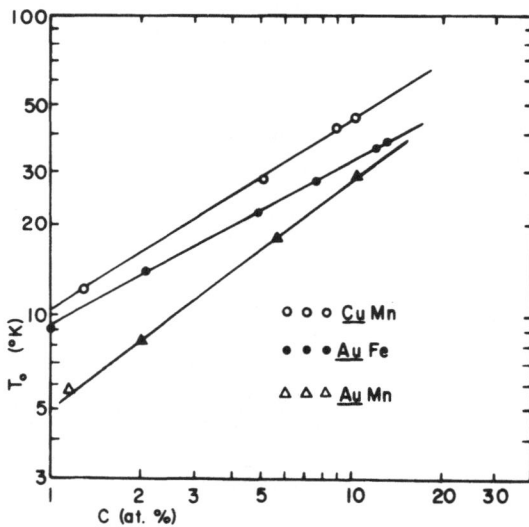

Figure 5. A log-log plot of T_o vs. C for <u>Cu</u>Mn, <u>Au</u>Mn, and
<u>Au</u>Fe where <u>Au</u>Fe data are taken from Ref. 5.

1 at.% $\leq$ C $\leq$ 8 at.%. There is evidence that both θ and p_{eff} depend sensitively upon the metallurgical factors which determine the amount of local atomic ordering in these alloys.

CONCLUSIONS

A comparison of these results for CuMn and AuMn alloys with our earlier work on AuFe shows clearly that the sharp peaks in the low field $\chi(T)$ are common to all of these spin-glass alloys. The Curie-Weiss paramagnetic behavior above T_0, the sharp transition at T_0, and the fall off in $\chi(T)$ to a non-zero intercept as T approaches zero indicate a cooperative antiferromagnetic type of transition, despite the conceptual difficulties involved in seeing how this occurs in random-type alloys. The broad maxima at temperatures directly proportional to C predicted by the theory of Klein[4] are clearly incompatible with our data. It would rather seem that a theoretical approach based on the spin density wave mechanism of Overhauser[10] would be more likely to explain the origin of these sharp cooperative transitions.

In order to understand the concentration dependences of θ and p_{eff} in CuMn and AuMn we are currently studying the effect of quenching on θ and p_{eff}. We are also extending the measurements of the low field $\chi(T)$ to dilute concentrations to determine the single magnetic atom limit for p_{eff}, and to more concentrated alloys to study the superparamagnetism at higher concentrations, especially in AuMn alloys which approach the ordered phase Au_4Mn. To obtain further insight into superparamagnetic clustering in these alloys we are performing computer simulations of alloys allowing local atomic ordering, similar to the simulations of random systems used to understand quenched AuFe alloys.[9]

REFERENCES

1. J. Owen, M.E. Browne, V. Arp, and A.F. Kip, J. Phys. Chem. Solids 2, 85 (1957).
2. J.S. Kouvel, J. Phys. Chem. Solids 21, 57 (1961); 23, 795 (1963).
3. O.S. Lutes and J.L. Schmit, Phys. Rev. 134, A676 (1964).

4. M.W. Klein, Phys. Rev. 136, A1156 (1964); 173, 552
    (1968); 188, 933 (1969).
5. V. Cannella and J.A. Mydosh, to be published in Phys.
    Rev. B, Dec., 1972.
6. See R.J. Borg, Phys. Rev. B 1, 349 (1970) where other
    references may be found.
7. E. Maxwell, Rev. Sci. Inst. 36, 553 (N) (1965).
8. V. Cannella, Ph.D. Dissertation, (Fordham Univ., 1971)
    (unpublished).
9. K. Duff and V. Cannella, article in this conference.
10. A.W. Overhauser, J. Phys. Chem. Solids 13, 71 (1960);
    Phys. Rev. 128, 1437 (1962); J. Appl. Phys. 34, 1019
    (1963).

*Work supported in part by the Air Force Office of Scientific Research AFSC under Grant No. AFOSR - 71 - 2002

DISCUSSION

S.C. Moss: How low in concentration are these sharp transitions found?

V. Cannella: For CuMn and AuMn we have not yet performed measurements on lower concentrations, but in AuFe we have found the sharp peaks as low as 0.1 at.%, and have indications that the phenomenon extends even lower in concentration. We expect to publish the low concentration AuFe data shortly, and we are currently doing the measurements on CuMn and AuMn.

Questioner: It seems intuitive that an oscillatory interaction is not going to line up spins into long range order. Perhaps a mean field would work, but an oscillatory RKKY interaction does not seem able to provide the mechanism for cooperative long-range order.

V. Cannella: I do not have the explanation for this behavior, I am simply presenting the data, which were taken at two different places with completely different sets of experimental apparatus. The a-c measurements were run at frequencies of 17, 155, and 500 Hz with no change in position or shape of the peak, likewise the curves may be taken at 5 gauss or .5 gauss with no changes. I agree that a rounded peak would be easier to explain, but the data shows that the peaks are sharp.

S. von Molnar: Wouldn't you expect that some other thermodynamic quantities would also give some sort of transition,
for example, has anyone measured the specific heat?

V. Cannella: The specific heat of the AuFe 1 to 8% alloys
has never been measured in the region of the ordering temperatures which we found, it has only been measured at lower temperatures. I am not familiar with the specific heat
measurements on the CuMn and AuMn systems.

J.A. Mydosh: May I also point out that Mössbauer measurements on AuFe alloys have also shown sharp ordering temperatures which agree extremely well with the susceptibility
measurements.

S. von Molnar: Then any microscopic measurement should
show this behavior?

B.R. Coles: I think that the remarks about the Mössbauer
work must be stated with a certain amount of caution. The
way in which you determine the transition temperature from
the Mössbauer data is not absolutely straightforward. In
the lower concentration specific heat data for CuMn, people
have measured C(T) through the transition temperature, and
there is no critical phenomenon. In the electrical resistivity, which is also a zero field measurement and normally
gives much sharper transitions than the susceptibility,
again, there is no sharp transition. I thought that a very
interesting feature of your results was that if you fit the
very high temperature susceptibility, where the moment is a
single atom moment, to plots of $1/\chi$ vs. T, then you see
deviations above the peak in $\chi(T)$. Therefore it is clear
that interaction effects are setting in above the temperature of the peak, so the peak does not mark the first onset
of the interactions.

V. Cannella: I would give a different interpretation to the
deviations from a linear $1/\chi$ vs. T plot at temperatures well
above the ordering temperatures. I think we must distinguish two types of interactions, a short-range ferromagnetic
interaction between moments in some specific local arrangement of magnetic atoms (e.g. near neighbors); and the longer
range oscillatory type interaction. There certainly exist
interactions far above the peak in $\chi(T)$, but these are the
short-range ferromagnetic interactions which provide the

mechanism for forming superparamagnetic clusters, not the oscillatory "antiferromagnetic" interactions which are characterized by the sharp susceptibility peaks. The deviations from a linear $1/\chi$ vs. T at higher temperatures are due to the breakup of the superparamagnetic clusters and correspond to a decrease in the effective moments, although our data does not go to high enough temperatures to find single atom values for the higher concentration alloys. May I refer you to the paper by K. Duff and myself later in this conference where we discuss the superparamagnetic clustering and the magnitude of χ for AuFe alloys.

N. Rivier: I don't think it is obvious that the susceptibility as a response to a static field is going to be the same as the response to a dynamic field. You said yours was an a-c measurement and there may be something like a mobility edge in the spin system to give this type of sharp behavior.

V. Cannella: I agree in principle, but we have done measurements from frequencies of 500 Hz to as low as 17 Hz with no appreciable change in behavior. It seems for a system of paramagnetic spins that 17 Hz is certainly approaching the static field limit.

P.A. Beck: I think that it is clear that these so-called transition temperatures have nothing to do with ordering in any long-range manner. Recalling the speaker's figure showing the elimination of the sharp peak due to cold-working I wonder whether we may be dealing here, in very small fields, with a selective response to the a-c fields of only the largest magnetic clusters. In that case the cold-working would cut up these clusters and destroy the sharp peaks.

V. Cannella: The effect of cold-working is found only in higher concentration samples. At low concentrations, say 1% and below, I have subjected AuFe samples to extensive cold-working, and find the same sharp peak as for those samples which have been annealed. As a result, I believe that in cold-working higher concentration samples the large superparamagnetic clusters become very strained, and this interferes with the long range oscillatory interactions.

T.A. Kaplan: A question which is apparently causing considerable difficulty for this audience is the following.
How can one possibly conceive of long-range antiferromagnetic ordering (and hence a sharp phase transition) in these
disordered alloys with long-range oscillatory exchange interactions? The difficulty arises from the almost certain
expectation of "misfits" or competing interactions (e.g.
two neighboring spins might both be neighbors of a third,
all with antiferromagnetic interactions). However, from
earlier experience with this situation for crystals one
knows that instead of necessarily getting no long-range
order, one usually obtains more complex states (e.g. spirals,
or worse) still maintaining the sharpness of the phase transition. It is clear that even for a very low-symmetry exchange $J(\vec{R}_i, \vec{R}_j)$, expected for a disordered or amorphous
alloy, there will be some state, i.e. some set of spins,
$\vec{S}_i^0$ (complex of course) which minimizes the energy -

$\sum J(\vec{R}_i, \vec{R}_j) \, \vec{S}_i \cdot \vec{S}_j$ (or more generally the free energy at

finite temperature). A priori, I see no reason to expect
the transition to be less sharp for a complex spin configuration than for a simple one. Such a set, $\vec{S}_i^0$, (essentially non-degenerate for a given $J(\vec{R}_i, \vec{R}_j)$) is then perhaps
the "ordered antiferromagnetic state" you seem to observe.

V. Cannella: I agree fully. I have not presented an explanation for the data, I have only presented experimental
evidence which indicates that a critical phenomenon is occurring, and as far as we can tell it must be cooperative
in order to explain the sharp transition, and it would seem
to be a magnetic transition since we are looking at a magnetic response. I really don't think I can draw any further conclusions except to emphasize the discrepancy between
this data and Klein's theoretical treatment of these alloys.
Klein's work predicts: 1) broad, rounded maxima in $\chi(T)$;
2) that these peaks should not be dependent upon fields
(for low fields); 3) that for low concentrations the magnitude of the peak should be concentration independent, and
4) that the temperature of the maximum should be directly
proportional to concentration. Our data violates all four
of these predictions.

I.S. Jacobs: I would like to come back to Professor Coles
comment. I think it is very important to recall that the
high temperature measurements give you average interaction
fields. You talk of superparamagnetic clusters which means
that there are a lot of magnetic atoms which are indeed co-
operating. Superparamagnetism means that there are net
moments for the clusters, and it is well known in superparamagnetism that these moments can freeze in due to magneto-
crystalline anisotropy, and this mechanism can also be
field dependent.

V. Cannella: I agree with your comments about superparamagnetism but I do not think that the anisotropy mechanism
can produce our susceptibility peaks. For low concentration alloys, say 1 at.% of magnetic impurities, about 90%
of the magnetic contributions come from single magnetic
atoms. For these low concentrations we find sharp peaks in
$\chi(T)$ and no appreciable deviation from a linear $1/\chi$ vs. T.
We cannot, therefore attribute a sharp peak at 1% to the
freezing-in of superparamagnetic clusters. In fact, the
"freezing-in" temperature of superparamagnetic clusters is
dependent upon the size of the cluster, so it seems impossible to get a sharp peak from this mechanism even at higher concentrations because of the distribution of cluster
sizes.

CALCULATIONS OF SUPERPARAMAGNETIC CLUSTERING IN RANDOM

ALLOYS OF MAGNETIC ATOMS IN NONMAGNETIC MATRICES*

K. J. Duff and V. Cannella

Wayne State University, Department of Physics

Detroit, Michigan 48202

ABSTRACT

Superparamagnetic clustering has been investigated in FCC and BCC random substitutional alloys of magnetic solutes in nonmagnetic matrices using a Monte Carlo type computer simulation technique. Two alternative definitions of "super-paramagnetic cluster" were used: (a) a set of impurity atoms each of which is coupled to at least one other by a ferromagnetic first near neighbor (1NN) interaction; (b) a set of impurity atoms each of which is coupled ferromagnetically to at least one other by either a first near neighbor interaction or a second near neighbor (2NN) interaction. The magnetic susceptibility calculated for clusters using definition b agrees quite well with the experimental susceptibility data for quenched lower concentration (1 at.% $\leq C \leq 12$ at.%) $\underline{Au}$Fe alloys in the paramagnetic temperature region. This implies ferromagnetic coupling between Fe-Fe first and second near neighbors in $\underline{Au}$Fe, and supports the validity of the present method in treating superparamagnetic clustering in alloys.

The problem of calculating the paramagnetic susceptibility of substitutional spin-glass alloys becomes quite complex when local ferromagnetic interactions between the magnetic atoms produce superparamagnetic clustering. The

*Work supported in part by the Air Force Office of Scientific Research AFSC under grant No. AFOSR-71-2002.

term "spin-glass alloy" is used to denote substitutional
alloy systems containing magnetic moment bearing atoms dissolved in a nonmagnetic metallic matrix, for example, CuMn,
AuFe, etc. Even under the most simplifying physical assumptions (a random distribution of magnetic atoms, low temperatures, and a cubic distribution of anisotropy axis orientations for single domain particles) the problem reduces
to finding the density $N_n(C)$ of superparamagnetic clusters
containing n magnetic atoms for a given concentration C, or
equivalently to finding the fraction $f_n(C)$ of magnetic atoms
in clusters containing n magnetic atoms. Explicitly we consider that at low temperatures (well below the ferromagnetic
Curie temperature of the superparamagnetic clusters) we may
write the magnetization M_n of a cluster containing n atoms
as

$$M_n = M_n(T=0) = ng \, \mu_B \, J \tag{1}$$

where g is the Landé factor, μ_B is the Bohr magneton, and J
is the spin on each magnetic atom. This assumes that at
low temperatures the magnetization of a cluster of n atoms
is independent of the shape of the cluster, or alternatively, that all cluster geometries are equivalent. The susceptibility of an array of such noninteracting superparamagnetic clusters may be written[1]

$$\chi = \sum_n N_n \, M_n^2 / 3kT \tag{2}$$

where k is the Boltzmann constant and T is the temperature
(°K), provided we have a statistical or a cubic distribution of magnetic anisotropy axis orientations. We may re-
write

$$M_n^2 = g^2 \, \mu_B^2 \, nJ(nJ+1) \quad \text{and} \quad \chi = \sum_n N_n \, g^2 \, \mu_B^2 \, nJ(nJ+1)/3kT \tag{3}$$

to allow for quantum effects in small clusters. Interactions between clusters may be accounted for by introducing
a paramagnetic Curie temperature θ, and we make use of the
equality $nN_n = Df_n$ where D = D(C) is the density of magnetic atoms in the alloy, to obtain

$$\chi = g^2 \, \mu_B^2 \, D \, J \sum_n f_n \, (nJ+1)/[3k(T-\theta)] \tag{4}$$

We have now isolated the cluster dependent factor which we will call $S(C)$

$$S(C) = \sum_n f_n(C) \; (nJ+1), \tag{5}$$

and this is, in turn, dependent upon $f_n(C)$, the fraction of magnetic atoms found in clusters containing n magnetic atoms for a given concentration.

The calculation of the values of $f_n(C)$ for various n and C depends upon the conditions imposed upon the cluster-forming interactions. Our present model assumes that only geometrical proximity conditions affect the formation of superparamagnetic clusters at low temperatures, namely, that cluster-forming ferromagnetic interactions occur either between first near neighbor (1NN) magnetic atoms only, or between both first and second near neighbor (1NN + 2NN) magnetic atoms. For a random distribution of substitutional solute atoms and only 1NN interactions it is possible to obtain approximate analytic solutions[2] of $f_n(C)$ for small n (n ≤ 7) and these may be extrapolated to account for virtually all magnetic atoms in lower concentration samples (C ≤ 8 at.%). However, for higher concentrations where larger clusters are important, or for more complex cluster formations (1NN + 2NN), even approximate analytic solutions for $f_n(C)$ become too difficult, and computer simulation becomes preferable.

We have devised Monte Carlo computer simulations of random substitutional binary alloys with either FCC or BCC lattice structure. The computer identified the clusters in the simulation and calculated $f_n(C)$ and $S(C)$ for various concentrations and cluster types. We could, in principle, create lattices of arbitrary size but in practice we used lattices of between 37,000 sites and 275,000 sites, applying cyclic boundary conditions to eliminate surface effects. To assure that our simulated lattices were large enough to give us reliable statistics we required that the statistical fluctuations in $S(C)$ should be small. Statistical fluctuations in $f_n(C)$ were always more severe for higher values of n above concentrations of a few at.%. But the fluctuations in $S(C)$ were quite small (a few %) for C ≤ 10% even with smaller lattices (40,000 sites), since for concentrations well away from the ferromagnetic (percolation) limit the size of the largest cluster was a function of the con-

centration, not of the size of the lattice. In Tables I-IV
we present values of $f_n(C)$, for n up to 40, taken from actual typical computer simulations. Tables I and II present
$f_n(C)$ for a FCC lattice with 1NN clusters and with 1NN+2NN
clusters respectively. Tables III and IV present similar
data for a BCC lattice. The concentrations were chosen
such that the maximum $n \leq 40$. For Table I the lattice size
was ∿119,000 sites; for Table II, ∿275,000 sites; for Table
III, ∿138,000; and for Table IV, ∿60,000. It is difficult
to tabulate the values of $f_n(C)$ for higher concentrations
where the cluster size n can be several hundred magnetic
atoms. Instead, we present in Table V the values of S(C)
found for various lattice types, cluster types, and concentrations. S(C) in Table V is as defined in Eq. 5. We used
the value J = 1.2 for all cases since this is appropriate
for single Fe atoms in the Au lattice.[2,3]

We have used these calculations of the superparamagnetic clusters in a random substitutional alloy to attempt to
understand the magnitudes of experimental magnetic susceptibility for quenched AuFe alloys.[2] For purposes of comparison we considered the Curie constant ξ defined by the
relation $\chi = \xi/(T-\theta)$. Experimental values ξ(exp) were determined from the slope of experimental curves of $1/\chi$ vs.
T.[4] Calculated values of ξ were found using

$$\xi(\text{calculated}) = g^2 \mu_B^2 \, D \, J \, S(C)/3k \qquad (6)$$

where S(C) determined the lattice and cluster type. In
Table VI we compare ξ(exp) for various concentrations of
AuFe with calculated values of ξ. ξ(single) was calculated
for single noninteracting magnetic atoms. ξ(1NN) and
ξ(1NN + 2NN) were calculated using S(C) for a FCC lattice
allowing 1NN clusters and 1NN + 2NN clusters respectively.
It is clear from Table VI that the experimental values for
the higher concentrations are much larger than can be explained using single magnetic atoms or 1NN clusters. The
reasonable agreement between ξ(exp) and ξ(1NN + 2NN) for
$C \leq 12$ at.% indicates that a ferromagnetic interaction probably exists between Fe-Fe first and second near neighbors
in AuFe alloys. The fact that ξ(1NN + 2NN) are smaller
than ξ(exp) suggests that even our quenched samples may
have some local atomic order which enhances the formation
of the superparamagnetic clusters. This tendency towards
the formation of local atomic order has been indicated by
measurements of the susceptibility (and thermopower) of
unquenched AuFe samples.[2]

Table I. Typical values for $f_n(C)$ for various C, $n \leq 40$, for a random FCC lattice with 1NN clusters.

n	C=1%	C=2%	C=5%	C=8%	C=12%	n	C=12%
1	.8990	.8091	.5420	.3700	.2176	16	.0146
2	.0876	.1464	.2364	.2178	.1527	17	.0095
3	.0134	.0380	.1152	.1370	.1151	18	.0088
4		.0065	.0508	.0894	.0885	19	.0080
5			.0267	.0577	.0707	20	.0056
6			.0150	.0363	.0508	21	.0044
7			.0047	.0355	.0476	22	.0031
8			.0040	.0173	.0342	24	.0034
9			.0015	.0146	.0271	25	.0070
10			.0017	.0086	.0224	26	.0018
11				.0059	.0216	27	.0019
12			.0020	.0039	.0193	34	.0024
13				.0014	.0220	36	.0025
14				.0030	.0167	37	.0026
15				.0016	.0168	39	.0027

Table II. Typical values of $f_n(C)$ for various C, $n \leq 40$, for a random FCC lattice with 1NN + 2NN clusters.

n	C=.5%	C=1%	C=2%	C=5%	C=8%	n	C=8%
1	.9104	.8439	.7035	.3823	.2253	20	.0018
2	.0854	.1241	.2052	.2249	.1707	21	.0077
3	.0042	.0249	.0631	.1371	.0994	22	.0030
4	*	.0072	.0194	.0843	.0859	23	.0052
5		*	.0045	.0553	.0656	24	.0022
6			.0043	.0330	.0528	25	.0034
7			*	.0195	.0498	26	.0012
8				.0183	.0368	27	.0012
9				.0148	.0324	28	.0026
10				.0064	.0310	29	.0026
11				.0031	.0246	30	.0014
12				.0051	.0224	31	.0042
13				.0037	.0130	33	.0030
14				.0050	.0109	34	.0016
15				.0021	.0137	35	.0016
16				.0023	.0080	37	.0034
17				*	.0124	38	.0017
18				.0013	.0082	39	.0018
19				.0014	.0069		

Table III. Typical values for $f_n(C)$ for various C, n $\leq$ 40, for a random BCC lattice with 1NN clusters.

n	C=2%	C=5%	C=8%	C=12%	C=15%	n	C=12%	C=15%
1	.8330	.6485	.5128	.3631	.2722	18	.0065	.0096
2	.1308	.2032	.1914	.1580	.1244	19	.0023	.0083
3	.1260	.0783	.1207	.1191	.0928	20	.0012	.0145
4	.0101	.0367	.0727	.0811	.0728	21		.0112
5		.0175	.0411	.0619	.0668	22		.0085
6		.0053	.0233	.0449	.0502	23	.0028	.0022
7		.0071	.0133	.0444	.0356	24	.0029	.0023
8			.0058	.0275	.0372	25	.0015	.0012
9		.0013	.0057	.0201	.0327	27		.0026
10			.0045	.0175	.0247	28		.0014
11			.0050	.0126	.0202	30		.0029
12			.0011	.0051	.0244	31		.0015
13			.0012	.0094	.0208	32		.0031
14		.0020		.0068	.0102	33		.0016
15				.0036	.0131	35		.0034
16			.0015	.0048	.0101	36		.0017
17				.0030	.0140	39		.0019

Table IV. Typical values of $f_n(C)$ for various C, n $\leq$ 40, for a random BCC lattice with 1NN + 2NN clusters.

n	C=1%	C=2%	C=5%	C=8%	C=10%	n	C=8%	C=10%
1	.8502	.7603	.4833	.3069	.2408	14	.0051	.0164
2	.1190	.1869	.2399	.1182	.1540	15	.0054	.0075
3	.0280	.0410	.1270	.1391	.1095	16	.0029	.0161
4	.0029	.0101	.0756	.1025	.0844	17		.0085
5		.0018	.0359	.0668	.0653	18		.0060
6			.0014	.0482	.0563	19		.0032
7			.0123	.0329	.0410	20	.0036	.0101
8			.0105	.0282	.0321	23	.0021	
9			.0026	.0227	.0301	27		.0045
10			.0015	.0108	.0385	29		.0097
11				.0099	.0147	31		.0104
12				.0065	.0181	32		.0054
13				.0117	.0174	38	.0034	

Table V.   Values of S(C) as defined in Eq. 5 for various
           lattice types, cluster types, and concentrations.

C (at.%)	S(C) for FCC Lattice		S(C) for BCC Lattice	
	1NN Clusters	1NN+2NN Clusters	1NN Clusters	1NN+2NN Clusters
1	2.35	2.4	2.3	2.4
2	2.5	2.8	2.5	2.6
5	3.2	4.4	3.0	3.5
8	4.6	$8.3 \pm .2$	3.7	5.5
10		18 ± 2		$7.7 \pm .2$
11		31 ± 2		
12	$8.3 \pm .2$	71 ± 12	5.5	12 ± 1
14				26 ± 2
15	18 ± 2		8.3	
17			12 ± 1	

Table VI.   Comparisons of the experimental Curie constant
            ξ(exp) for various concentrations of quenched
            AuFe alloys with calculated values of ξ(single),
            ξ(1NN), and ξ(1NN + 2NN) for a FCC lattice
            structure.

C (at.%)	ξ(single) (emu °K/g)	ξ(1NN) (emu °K/g)	ξ(1NN+2NN) (emu °K/g)	ξ(exp) (emu °K/g)
1	6.7×10^{-5}	7.2×10^{-5}	7.3×10^{-5}	7.3×10^{-5}
2	1.4×10^{-4}	1.5×10^{-4}	1.7×10^{-5}	1.7×10^{-5}
5	3.5×10^{-4}	5.3×10^{-4}	7.0×10^{-4}	7.7×10^{-4}
8	5.7×10^{-4}	1.2×10^{-3}	2.2×10^{-3}	2.7×10^{-3}
12	8.6×10^{-4}	3.4×10^{-3}	2.8×10^{-2}	3.7×10^{-2}

These computer simulations have been extended to higher concentrations in order to study the onset of ferromagnetic order in random substitutional binary alloys. A comparison of this work with the experimental data for quenched ferromagnetic AuFe alloys will be published shortly. We are currently using computer simulations which allow local atomic ordering to study the superparamagnetism of annealed AuMn and CuMn.

REFERENCES

1. E. Kneller, in _Magnetism and Metallurgy_, edited by A.E. Berkowitz and E. Kneller (Academic, New York, 1969) Vol. 1, pp. 393-409.

2. V. Cannella and J.A. Mydosh, to be published in Phys. Rev. B, Dec. 1972.

3. J.L. Tholence and R. Tournier, J. Phys. (Paris) _32_, C1-211 (1971).

4. It should be mentioned that the value of $\xi(\exp)$ for $C = 12\%$ (and 13%) in Ref. 2 was determined from a portion of the $1/\chi$ vs. T curve rather far from T_0. In this paper the value of $\xi(\exp)$ for $C = 12\%$ was determined from the portion of the curve immediately above T_0 which is consistent with the determination of $\xi(\exp)$ for the other concentrations.

LAW AND ORDER IN MAGNETIC GLASSES

N. Rivier and K. Adkins

Department of Physics, Imperial College

London SW7 2BZ, U.K.

I.   A magnetic "glass" can be described as a metal pollut-
ed by several magnetic impurities located at random, and interacting with each other. There is no recognizable long-
range order, but the short-range order manifests itself as
a peak in the specific heat, and the magnetic nature of this
order can be inferred from the quenching by an external magnetic field of the corresponding peak in the magnetic susceptibility.

Generally, the interaction is of the Ruderman-Kittel-
Kasuya-Yosida (RKKY) type, that is, oscillatory and long-
range. The oscillatory character of the interaction together with the randomness in the position of the impurities is responsible for the absence of long-range order.*
The slow (r^{-3}) spatial dependence of the RKKY interaction
gives rise to very simple scaling laws,[1]   obeyed by these
materials whenever the wiggly modulation can be averaged.
This will allow us to regard <u>the magnetic glass as a well-
defined state of magnetic matter</u> in a region of the (T,c)
plane, like ferromagnetism, Langevin paramagnetism, etc.

One can see two justifications for the name "magnetic
glass". First, the suggestion of Anderson[4]   that the

---

*Two competing interactions with different ranges would do
as well, as in <u>PdMn</u> at intermediary (4-5%) concentrations,
discussed in these Proceedings by Professor Coles.[2,3]

short-range magnetic order could be described as a spin
density wave function that is localized, like the electronic wave function of some amorphous semiconductors. Second,
the specific heat, linear in the temperature and independent of the concentration at low temperatures (Zimmerman
and Hoare[5]) has the universal T-law recently found in
several glasses by Pohl.[6]

II. We are going to describe the magnetic glass by a local
molecular field, following Marshall[7] and Klein and Brout.[8]
The quantity investigated is the effective magnetic field
seen by one impurity at $\vec{r}$ due to all the other impurities,
in addition to a possible external magnetic field. Since
the impurities are located at random, their contribution
to the effective field $\vec{H}_i = \vec{S}_i J(\vec{r} - \vec{R}_i)$ is a random variable, whether there is magnetic order or not. The effec-
tive field can therefore be described by a probability distribution $P(\vec{H})$ which is continuous since J is a continuous
function of the distance. [The difference between a crystalline and an amorphous substitutional alloy is only relevant for the shells of atoms nearest to the impurity in
$\vec{r}$, and affects therefore only the high field contributions
to $P(\vec{H})$]. The question of the localization of the spin
density wave function, and generally the nature of the
correlation between spins arises in the next member of the
hierarchy, the joint probability $P[\vec{H}(\vec{r}), \vec{H}'(\vec{r}')]$.

The interest of this approach is twofold: first it
introduces the randomness of the system (responsible for
the localized nature of the order) in an elementary way -
the contributions of each impurity $\vec{H}_i$ are independent, and we
we have a one-dimensional random walk problem for an Ising
RKKY interaction and a three-dimensional random walk problem for a Heisenberg RKKY interaction. An ordinary glass
where the electronic paths are not self-avoiding is already
a more complicated random system. Second, the effective
field is precisely the auxiliary field coupled to the impurity spin in a functional integral solution of the
Hamiltonian $H = \sum_{ij} J_{ij} \vec{S}_i \cdot \vec{S}_j$ (i, j random positions). It
is therefore possible in principle to obtain the elementary
excitations and all the dynamics of the system by studying
the Green functions of the effective field $P[\vec{H}(\vec{r},t); \vec{H}'(\vec{r},t')]$ (c.f., e.g., Ref. 9).

The calculation of $P(\vec{H})$ follows a general method of Markoff, first applied to the spin glass problem by Anderson (c.f., Ref. 8). The effective field seen by the impurity at the origin and due to N other impurities is the sum of N independent variables $\vec{H}_i$ described by the same probability distribution $w(\vec{H}_i)$, therefore,

$$P_N(\vec{H}) = \int d\vec{H}_1 \cdots \int d\vec{H}_N \; w(\vec{H}) \cdots w(\vec{H}_N)$$

$$\times \; \delta[\vec{H} - \sum_i^N \vec{H}_i] = \int \frac{d^3k}{(2\pi)^n} \; e^{-i\vec{k}\cdot\vec{H}}[w(\vec{k})]^N, \qquad (1)$$

where n is the dimensionality of the "random walk" (n = 1 for an Ising, n = 3 for a Heisenberg, RKKY interaction, n = ∞ for a spherical model), and $w(\vec{k})$ is the Fourier transform of $w(\vec{H}_i)$. Specifically $\vec{H}_i = \vec{S}J(\vec{R}_i)$, where $J(\vec{R})$ is the RKKY interaction. The i-th impurity can be anywhere within the volume Ω, with an orientation S_e given by the probability distribution $p_{\vec{R},\vec{H}}(\vec{S}_e)$ which is a function of its position as well as of the resulting total magnetic field, and describes the correlation between the i-th impurity and that at the origin, so that

$$w(\vec{H}_i) = \frac{1}{\Omega} \int d^3R \sum_e p_{\vec{R},\vec{H}} (\vec{S}_e) \; \delta \; [\vec{H}_i - \vec{S}J(\vec{R})] \qquad (2)$$

is an implicit function of the resulting $\vec{H}$ itself. Writing $\Omega = Na^3/c$, where c is the concentration of impurities, with the identity $\lim_{N\to\infty}(1-x/N)^N = (\exp{-x})$ one obtains the general equation

$$P(\vec{H}) = \frac{A}{(2\pi)^n} \int d^n k e^{-i\vec{k}\cdot\vec{H}} e^{-\frac{c}{a^3} \int d\vec{R} \sum p_{\vec{R},\vec{H}}(S_e)}$$

$$\times \; [1 - e^{i\vec{k}\cdot\vec{S}J(\vec{R})}]. \qquad (3)$$

A is a normalization constant. The only unknown function is $p_{\vec{R},\vec{H}}(S_e)$ which describes the correlation between spins (clusters) and can be obtained from the joint probability $P[\vec{H}, \vec{H}']$ in the next step of the hierarchy. $p_{\vec{R},\vec{H}}(S_e)$ is also the only function in Eq. 3 which depends on the temperature and on an applied magnetic field. Equation 3 is new, but the derivation should be familiar to the reader of Klein and Brout, or generally of the literature on the Central Limit Theorem.

The distributions $P(\vec{H})$ in the simplest cases are given in Table I. Detailed calculation will appear elsewhere (e.g., Ref. 10). The following general features can be noticed:

1. $P(\vec{H})$ is an isotropic (symmetric in the Ising case) function of $\vec{H}$: the clusters are free to rotate.

2. The __broadening__ of the distribution is given by the real part of the exponent in Eq. 3. For spin 1/2, it is independent of the presence and of the nature of the short range order (SRO). It is also independent of the temperature.

3. The __shift__ in the distribution is temperature-dependent. For example, the Ising distribution exhibits three peaks at temperature $kT < 1/2H_0$, that coalesce into one peak at $kT \geq 1/2H_0$, indicating that the short-range order disappears at a local Curie temperature. The local "Weiss" field decreases with temperature.*

4. The distribution $P(H_z) = [2\pi \int dH_\perp \, H_\perp \, P(\vec{H})]$ is independent (in the absence of SRO), or nearly independent (in the presence of SRO), of the dimensionality of the interaction. That provides __model__ __independent__ expressions for static quantities like the magnetic susceptibility or the specific heat.

An objection to the Marshall-Klein and Brout theory has often been made whereby the linearity in temperature of the specific heat is an artifical result of the Ising interaction. A Heisenberg interaction between the impurities gives strictly $P(H = 0) = 0$ (see Table I) since $H = 0$ is only a point in a three dimensional space. The relevant quantity, however, is the __finite__ $P(H_z = 0)$ rather than the vanishing $P(H = 0)$, because the local magnetic energy $\vec{S} \cdot \vec{H}$ involves a scalar product, hence a preferred direction

---

*As expected physically, but Klein predicts the opposite.[8]

TABLE I $P(\vec{H})$ distributions for spin 1/2

<u>Ising</u>	no SRO	$P(H) = \Delta/\pi \ (H_z^2 + \Delta^2)$
	SRO	$P(H_z) = \Delta/\pi A[(H_z - H_o \tan h1/2\beta H_z)^2 + \Delta^2]$

<u>Heisenberg</u>	no SRO	$P(\vec{H}) = \Delta/\pi^2 (H^2 + \Delta^2)^2$
		$P(H) = 4\pi \ H^2 P(H)$
		$P(H_z) = 2\pi \int dH_\perp H_\perp P(\vec{H}) = \Delta/\pi(H_z^2 + \Delta^2)$
	SRO, T=0	$P(\vec{H}) = \Delta/\pi^2 B[(H - H_o)^2 + \Delta^2]^2$
		$P(H_z)$ similar to Ising

A, B normalization constants Δ, H_0 are temperature independent and proportional to the concentration. For reference,

$$\Delta = 2|k|^{-1} \int d\vec{R} \ \sin^2 \frac{1}{4} \ kJ(\vec{R}) \simeq \frac{1}{3} \ \pi^2 \ [\alpha/(2k_F a)^3] \ c$$

(that of $\vec{S}$).*

    Regarding the susceptibility, remember that the spins
uncorrelated to that at the origin form a polarizable medium, and give rise to a susceptibility of the form
$\chi = \chi_0/(1 - a\chi_0)$, where $a < S^z >$ is the Weiss field due
to the distant impurities obtained self-consistently. This
allows for the occurrence of long-range order, whether
ferro- or antiferromagnetic, in some glasses.

5.   The general shape of $P(\vec{H})$ is a <u>Lorentzian.</u> When averaging over the RKKY wiggles $J \to < J > = \alpha/(2k_F R)^3$, one obtains by direct change of variable and use of the Boltzmann
distribution, the scaling laws of Souletie[1].

$$P(H_z, T) = c^{-1} f (H_z/c, T/c) \tag{5}$$

*The proof is straightforward: the magnetic energy per spin
is $\quad E = \int d\vec{H} \ P(\vec{H}) \ \vec{S} \cdot \vec{H} = \int dH_z B(H_z, T) H_z [2\pi \int dH_\perp H_\perp P(\vec{H})]$,
where $B(H_z, T)$ is the Brillouin function for the spin $\vec{S}$ in a
field H_z. Thus
$$E = \int dH_z \ B(H_z, T) \ H_z \ P(H_z), \tag{4}$$
by definition of $P(H_z)$. This is Marshall's well known expression, generalized here to any interaction.

leading to universal expressions for the physical quantities (χ, C_{sp}) as functions of the reduced variables T/c.

One would have expected a Gaussian $P(\vec{H})$ since the calculation of Eq. (3) is identical to the proof of the Central
Limit Theorem (cf., e.g., Ref. 11). The CLT implies some
assumptions on the analytic behavior of the probability
distribution of each step, assumptions that are not obeyed
by $w(H_i)$ because of the slow, oscillatory nature of the interaction (whether averaged RKKY or not). Precisely, the
derivative $w'(H_i)$ is neither continuous nor absolutely in-
tegrable. This affects the large k, hence the small H contribution to $P(\vec{H})$ which interests us here.*

Incidentally, the concentration independent, linear
specific heat[5]   of spin glasses can be seen as a direct
experimental proof of the inapplicability of the CLT. After a Sommerfeld expansion of Eq. 4 one obtains the specific
heat

$$C_{sp} = (cst) \times c\ P(H_z = 0)\ kT, \tag{6}$$

regardless of the dimensionality of the interaction. A
Gaussian distribution, as predicted by the CLT, gives
$P(H_z = 0) \propto c^{-1/2}$ and $C_{sp} \propto c^{1/2}\ kT$, in direct contradiction with the experimental results.

The concentration independence follows from a distribution that obeys Blandin's scaling laws, Eq. 5, i.e. from
a r^{-3} interaction between spins. This is also the case of
the interaction via phonons between the localized structures
that could explain the low temperature behavior of Pohl's
glasses[12] (dipole-dipole interaction between local ionic
displacements). Moreover, since the RKKY or Caroli[13] coupling constant α is more or less constant for the whole
series of Transition-Transition or Noble-Transition metal
alloys, the linear term in the specific heat should be
quantitatively the same in all these glasses. The unfor-
tunate fact that C_{sp} in Eq. 6 samples the spins or cluster
that are free to rotate, and gives no indication on the
nature of the order, is certainly related to this universality.

---

*It is easy to give Gaussian wings to $P(H_z)$ by keeping the
impurities on a discrete lattice or by taking a lower cut-
off for the distance between two impurities. However, P(0)
will still go as c^{-1}.

The question of the localization of the spin density wave function is related to that of the local field distribution and must be discussed with the joint probability

$$D(\vec{H}, \vec{H}'; \vec{R}) = P[\vec{H}(\vec{r}), \vec{H}'(\vec{r}+\vec{R})] - P[\vec{H}(\vec{r})] \, P[\vec{H}'(\vec{r}+\vec{R})]$$

Following the steps leading to Eq. 3 one obtains easily,

$$P[\vec{H}(\vec{r}), \vec{H}'(\vec{r}+\vec{R})] = \int d^3k \; e^{-i\vec{k}'\cdot\vec{H}'} \; e^{-cf(\vec{k},\vec{k}';\vec{R})} \qquad (7)$$

where both the shift Im f and the broadening Re f are linear in k and k' to a good approximation. For spin 1/2 the broadening is independent of the function $p_{\vec{R},\vec{H}}(S_e)$ (which must be obtained self-consistently here), and is a good measure of the amount of correlation. Re f goes as $|\vec{k}|+|\vec{k}'|$ in the absence of correlation, and as $|\vec{k} + \vec{k}'|$ when R=0 (full correlation), in complete anology with an interference problem.

Averaging over the RKKY oscillations yields exactly the scaling law

$$P(H,H';\vec{R}) = c^{-2} \, g \, (H/c, \, H'/c; \, \vec{R}/c^{1/3}) \qquad (8)$$

Consequently, there is a relation between the average radius of the clusters, R_o, and the concentration

$$R_o \propto c^{-1/3} \qquad (9)$$

which implies magnetic screening in the Blandin regime. (This is in fact an old result of Blandin and Klein and Brout).

A rough evaluation of Eq. 7 yields $D \sim e^{-KR^3} \times f(\vec{H},\vec{H}')$. The constant of proportionality in Eq. 9 can be evaluated to be a/1.5, i.e. R_o = a for a concentration of 30%. That corresponds to 2.4 impurities per cluster on the average (including the impurity at the origin) <u>independently of the strength of the interaction</u>. This number, once again, is due to Blandin and Klein and Brout.

III.   It should have become clear at this stage that the
magnetic glass exhibits some very general, collective, properties.   These properties take a simple form as functions
of reduced variables in the Blandin regime, i.e., whenever
the RKKY oscillations can be averaged.   This is only possible at low concentrations: nearest neighbor impurities
hold much more tightly together than the clusters discussed
above, the clusters become very large and the localized
magnons very hard.   One estimates the highest concentration
c_2 for the Blandin regime as that for which the probability
that an impurity has no nearest neighbor within a radius
$\eta^{1/2}\pi/2k_F$ (corresponding to the first zero of the RKKY oscillations; η is the Stoner enhancement of the host metal*)
becomes less than 1/2.   One obtains (with a Poisson distribution of impurities),

$$c_2 = (3 \ln 2/4\pi) \; (2k_F \; a/\pi)^3 \; \eta^{-3/2} \qquad\qquad (10)$$

i.e., $c_2 \sim 10\%$ for a noble host metal.   This upper bound is
again independent of the strength of the interaction.   As
expected, c_2 is strongly reduced in a polarizable host (e.g.,
Pd, where $\eta = 10$).   For $c > c_2$ one expects mictomagnetic behavior of the type discussed by Professor Beck in these
Proceedings.

The Blandin regime is limited at low concentrations as
well, i.e., there is a concentration c_1 below which one observes single impurity behavior such as the Kondo effect,
(see, e.g., Ref. 15).   The existence of a lowered bound is
somewhat surprising for a long-range interaction like RKKY.
The proof is based on the fundamental theorem of Information
Theory (see, e.g., Ref. 16).   The question is whether one
can obtain by some experiment any information on the state
of magnetization of an isolated impurity, which, if it has
spin 1/2, can be seen as a binary symmetric source.   This,
in the presence of noise, is described by the local field
distribution P(H).

A spin 1/2 impurity is capable of flipping its spin up
and down (or, alternatively, an impurity state can be occupied at any instant by an up or down spin electron).   One

---

*It is a simple matter to show that a host susceptibility,
 enhanced by η, has a width in reciprocal space reduced by
 $\eta^{-1/2}$.   See, alternatively, Ref. 14.

can define a correlation function $< S^z(t)\ S^z(0) >$ and therefore a rate of loss of correlation or of spin memory τ_{sf}^{-1}
(independently of the Kondo- or spin fluctuation-model).
This is a well defined limit, characteristic of the impurity
and of the Fermi gas with which it interacts. It is a consequence of the fact that every spin flip produces an infrared catastrophe in the response of the Fermi gas. In terms
of information theory, τ_{sf}^{-1} is the noiseless channel capacity of the system, and indicates the maximum possible rate
of transmission of information.

The presence of other impurities affects the binary
signal (up-down) given by one impurity. Generally (and
strictly so in the absence of SRO), the effect can be described as a noise, of rate Δ (equivocation rate).*

The maximum rate of information that can be transmitted through the noisy channel is

$$C = \tau_{sf}^{-1} - \Delta \tag{11}$$

It is picked up by the measurement apparatus in an experiment which we shall assume static (e.g., resistivity or
specific heat measurement so that the information about the
impurity can be transmitted at an arbitrarily slow rate H.
Shannon's fundamental theorem states that full information
can be obtained on the instantaneous state of the impurity
by itself if and only if H < C, by a suitable coding, i.e.,
a suitable experiment. In this condition, one can observe
an individual impurity if $\tau_{sf}^{-1} > \Delta$, and it is impossible
to do so if $\tau_{sf}^{-1} < \Delta$.

A similar application of Shannon's theorem to a single
impurity in thermal noise (of equivocation rate $k_B T$) yields
the well-known phenomenological result $\tau_{sf}^{-1} = k_B T_k$, where
T_k is the Kondo temperature of the single impurity alloy.[17]

Thus, the critical concentration c_1 for observing single impurity behavior in an alloy is given by the identity

$$k_B T_k = \Delta \tag{12}$$

---

*Since the impurities are free to rotate, and the noise has
 a symmetrical distribution, one has here perfect equivocation, like that of an ideal liar who lies on the average
50% of the time.

Introducing the value quoted for Δ in Table I, one obtains,

$$c_1 = (3k_B/\pi^2\alpha)(2k_Fa)^3 T_k \tag{13}$$

For transition or noble metal hosts, $\alpha \simeq 1$ eV[13] and $2k_Fa \simeq 1$ so that

$$c_1 \simeq 10^{-4} T_k \simeq T_k/T_F \tag{14}$$

an empirical relation due to Star,[15] probably known to other experimental groups (Grenoble, Sussex). Variation in the coupling between impurity and conduction electrons over different alloys is negligible in α (where it appears quadratically) compared to its effect on T_k (where it appears exponentially).

In the spin glass regime, i.e., $\Delta > k_B T_k$, the impurity is tightly coupled to the "noise" due to all the other impurities. This defines a new source, and a new channel capacity (the rate of loss of correlation). One expects therefore Langevin paramagnetism whenever the thermal noise dominates the spin glass capacity, i.e., if $k_B T > \Delta$.

There is a very strong analogy between loss of correlation due to infrared catastrophe in the electron sea (τ_{sf}) and that due to interaction with other impurities at random. Indeed, the susceptibility and the specific heat are qualitatively similar functions of the temperature in the Kondo and the spin glass (without SRO) regimes. However, their scaling with concentration is very different: $\chi = c/T_k f(T/T_k)$ in the Kondo regime, whereas $\chi = c/\Delta f(T/\Delta) = g(T/c)$ in the spin glass regime. Similarly, the initial linear specific heat has a slope proportional to c/T_k in the Kondo regime, and to c/Δ (concentration independent) in the spin glass regime. This, incidentally, makes Eq. 12 totally straightforward.

In conclusion, the Blandin regime of a magnetic glass is valid for concentrations $c_1 < c < c_2$. Neither c_1 nor c_2 are sharp critical points in the sense of thermodynamics, since they mark the onset of a local order, involving a finite number of degrees of freedom only. The transition between the magnetic glass regime and single impurity behavior on the one hand, mictomagnetic regime on the other hand is smooth. The Blandin scaling laws are a consequence of the r^{-3} dependence of the interaction only, and the lower and upper bounds are themselves dependent only on general parameters of the host metal and of the dilute alloy.

The Blandin regime of magnetic glass is therefore a very general state of magnetic matter, involving <u>collective</u>, and <u>random</u> distribution of impurities in a metal.

REFERENCES

1. A. Blandin, Thesis, Paris (1961); A. Souletie, Thesis, Grenoble (1968); A. Souletie, and R. Tournier, J. Low Temp. Phys. <u>1</u>, 95 (1969).
2. B.R. Coles, these Proceedings; H. Jamieson, R.H. Taylor, and B.R. Coles, to be published.
3. R.H. Taylor, Thesis, London (1972).
4. P.W. Anderson, Mat. Res. Bull. <u>5</u>, 549 (1970).
5. J.E. Zimmerman and F.E. Hoare, J. Phys. Chem. Solids <u>17</u>, 52 (1960).
6. R.C. Zeller and R.O. Pohl, Phys. Rev. <u>5</u>, 2029 (1971).
7. W. Marshall, Phys. Rev. <u>118</u>, 1519 (1960; M.W. Klein, and R. Brout, Phys. Rev. <u>132</u>, 2412 (1963); M.W. Klein, Phys. Rev. <u>136</u>, 1156 (1964); and subsequent publications.
8. D. Sherrington, J. Phys. C. <u>4</u>, 401 (1971).
9. K.J. Adkins, Thesis, London (1973).
10. A.I. Khinchin, <u>Mathematical Foundations of Statistical Mechanics</u>, Dover (1949), appendix.
11. P.W. Anderson, B. Halperin, and C.M. Varma, Phil. Mag. <u>25</u>, 1 (1972; W.A. Phillips, J. Low Temp. Phys. <u>7</u>, 351 (1972).
12. A. Blandin, in <u>Theory of Magnetism in Transition Metals</u>, W. Marshall (ed.), Academic Press (1967), p. 393.
13. T. Moriya, ibid., p. 234.
14. W.M. Star, Thesis, Leiden (1971).
15. E. Roubine, <u>Introduction à la Theorie de la Communication</u>, Vol. 3, Masson (1970).
16. N. Rivier and M.J. Zuckermann, Phys. Rev. Letters <u>21</u>, 904 (1968).

DISCUSSION

R. Hasegawa: A few weeks ago you told me that you have
some difficulties with the susceptibility calculations.
Could you comment on this, or have you resolved this?

N. Rivier: The difficulties are practical ones and concern
the relationship between the computer and my student or myself. Formally, one can see that the susceptibility is
giant and why it is so. The question is whether the quenching by an external magnetic field is as drastic as it should
be. The non-linear effects are there: the shift in the
$P(H)$ distribution due to the impurities within the cluster
is dependent on the magnetic field. Furthermore, the impurities outside a cluster form a polarizable medium, leading to an additional shift in the distribution as a function
of the field and to an enhanced susceptibility.

S. Kirkpatrick: You discussed the spatial distribution function of the static field, what does that tell you about the
localization of spin waves?

N. Rivier: About the localization of spin waves, nothing,
but about the localization of the spin density wave function,
the fact that you get a correlation function which decays
exponentially in space is already an indication.

S. Kirkpatrick: The question is really whether the spin ex-
citation tries to follow the slowly varying magnetic field.
Does it in consequence of these variations get localized?
Has that been looked at?

N. Rivier: That is a very interesting question. I don't
know. I think that the elementary excitations in the
Blandin regime are localized magnons: the cluster is going
to wobble, whereas it rotates rigidly in the mictomagnetic
regime. To obtain a theoretical answer, you must follow
through the functional integral approach that I only sketched. Experimentally, the specific heat is <u>not</u> telling you
anything about the inside of the cluster, since its major
contribution - the linear law - is due to spins or clusters
that are free to rotate. The resistivity is a better can-
didate.

Question: Is the slope of the linear term in the specific heat the same for a material with spin 5/2-impurities as for spin 3/2 or 1/2, etc?

N. Rivier: I have not made the calculation for spins other than 1/2. Yours is an interesting point related to the universality found by Pohl in ordinary glasses. We can show that the linear term of the specific heat is independent of the concentration and that it depends on the particular alloy only through Caroli's coupling constant for spin 1/2 impurities, that is small compared to T_k. I don't know of enough experimental evidence regarding this question.

T.A. Kaplan: Do you make a molecular field approximation, say within the functional integral scheme?

N. Rivier: Yes. In the present state of the theory, the functional integral scheme is just a fancy name used to justify the molecular field approach. It is nevertheless a consistent scheme that can give you a complete solution of the problem, with the correlations between impurities and the dynamics included.

P.A. Beck: I think this linear contribution to the low temperature specific heat is just a very rough first approximation in a narrow temperature range. If you look at the thing over a wider temperature range it turns out to be much more like a Schottky specific heat.

N. Rivier: Is the linear law breaking down at low temperatures as well?

P.A. Beck: Yes. It becomes a Schottky anomaly.

N. Rivier: Does it rise exponentially from T = 0?

P.A. Beck: I am referring to the recent paper by W.A. Phillips.

N. Rivier: Does it become a Schottky anomaly theoretically?

P.A. Beck: No, emperically, actual experimental data show this.

SPIN ARRANGEMENTS IN DILUTE ALLOYS

B. Window

Physics Dept., Carnegie-Mellon University

Pittsburgh, Pennsylvania 15213

ABSTRACT

Mössbauer results for gold iron and copper iron alloys at 300 K and 4.2 K are analyzed to show that the spins in such alloys point in directions determined by the local environment. For clusters of three or more iron atoms linked as near neighbours, the spins point in directions consistent with a large pseudodipolar anisotropic exchange interaction between near neighbours. For pairs of atoms single ion anisotropy appears to dominate the pseudodipolar anisotropy. The unifying feature of the numerous dilute alloy systems which behave like gold iron alloys is probably that the intracluster anisotropy is comparable to or greater than the intercluster or further near neighbour exchange.

The unusual magnetic ordering in alloys of dilute magnetic impurities in non magnetic hosts has been investigated by many techniques, but one experimental method which has pointed the way for other measurements is the Mössbauer effect; it was the first technique to measure the sharp ordering temperatures characteristic of such alloys. The recent small field susceptibility measurements on gold alloys by Cannella and Mydosh[1] have shown that the essential condition for a sharp anomaly is the use of small fields, and the ordering temperatures they observe coincide with those measured in the Mössbauer experiments of Violet and Borg[2].

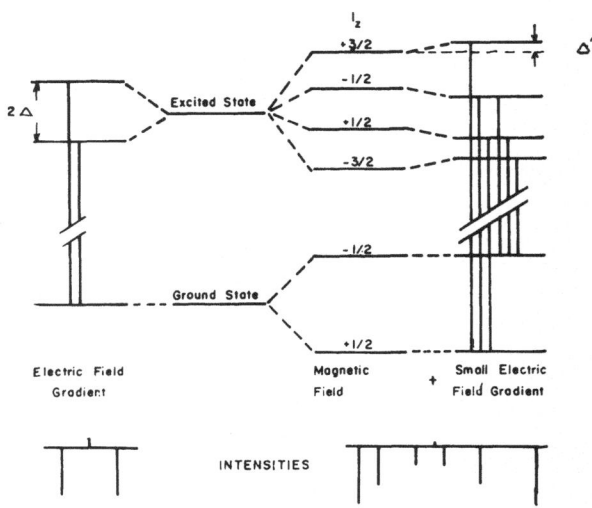

Figure 1 Energy level diagrams for the ground and excited
states of the ^{57}Fe nucleus are shown for a small electric
field gradient (a) and a large magnetic field with a small
electric field gradient (b).

Another aspect of the technique is the ability in certain favorable cases to determine where the spins point
relative to the near neighbour impurities. This determination relies on there being observable electric field
gradients (efg) due to impurity near neighbours in the
otherwise cubic environment, and hence is applicable only
to iron in copper or gold alloys.[3]

The non magnetic 300K spectra of gold iron alloys can
be fitted as a function of iron concentration to give a
point charge per neighbor (q_n) and an isomer shift change
as a function of the number of iron neighbours (n). We
express the point charge q_n in units of the quadrupole-
splitting (Δ) expected in the Mössbauer spectrum for one n
neighbour by itself. (See ref. 3 for a discussion of the
fitting) For a configuration with n neighbours, the quadrupole splitting (Δ) (Figure 1) can be found by evaluating
V_{xx}, V_{yy}, V_{zz} in the principal axis system of the efg (x,y,
z), e.g.

$$V_{xx} = \frac{1}{2} q_n \sum_{i=1,n} (3 \cos^2\theta_{xi} - 1)$$

where θ_{xi} is the angle subtended at the origin by the x axis and i th neighbour. Then,

$$\Delta = V_{zz} (1 + \eta^2/3)^{1/2} \text{ where } \eta = |(V_{xx} - V_{yy})/V_{zz}|$$

In the presence of a large magnetic field, the efg can be treated as a small perturbation, and the resultant observed efg (Δ^1) (Figure 1) is given by

$$\Delta^1 = 1/2 \ q_n \sum_{i=1,n} (3 \cos^2\theta_{z'i} - 1)$$

where the magnetic field direction defines the z' axis, and $\theta_{z'i}$ is the angle subtended at the origin by this axis and i th impurity. Hence, if q_n is obtained from the 300 K data, and Δ^1 from the 4.2 K data as a function of n, values of

$$\varepsilon_n = 1/(2n) \sum_{i=1,n} (3 \cos^2\theta_{z'i} - 1) \quad \text{can be obtained.}$$

These values will be averages over all the possible arrangements of n neighbours, and will be zero when long range magnetic ordering does occur. The pronounced asymmetry of the spectra of AuFe alloys at 4.2 K (Figure 2) immediately

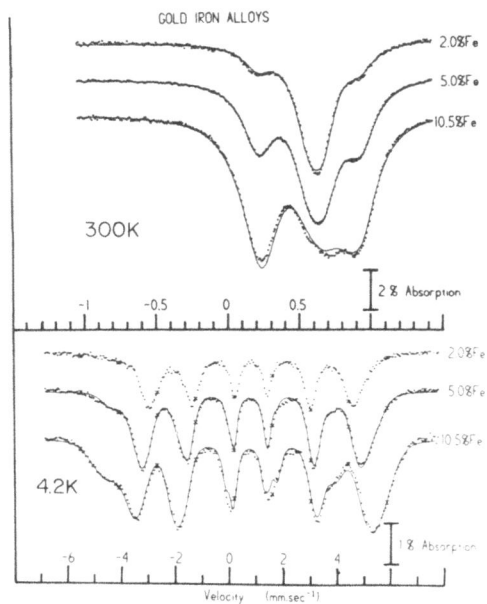

Figure 2 Mössbauer spectra of gold iron alloys at 300K and 4.2K together with fitted curves (The fitting is described in reference 3)

shows that the spin directions are related to where the iron near neighbours are situated. These spectra have been fitted (see reference 3) by considering n=0,1 and $\geq$ 2 iron neighbour configurations to give $\varepsilon_1 \approx \pm 1/2$ and $\varepsilon_2 \approx \mp 1/2$, the sign depending on the sign of q_n. There is some evidence for a negative charge q_n, and so $\varepsilon_1 \approx -1/2$ and $\varepsilon_2 \approx +1/2$.

We now consider some facets of exchange anisotropy in metals and alloys that have been overlooked in most work on these alloys. Anisotropy arises from spin orbit coupling[4], and can be divided into single ion terms, most commonly represented by a DS_z^2 term in the Hamiltonian for a crystal field of axial symmetry, and terms due to the interaction between neighbouring atoms (called exchange anisotropy)[4,5]. The latter is more important in magnetic metals, and was used by Van Vleck[4] to explain the well known anisotropy of ferromagnetic iron and nickel. Treating the spin orbit coupling as a perturbation, the leading anisotropic exchange contribution is a pseudipolar interaction

$$J(\lambda/\Delta\varepsilon)^2 \; [\vec{S}_i \cdot \vec{S}_j - 3(\vec{S}_i \cdot \vec{r}_{ij})(\vec{S}_j \cdot \vec{r}_{ij})/r_{ij}^2 \;]$$

where J is the exchange interaction, λ is the spin orbit coupling constant, and $\Delta\varepsilon$ is the energy difference between the ground state and the next excited orbital state. This term vanishes in cubic symmetry. The anisotropy of fcc nickel and bcc iron comes from higher order terms of order

$$K_1 \sim J(\lambda/\Delta\varepsilon)^4 \sim 10^{-4} \; J \qquad \text{from experiment}$$

In the dilute magnetic alloys with no long range magnetic order the pseudo dipolar terms will not vanish within a particular cluster of atoms, and will be of order

$$K_1 \sim J(\lambda/\Delta\varepsilon)^2 \sim 10^{-2} \; J$$

Hence it is feasible that K_1 could be 0.5 - 10% of J.

Some numerical calculations have been made on a random lattice set up on a computer, where clusters linked by near neighbours were sorted out, their easy directions calculated using dipole-dipole interactions, and the distributions of ε_n calculated assuming the magnetization pointed in this direction. The results are shown in Figure 3 for a 5% and 10% iron sample, and ε_n for $n \geq 2$ has a mean of about 1/2, as observed experimentally. Clusters of a pair of iron atoms have been excluded in the calculation, as the dipole-dipole interaction would make the spins line up along the

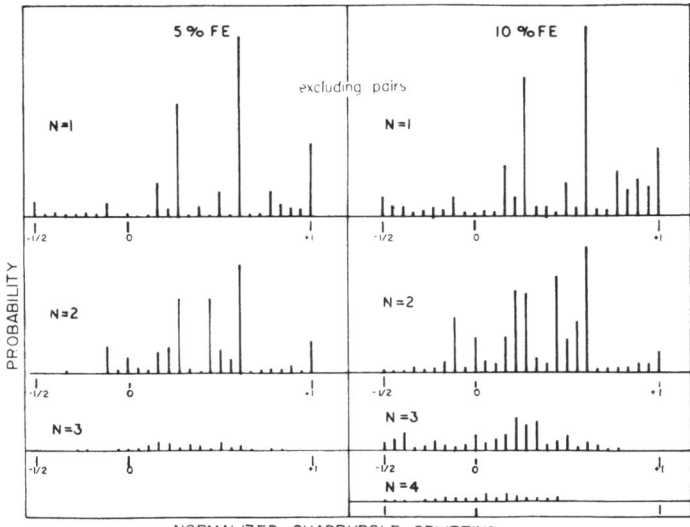

Figure 3   Distribution of ε_n, proportional to quadrupole splitting in the magnetic spectra for iron atoms with n iron near neighbours, calculated assuming the spins in clusters point along easy axes determined by pseudo dipolar interactions.

Fe-Fe axis, giving $\varepsilon_1 \sim +1$, which is not observed. In this situation, we propose that a DS_z^2 term in the Hamiltonian is responsible for the spin pointing perpendicular to the Fe-Fe axis. In pairs, the axial distortion of the iron screening charge is greatest, and one expects the largest $|D|$ for such configurations. In similar alloys such as iron in rhodium, the Mössbauer effect shows small quadrupole interactions and hence D should be small, and this source of anisotropy is probably not necessary for the unusual properties of these alloys. In reference 3, the fitting was interpreted as indicating that the spin pointed along a <111> direction so as to maximize ε_n, but the fit does not justify such a restrictive interpretation. In the high concentration ferromagnets where the first order terms average to zero over the large clusters, higher orders will determine the easy direction, possibly a <111> direction.

We propose a Hamiltonian for the spins in clusters given by

$$\mathcal{H}_{cluster} = -J\Sigma_{pairs} \vec{S}_i \cdot \vec{S}_j + K_1 \Sigma_{pairs} [\vec{S}_i \cdot \vec{S}_j - \frac{3(\vec{S}_i \cdot \vec{r}_{ij})(\vec{S}_j \cdot \vec{r}_{ij})}{r_{ij}^2}]$$
$$-g\beta\Sigma_{i=1,n} \vec{H}_i \cdot \vec{S}_i$$

where $\vec{H}_i$ includes the exchange interactions on the i th spin in the cluster due to other clusters and any applied magnetic field. The first term causes the spins to couple parallel well above the ordering temperature, while the second term forces the spins to point along a single direction (up or down the easy axis) when the third makes the sample order magnetically. The alignment of the spins well above the bulk ordering temperature will give rise to large specific heat anomalies at the higher temperatures[6] and cause the large values of the effective moment observed by Cannella and Mydosh[1] just above the ordering temperature. Calculations of other properties would follow the treatment of Kouvel[7].

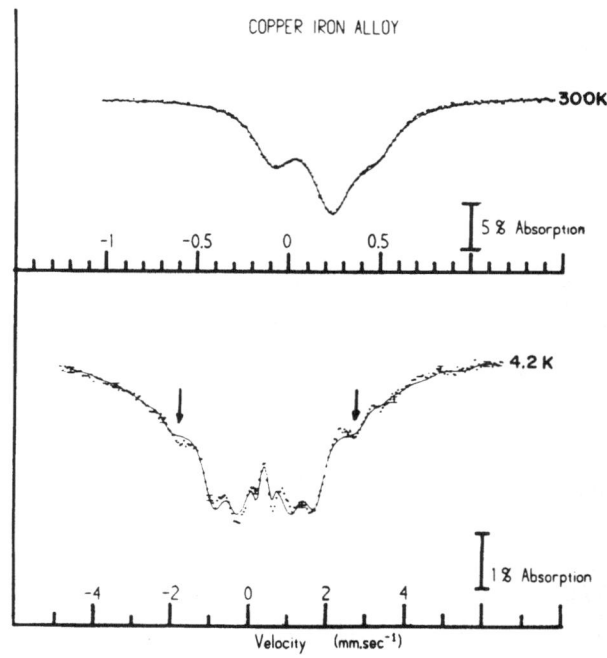

Figure 4 Mössbauer spectra of a quenched copper 1% iron alloy at 300K and 4.2K. The displacement of the peaks, marked with arrows is consistent with pairs of iron atoms pointing normal to the Fe-Fe axis.

In Figure 4 the Mössbauer spectra of a quenched copper one atomic percent iron alloy are shown. The displacement of the peaks marked with arrows are consistent with the spins of iron atoms in pairs pointing normal to the Fe-Fe axis. (8)

REFERENCES

1. V. Cannella and J. Mydosh, Phys. Rev. to be published (1972).
2. C. E. Violet and R. J. Borg, Phys. Rev. 149, 540 (1966)
3. B. Window, Phys. Rev., to be published (1972)
4. J. H. Van Vleck, Phys. Rev. 52, 1178 (1937)
5. K. Yosida, J. Appl. Phys. 39, 511 (1968)
6. J. E. Zimmerman and F. E. Hoare, J. Phys. Chem. Solids 17, 52 (1960)
7. J. Kouvel, J. Phys. Chem. Solids 24, 795 (1963)
8. B. Window, Phil. Mag., to be published.

DISCUSSION

W. Buyers: How does the magnitude of the pseudo-dipolar anisotropy compare with that of direct dipole-dipole aniso-tropy when you have got two iron atoms adjacent?

B. Window: I haven't calculated it, but usually just pure dipole-dipole anisotropy is not very significant.

W. Buyers: In some systems it is in fact the dominant source of the anisotropy. For example, even in MnF_2, one of the simplest antiferromagnets, the energy gap is 95% magnetic dipole interaction.

B. Window: You are sure it isn't pseudo dipolar?

W. Buyers: I think that in this case you can calculate what the dipole interaction is, you know what all the num-bers are, and that represents all of the observed energy gap.

B. Window: The near neighbour iron atoms in the alloy are very close together, and interact strongly by direct ex-change. In this case, the pseudo dipole terms should dominate the bare dipole interaction.

R. Bukrey: You have a reasonable quadrupole interaction for dimer iron-iron at room temperature. Do you see that same size quadrupole interaction when you lower the temperature?

B. Window: You see about half of it.

ELECTRICAL RESISTIVITY OF A SPIN-GLASS RANDOM ALLOY: AuFe

J. A. Mydosh

Institut für Festkörperforschung, Kernforschungs

anlage-Jülich, Germany

and

P. J. Ford

Institute of Physics, The University of Zagreb,

Yugoslavia

ABSTRACT

The AuFe system may be considered as a typical example
of a spin glass[1] or mictomagnet,[2] except in the most dilute
concentrations, (below ≈ 0.05 at.% Fe), where the Kondo effect
becomes dominant. This system has long been studied,[3] but
only recently have the concepts of a spin glass been applied
to explain its rather strange magnetic behavior.[3,5] Although
there has been this renewed interest,[6] mainly in the magnetic properties of spin-glass alloys, there has as yet been no
recent study of the electrical resistivity of such a system.[7]
This would be useful, firstly to examine the high temperature cluster formation, secondly to study the behavior a-
round the ordering or "freezing" temperature, T_0, and finally
to look at the low temperature spin-wave excitations.

A simple way to introduce the behavior of the AuFe system is to plot the ordering temperature, T_0, as a function
of the iron concentration. This is shown in Fig. 1 where it
can be seen that for $c < 12$ at.% Fe, T_0 is varying in a slow
and nonlinear manner, and that this persists down to

237

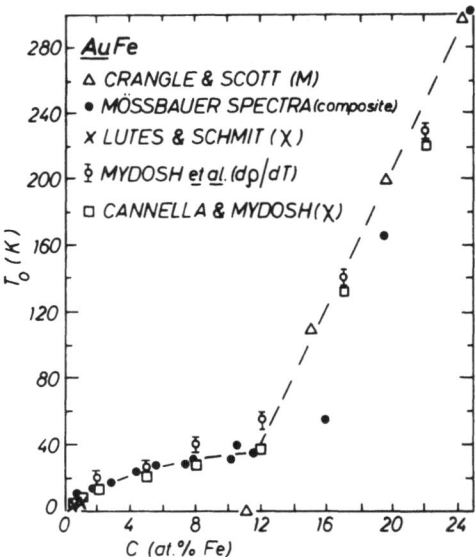

Figure 1:   Determination of the ordering temperature, T_0,
            for AuFe alloys by a variety of experimental
            techniques.   (See Ref. 3 for a more complete
            description of this figure.)

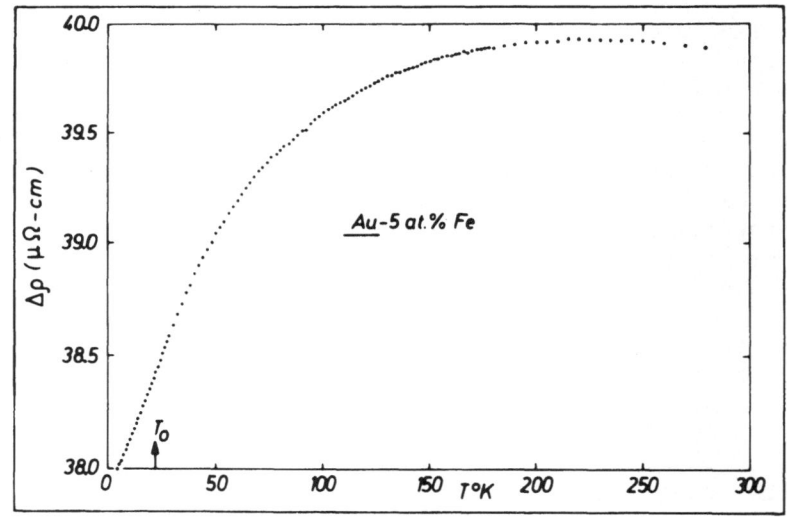

Figure 2:   Δρ versus T for a 5 at.% Fe alloy.

$c \simeq 0.05$ at.% Fe.[8] At T_0 the "superparamagnetic" clusters would no longer be able to freely respond to a small external field, but would become "frozen" in random directions, so that there would be no long-range order. For $c > 12$ at.% Fe there is an overall ferromagnetic behavior although this is unlike a classical ferromagnet since there are regions of mixed magnetism, which gives a rather inhomogeneous spin polarization.

We have measured the electrical resistivity, ρ, of a series of $\underline{Au}$Fe alloys with concentrations of 0.5 to 22 at.% Fe, in the temperature range 0.5 to 300°K. By subtracting the phonon contribution of the Au host, we have obtained

$$\Delta\rho(T, c) = \rho_{alloy}(T, c) - \rho_{gold}(T).$$

This is directly related to the spin scattering, particularly since the temperature dependences are large, so that deviations from Matthiessen's rule are small by comparison.[9]

In Fig. 2 we show the temperature variation of $\Delta\rho$ up to 300°K for a 5 at.% Fe alloy, which is typical of $\underline{Au}$Fe alloys in the spin glass regime. At the lowest temperatures, (below $\simeq15$°K), the resistivity is varying faster than T, and it will be shown later follows a $T^{3/2}$ dependence. Above this temperature, the resistivity is varying linearly with T. This region corresponds to the ordering temperature, (see Fig. 1), where the temperature dependence of $d\rho/dT$ shows a maximum, (Ref. 7, Mydosh $\underline{et}$ $\underline{al}$.), which is characteristic of that ordering is taking place. Above 50°K, the resistivity begins to vary slower than T and becomes approximately temperature independent above about 150°K.

In order to understand better the initial temperature behavior of $\Delta\rho$, we have plotted $\Delta\rho$ against various powers of T. We find that the power law which fits the data is $T^{3/2}$, and this is followed over the entire concentration range of our measurements. A typical example of such a plot is shown in Fig. 3 for a 2 at.% Fe sample. From these plots we have determined the coefficient of the $T^{3/2}$ term and the temperature range of this dependence, and these are summarized in Table 1. For the lowest concentrations, the $T^{3/2}$ dependence is only valid to $\simeq0.2$ T_0, but as the concentration is increased into the middle of the spin-glass

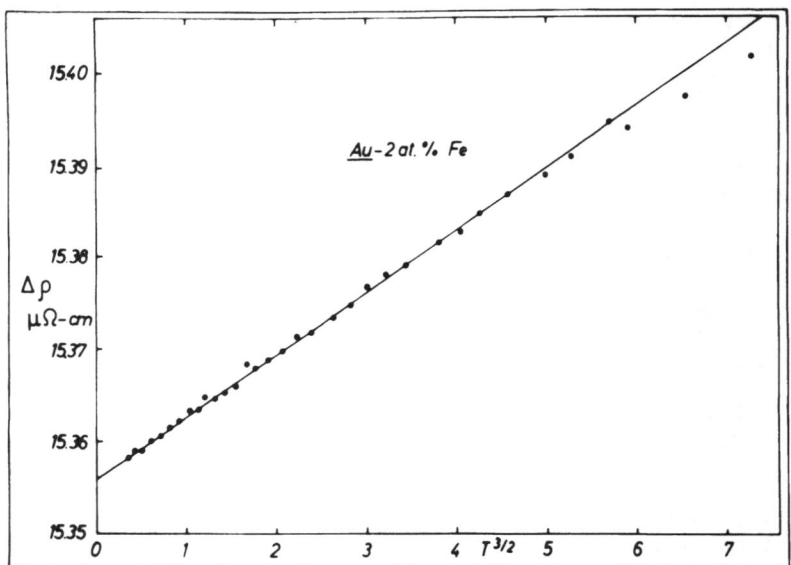

Figure 3: $\Delta\rho$ versus $T^{3/2}$ for a 2 at.% Fe alloy.   (The
range of fit is to $\approx 3°K$.)

Conc. (at.% Fe)	Ordering Temp. T_0 (K)	Range of $T^{3/2}$ Behavior ($T_{3/2}$ / T_0)	Coefficient of $T^{3/2}$ Term ($10^{-3}\mu\Omega$-cm/$K^{3/2}$)
0.5	5.0	0.23	9.6
0.8	7.0	0.17	7.8
1.0	8.5	0.19	7.7
1.5	10.5	0.22	7.0
2.0	13.9	0.21	6.8
5.0	22.2	0.53	5.5
8.0	27.9	0.65	5.1
12.0	36.0	0.97	7.5
17.0	132	0.25	17
22.0	218	0.40	22

Table 1:   Initial $\Delta\rho$ versus $T^{3/2}$ behavior for AuFe alloys.

regime, the coefficient of the $T^{3/2}$ becomes slightly smaller, but the temperature range of this behavior becomes larger, and almost reaches the ordering temperature for the
12 at.% Fe concentration. The $T^{3/2}$ dependence also persists in the ferromagnetic concentrations but with a much
greater coefficient. A similar initial-temperature behavior
has been found in the thermoelectric power of AuFe in the
same concentration range.[3,10]

At the present time, there seems to exist no theories
which are directly applicable to the spin-glass resistivity.
In an attempt to phenomenologically interpret our resistivity results in a way which is consistent with the various
other measurements made on AuFe, we propose the following
model. At high temperatures, ($\approx 300°K$), magnetic clusters
are formed, i.e. there are correlations between the magnetic impurities, which are probably mainly ferromagnetic but
also may include some antiferromagnetic alignments. These
regions of short-range order grow in size and/or number and
slowly reduce the spin disorder scattering. This reduction
in the scattering becomes stronger as the temperature is
lowered, due to interactions between the clusters, until
the clusters finally order, (or freeze), at T_0. The exact
nature of this behavior is presently unclear. However, in
small external fields, there seems to be a well-defined T_0,[3,5]
but with no long-range spatial order, since the clusters
become "locked" into an overall random alignment, i.e., form
a spin-glass. For the larger concentrations, (c $\gtrsim$ 12 at.%
Fe), the ordering is into an inhomogeneous ferromagnet. At
the lowest temperatures, spin-wave excitations probably
cause the initial $T^{3/2}$ dependence for both the spin glass
and the ferromagnetic regimes.[11] In the spin-glass case,
these excitations could be localized within the various
clusters. As the temperature approaches T_0, the resistivity
is no longer dominated by spin waves, and the linear T dependence observed in this region, can probably be accounted
for by a molecular field treatment.

A more detailed account of this study will shortly be
forthcoming.

REFERENCES

1. B.R. Coles, as quoted by P.W. Anderson, Mater. Res. Bull. $\underline{5}$, 549 (1970).
2. P.A. Beck, Met. Trans. $\underline{2}$, 2015 (1971).
3. See for example V. Cannella and J.A. Mydosh, Phys. Rev. B. Nov. 1972, and references cited therein.
4. B. de Mayo, in Magnetism and Magnetic Materials - 1971, AIP Conference Proceedings No. 5 edited by C.D. Graham, Jr. and J.J. Rhyne (American Institute of Physics, New York, 1972), p. 492.
5. R.J. Borg and T.A. Kitchens, J. Phys. Chem. Solids (to be published).
6. See the variety of papers in this symposium Proceedings directly related to spin-glasses and the AuFe alloy system.
7. Previous resistivity studies of AuFe, but from other points of view include: R.C. Sundahl et al. J. Appl. Phys. $\underline{37}$, 1024 (1966); J.A. Mydosh et al. in Proceedings of the Eleventh Conference on Low Temperature Physics (St. Andrews University, Printing Department, St. Andrews, Scotland, (1969), Vol. 2, p. 1324; and P.J. Ford, T.E. Whall and J.W. Loram, Phys. Rev. $\underline{B2}$, 1547 (1970).
8. See Ref. 5, and also V. Cannella and J.A. Mydosh, to be published.
9. T.E. Whall, P.J. Ford, and J.W. Loram, Phys. Rev. B, October, 1972.
10. V. Cannella, J.A. Mydosh, and M.P. Kawatra, J. Appl. Phys. $\underline{41}$, 1421 (1970).
11. A recent theory of P.D. Long and R.E. Turner (J. Phys. C $\underline{2}$, S$\underline{127}$, (1970), invoking the lack of translational invariance, was used to explain the initial $T^{3/2}$ dependence found in dilute PdFe alloys.

DISCUSSION

S. Von Molnar: In your discussion of Figure 2, you say that you think you might be forming clusters. Would you expect an increase in the scattering cross section and some increase in resistivity?

J.A. Mydosh: I would tend to say no, because at low temperatures this clustering takes out the independent scatterers by correlating them such that there would be some reduction in the overall spin disorder scattering. However at high temperatures, where the clusters are just beginning to form, I would expect some increase in the resistivity.

C.W. Rector: I would think that in your method of subtracting out the host conductivity, that it would depend rather critically on the details of the particular model you use. Would you explain how, in detail, did you subtract out the host conductivity?

J.A. Mydosh: It is really very simple. You just measure pure gold and subtract out this resistivity from the resistivity of the alloy. You want to try to get rid of phonons as best you can, although granted for 22%, this procedure is extremely crude.

N. Rivier: This comment is about Fig. 2. It looks very similar to the resistivity due to any sort of scattering by some local dynamical structure. In other words, it looks very similar to the low concentration data of Coles in $\underline{Rh}Fe$ or those of Sarachick in IrFe, except for the absence of T^2 law at low temperatures. I wonder whether you really have gone far enough down in temperatures to ascertain whether this $T^{3/2}$ is not a kind of a transition from a T^2 law to a T law, or whether it is something different.

J.A. Mydosh: Let me just say that our lowest temperatures were of order 0.4 to 0.5 degrees. We have tried T^2 plots starting from these temperatures and it simply doesn't work with T^2.

N. Rivier: Can I make a second point in answer to the gentleman who was asking about whether or not clusters increase the resistivity? One must understand that in the dilute limit of $\underline{Au}Fe$ you have a virtual bound state at the Fermi level, and therefore you are in the unitarity limit, and anything like clustering means some sort of resonance from one virtual bound state to the other which is going to damp off the bound state, put it off the unitarity limit, so it is going to decrease the scattering.

V. Cannella: A comment which might substantiate the paper
and might answer Rivier's question as to whether or not the
$T^{3/2}$ is really a transition from a T^2 law to a T law. In
the thermoelectric power of these alloys, which we have measured over the same concentration and temperature region, we
have found a $T^{3/2}$ and a small $T^{5/2}$ dependence. That is, below the ordering temperatures we find a $T^{3/2}$ and a $T^{5/2}$ dependence in the thermoelectric power, which was very surprising. We didn't expect that.

J.A. Mydosh: This is an interesting point. Let me ask you.
Are the coefficients and ranges of $T^{3/2}$ behavior found in
the thermoelectric power similar to those found in the resistivity?

V. Cannella: The temperature range of fit was very similar
in the sense that by the time we get to the 12% it goes all
the way up to the ordering temperature. The coefficients I
think look quantitatively similar, but I would have to refer to the data to be certain.

NUCLEAR RESONANCE AND MAGNETIC ORDERING IN A RANDOM SPIN SYSTEM: (LaGd)Al$_2$[*]

D. E. MacLaughlin and M. Daugherty [†]

Department of Physics

University of California, Riverside, Ca. 92502

The results of Al^{27} zero-field nuclear quadrupole resonance (NQR) and spin-lattice relaxation measurements in the (LaGd)Al$_2$ magnetic alloy system are examined for features dependent on the spatial disorder of the Gd impurity spin system. Persistence of the NQR signal well below the magnetic ordering temperature T_0 indicates that the Al^{27} static hyperfine field is zero at an appreciable fraction of Al sites. This result is attributed to inhomogeneity in the Gd spontaneous magnetization. No clear evidence was obtained for critical-point behavior of the relaxation time near T_0. The rapid relaxation observed below T_0 indicates the presence of a large fluctuating Al^{27} hyperfine field. Thus the observed nuclei must be near regions of large Gd spin correlation, which in turn suggests that the Gd magnetization inhomogeneities are microscopic in size. For $T \gg T_0$ the relaxation times, appropriately scaled, are comparable to those observed in the intermetallic compound PrAl$_2$ in both magnitude and temperature dependence. Hence any major effect of spatial disorder on the spin system dynamics disappears in this temperature range, and the temperature-dependent contribution to relaxation in both random and periodic spin systems is dominated by short-range interspin correlation.

[*]Work supported by the National Science Foundation.

[†]Present address: Department of Physics, California State College, Dominguez Hills, Ca. 90246.

The problem of magnetic order in dilute alloys has attracted considerable theoretical and experimental attention.[1] In his discussion of the present status of the subject at this symposium, Professor Coles has pointed out the complexity of the ordering process.[2] We wish to comment in this paper on some of the kinds of information which can be obtained from nuclear resonance and relaxation studies of such alloys near the magnetic ordering temperature T_0. The difficulty of theoretical treatments in the not-too-dilute concentration range (magnetic ion concentration c = 0.01 - 0.1, say)[3] underlines the desirability of as much experimental information as possible, and it is useful to consider what can (and cannot) be learned from nuclear resonance.

Our measurements in the pseudobinary alloy system (LaGd)Al$_2$ make use of the zero-field Al^{27} nuclear quadrupole resonance (NQR) spectrum[4] to obtain resonance and relaxation data in the absence of a static magnetic field. In zero applied field the magnetic behavior of the Gd spin system is dominated by its internal spin-spin interactions. Furthermore, the absence of a field-induced Gd spin polarization ensures that static and fluctuating Al^{27} hyperfine fields due to Gd spins will be as sensitive as possible to details of the ordering process: indeed, in high applied field no effect at all attributable to magnetic ordering is observed in nuclear resonance experiments on the (LaGd)Al$_2$ system.[5] Our results consist of NQR spin-echo amplitude and spin-lattice relaxation data as a function of temperature and Gd concentration c, for two samples in which c = 0.02 and 0.05 respectively.

The temperature dependence of the NQR spin-echo signal amplitude S_{se} is given in Fig. 1 for the two samples. At high temperatures the expected Curie-like form $S_{se}T$ = const. is observed. Data for the two samples tend to follow a universal dependence on reduced temperature T/T_0, where T_0 is taken from the magnetization measurements of Maple on the same system.[6] The loss of NQR signal for $T/T_0 \lesssim 1$ is attributed to inhomogeneous broadening caused by the onset of spontaneous magnetization in the Gd spin system.[4] We wish to point out that this process is quite gradual, and that appreciable NQR signal remains well below the ordering temperature. Al^{27} nuclei in non-zero static hyperfine field are apparently wiped completely out of the NQR line, since no effect of either impurity concentration or ordering was

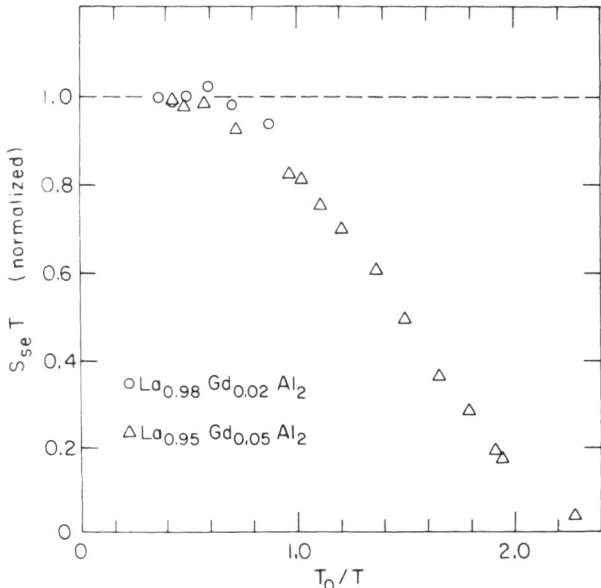

FIG. 1. Dependence on inverse reduced temperature T_0/T of the Al^{27} NQR spin-echo amplitude S_{se} times temperature T, normalized to unity at high temperature, in two $(\underline{La}Gd)Al_2$ samples. T_0 is the magnetic ordering temperature taken from magnetization measurements (Reference 6).

observed in the linewidth.[4] Hence the observed nuclei reside at sites for which the static hyperfine field is much less than the NQR linewidth $\Delta H \approx 20$ Oe. The NQR signal vanishes, however, for $T \ll T_0$. In this limit a non-zero static hyperfine field is present at $(100 \pm 5)\%$ of Al^{27} sites.

We take this behavior to indicate that in magnetically ordered $(\underline{La}Gd)Al_2$ the Gd magnetization is inhomogeneous. An alternative explanation might invoke the oscillatory RKKY interaction between Al^{27} nuclei and a homogeneous Gd spin system to explain the vanishing static hyperfine field at some Al^{27} sites. Although not ruled out by experiment this possibility seems unlikely, since in the concentration range $c = 0.01 - 0.1$ the probability that a given nucleus interacts significantly with more than one Gd spin is quite high. But zeroes of the RKKY function for several Gd spins would not be expected to fall at the same Al^{27} site for a large

number of such sites. It seems more reasonable that the
observed nuclei are in regions of relatively low Gd concentration, whereas appreciable Gd magnetization is on the
whole confined to "clusters" of high Gd concentration.

It should be noted that the loss of NQR signal cannot
be compared directly with the effective-field distribution
function $P(H,T)$ of Marshall,[7] Klein and Brout,[8] and others[3]
via the identification $S_{se}T \propto P(H=0,T)$, since the effective-
field distribution refers to the exchange field at magnetic
ion sites and not to the hyperfine field distribution at
nuclear sites. Nevertheless the decrease of $S_{se}T$ with decreasing temperature is consistent in sign with the behavior
of $P(H=0,T)$ as measured in the CuFe system using the Mossbauer effect.[9]

Spin-lattice relaxation times are given in Fig. 2 for

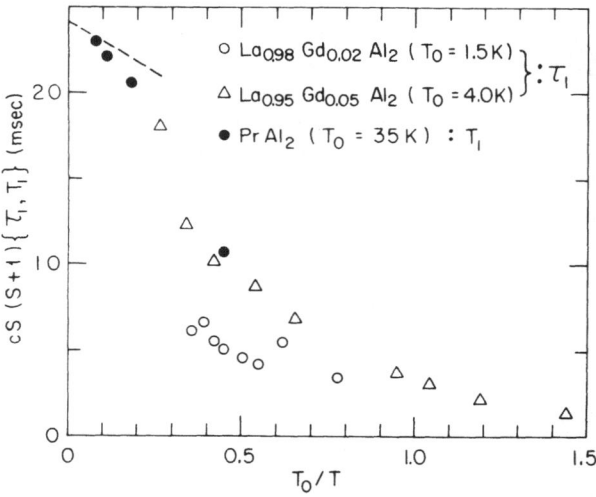

FIG. 2. Dependence of the spin-lattice relaxation times on
inverse reduced temperature T_0/T and magnetic ion concentration c in (LaGd)Al$_2$ and PrAl$_2$. Definitions of the relaxation times T_1 and τ_1, and the appropriateness of the
scale factor $cS(S+1)$, are discussed in the text. The
dashed curve gives the result of a pair-correlation calculation (Ref. 10).

our two (LaGd)Al$_2$ samples, as well as for the isostructural
intermetallic compound PrAl$_2$ as measured by Silbernagel et
al.[10] In the latter system the Pr spins occupy 100% of
lanthanide sites and the relaxation time T$_1$ is the same for
all Al^{27} sites; whereas in (LaGd)Al$_2$ a spatial average of
the position-dependent T$_1(\vec{r})$, denoted τ_1 in Fig. 2, was
obtained from our relaxation measurements.[4,5] If, as is
reasonable in both systems,[5,10] relaxation rates are taken
to be proportional to the factor cS(S+1) (c = 1 for PrAl$_2$),
a spin- and concentration-independent measure of the relaxation is obtained by multiplying T$_1$ and τ_1 by this factor
as shown in Fig. 2.

Several features of these data are noteworthy. No
anomaly or change of regime is found in the behavior of τ_1
near T/T$_0$ = 1. (More data in the PrAl$_2$ system are required
near T$_0$ to establish the situation there.) We conclude that
no unambiguous evidence of critical-point reduction of the
Gd spin fluctuation rate[11] is observed in the alloy, although such an effect could contribute to the observed general decrease of τ_1 with decreasing temperature.[4]

The ordering process in (LaGd)Al$_2$ appears to be spread
out over a considerable range in temperature, and hence is
probably not a cooperative phase transition. This by itself
suggests that the size of the Gd magnetization inhomogeneities is microscopic, and the conclusion is reinforced by the
observation that below T$_0$ the relaxation times become very
short. Fast relaxation of nuclei in zero static hyperfine
field implies a large fluctuating hyperfine field, hence
proximity of the observed nuclei to regions of Gd magnetization. That is, if the nuclei were in zero static field by
virtue of a macroscopic separation between them and regions
of Gd magnetization, one would also expect zero fluctuating
field contrary to observation. This qualitative argument
does not suffice to yield an estimate of the characteristic
size of the inhomogeneities; nor does it allow one to conclude that the impurity system is non-random, since clustering is also expected in a truly random disordered lattice.

Understanding of the temperature-dependent relaxation
mechanism is aided by comparison between the (LaGd)Al$_2$ and
PrAl$_2$ data. For both systems the high-temperature relaxation behavior (say, for T $\gtrsim$ 2T$_0$) is similar in both magnitude and temperature dependence.[12] Hence the same mechanism
is probably at work. Silbernagel et al.[10] have attributed

the $PrAl_2$ results to pair and higher-order spin correlation
in the paramagnetic state. For ferromagnetic pair correlation $T_1 \propto 1 - T_0/2T$, shown by the dashed curve in Fig. 2.
Higher-order correlations are expected to decrease T_1 below
the pair-correlation prediction, in agreement with experiment. It appears that such a description fits both periodic
and random systems, and that in the high-temperature region
the relaxations is insensitive to the spatial distribution
of the spins. This result confirms the view that in the
paramagnetic state short-range correlations dominate the
spin system dynamics, and that long-range periodicity does
not play an important role.

It is at temperatures in the region of magnetic ordering, then, that nuclear resonance and relaxation data appear
to be most sensitive to the disordered nature of the spin
system in dilute magnetic alloys. A full theory of this
region must take into account both disorder and high-order
spin correlations, and has not to our knowledge appeared.
The empirical conclusions arrived at above lend substance to
the hope that a better understanding of nuclear relaxation
processes in magnetic alloys will lead to a more complete
characterization of magnetic order in these materials.

We acknowledge with thanks useful discussions with
N. Y. Rivier, B. G. Silbernagel, and E. Šimánek.

REFERENCES AND FOOTNOTES

1. See, for example, M. W. Klein, Phys. Rev. 136, A1156
    (1964), and references cited therein.
2. B. R. Coles, in Proc. of the Symposium on Amorphous
    Magnetism, Wayne State University, Detroit, Michigan,
    1972.
3. S. H. Liu, Phys. Rev. 157, 411 (1967); K. H. Bennemann,
    J. W. Garland, and F. M. Mueller, Phys. Rev. Letters
    23, 1503 (1969).
4. See D. E. MacLaughlin and M. Daugherty, Phys. Rev. B
    (to be published) for a preliminary account of this
    work.
5. M. R. McHenry, B. G. Silbernagel, and J. H. Wernick,
    Phys. Rev. Letters 27, 426 (1971); Phys. Rev. B 5,
    2958 (1972).
6. M. B. Maple, thesis, University of California, San
    Diego, 1969 (unpublished).

7. W. Marshall, Phys. Rev. 118, 1519 (1960).

8. M. W. Klein and R. Brout, Phys. Rev. 132, 2412 (1963).

9. U. Gonser, R. W. Grant, C. J. Meechan, A. H. Muir, and
    H. Wiedersich, J. Appl. Phys. 36, 2124 (1965).

10. B. G. Silbernagel, V. Jaccarino, P. Pincus, and J. H.
    Wernick, Phys. Rev. Letters 20, 1091 (1968). See also
    F. Y. Fradin, J. Phys. Chem. Solids 31, 2715 (1970)
    for a related theoretical treatment.

11. T. Moriya, Progr. Theoret. Phys. (Kyoto) 28, 371 (1962).

12. The nearly perfect agreement between data for PrAl$_2$ and
    La$_{0.95}$Gd$_{0.05}$Al$_2$ is surely fortuitous for two reasons:
    first, no such agreement is found between the PrAl$_2$ and
    La$_{0.98}$Gd$_{0.02}$Al$_2$ data; second, the numerical factor
    which relates T_1 and τ_1 is unknown but probably not
    unity.

DISCUSSION

J. Souletie: Have you made such experiments in the superconducting state?

D.E. MacLaughlin: Yes, we have preliminary results in the
superconducting state which are very strange indeed. I'll
talk to you later about them.

J. Souletie: One of the interesting possibilities that you
could investigate is that of magnetic order in the superconducting state.

D.E. MacLaughlin: The phase diagram that I showed you (not
included in the text; see Ref. 6) indicated that at least
for this system the spin glass transition temperature goes
to zero before the onset of the superconducting state at
low concentrations of Gd. So maybe (LaGd)Al$_2$ is not the
system to use for that problem.

N. Rivier: Do you think that the correlation time of the
fluctuations of the random hyperfine field, which you use
in the formula for the spin-lattice relaxation rate, is the
time fluctuation of this hyperfine field; or do you think
it is equivalent to an ensemble average of the hyperfine
field? In other words, would you believe that this correlation time is in fact the inverse width of the local field
distribution?

D.E. MacLaughlin: No, I don't think that it is: I think
it is a real, honest-to-goodness correlation time. The
nuclei are surrounded by a magnetic environment, which produces a net hyperfine field. That field fluctuates in a
random way in time at the nuclear site. The correlation
time is not some kind of ensemble average, because the
nuclei are delta-function entities at fixed sites in the
material. A sample average is involved in the definition
of τ_1, however.

S.G. Bishop: You said that no effects of ordering have been
seen in high-field measurements. Did you mean by that that
Al^{27} nuclear magnetic resonance has been done, and no effects of the ordering have been seen?

D.E. MacLaughlin: That is exactly what I mean (see Ref. 5).
This is a very bizarre result, which is not yet understood.
Our results are much more conventional, if you wish.

H. Alloul: I should like to point out that similar things appear in CuMn. We have made nuclear relaxation measurements
in the ordered state of CuMn by conventional NMR from, say,
1 to 10 MHz. It appears that the low-c-end transition from
the ordered to the non-ordered state is enhanced in low
fields. At 1 MHz we see a discrete change from ordered to
non-ordered behavior in both relaxation time and signal
strength. In high field we don't see such a change any
more. This seems to relate very nicely to what MacLaughlin
has said.

OCCURRENCE OF LOCAL MOMENTS AND FERROMAGNETISM IN DILUTE

ALLOYS

D. J. Kim

Department of Physics, Aoyama Gakuin Univer-

sity*, Chitosedai, Setagaya-ku, Tokyo, Japan

and Francis Bitter National Magnet Laboratory[+],

Massachusetts Institute of Technology, Cam-

bridge, Massachusetts 02139

ABSTRACT

In this paper we discuss some basic theoretical aspects
of the magnetism of random dilute alloys, starting with the
Anderson model. First, we determine how an impurity state
is modified by the presence of the other impurities in its
neighborhood. Even if we consider only one species of impurity atoms, the electronic state at each impurity site is
different depending upon the distribution of other impurities in its local environment. Second, we discuss a general relation between the occurrence of a local moment on
an individual impurity and the ferromagnetic ordering of
the total impurity spins. Finally, we discuss the magnetic
interaction between the impurities. Especially in the case
of direct transfer interaction, we show that the interaction
between impurities can be either ferromagnetic or antiferromagnetic depending upon the fractional occupation of the
impurity states.

*Present address.
[+]Supported by the National Science Foundation.

A dilute alloy may be visualized as a random distribution of magnetic impurity atoms in an otherwise pure metal. In this paper we study the magnetism of the impurities although the modification of the magnetic property of the host metal by the presence of the impurities is equally important.[1,2] The problem may be divided into two: (I) What is the electronic structure of individual impurity atoms? (II) What is the effect of the exchange interaction between impurity atoms? In the simplified models the problem (I) is often neglected by assuming that the impurities have spins of the same length. According to the experiments, however, the magnetization of each impurity is different depending upon the distribution of other impurities in its local environment.[3] We study this local environment effect, or the Jaccarino-Walker effect, starting with the Anderson model[4] extended to the many impurity case. In our formulation the explicit form of the interaction between impurities is determined from the band structure of the host metal and the fractional occupation of the impurity states.

According to the Friedel-Anderson theory, whether an impurity has a local moment or not is determined by whether the magnetic susceptibility of the impurity diverges or not. Similarly, the criterion for the ferromagnetic ordering of the impurity spins is given by the divergence of the susceptibility of the total impurity system. Thus, in the present paper we calculate the magnetic susceptibilities of the individual spins in the Hartree-Fock approximation.

The Anderson Hamiltonian extended to the many impurity case is given as

$$H = \sum_{k,\sigma} \varepsilon_{k\sigma} c^{\dagger}_{k\sigma} c_{k\sigma} + \sum_{i,\sigma} E_{\sigma} d^{\dagger}_{i\sigma} d_{i\sigma} + \sum_{i} U d^{\dagger}_{i+} d_{i+} d^{\dagger}_{i-} d_{i-}$$

$$+ \sum_{i,k,\sigma} V e^{ikR_i} [d^{\dagger}_{i\sigma} c_{k\sigma} + c^{\dagger}_{k\sigma} d_{i\sigma}] + \sum_{i \neq j,\sigma} T_{ij} d^{\dagger}_{i\sigma} d_{i\sigma}.$$

$$(1)$$

The first term in the Hamiltonian is the energy of the conduction electrons, where $\varepsilon_{k\pm} = \varepsilon_k \pm \mu H$, H is the external magnetic field applied in the direction of the z-axis and $c_{k\sigma}^{\dagger}$ is the creation operator of an electron with momentum k and spin σ (= + or -). The second term is the energy of the unperturbed impurity level, where $E_{\pm} = E \pm \mu H$ and $d_{i\sigma}^{\dagger}$ is the creation operator of an electron at the i-th impurity site R_i. The third term is the Coulomb repulsion between electrons of opposite spins at each impurity site. The fourth term is the s-d mixing interaction between the host metal conduction electrons and the impurity electrons, and the last term is the direct transfer interaction between neighboring impurities.

The magnetic susceptibility can be most simply obtained by the Green's function method. The double time Green's function of the i-th impurity, $G_{i\pm}(\omega)$, in the Hartree-Fock approximation is easily calculated as

$$G_{i\pm}(\omega) = [\omega - E_{\pm} - UN_{i\mp} - V^2 F_{\pm}(\omega) - \Sigma_{i\pm}(\omega)]^{-1} \qquad (2)$$

$$\Sigma_{i\pm}(\omega) = \sum_{j(\neq i)} [T_{ij} + V^2 F_{ij\pm}(\omega)]^2 \ G_{j\pm}^{o}(\omega), \qquad (3)$$

where $N_{i\pm}$ is the expectation value of the number of the localized electrons at the i-th impurity which is to be calculated self-consistently from the Green's function as[5]

$$N_{i\pm} = -\frac{1}{\pi} \int_{-\infty}^{\infty} d\omega \ f(\omega) \ \text{Im} \ G_{i\pm}(\omega), \qquad (4)$$

where $f(\omega)$ is the Fermi distribution function, $G_{i\pm}^{o}(\omega)$ is the impurity Green's function in the single impurity limit, which can be obtained by putting $\Sigma_{i\pm}(\omega) = 0$ in Eq. (2), $F_{ij\pm}(\omega) = \sum_k \exp [ik (R_i - R_j)][\omega - \varepsilon_{k\pm} + io^{+}]^{-1}$ and $F_{\pm}(\omega) = \sum_k [\omega - \varepsilon_{k\pm} + io^{+}]^{-1}$.

The Green's function of the form of Eq. (2) was earlier used to discuss the effective exchange interaction between impurities by assuming from the beginning that the impurities have local moments of equal length.[6-10] In this paper we calculate the magnetic susceptibilities of the individual impurities by assuming that the impurities do not have local moments. The magnetic susceptibility of the i-th impurity is defined in terms of N_{i+}, Eq. (4), as

$$\chi_i = \frac{\mu(N_{i-} - N_{i+})}{H} = \frac{2\mu\Delta N_i}{H} \tag{5}$$

where we put $N_{i\pm} = N_i \pm \Delta N_i$. As in the calculation of the Pauli paramagnetic susceptibility of the conduction electrons, ΔN_i is given by

$$\Delta N_i = \rho_i(0)\ \Delta E_i \tag{6}$$

where $\rho_i(0)$ is the density of the states of the i-th impurity level at the chemical potential [which we take as the origin of measuring energy] and ΔE_i is the shift of the impurity level due to the external magnetic field. The many impurity effects appear through the modification of the impurity density of states from $\rho_i^0(0)$ to $\rho_i(0)$ [effect (I)] and the modification of the Zeeman shift of the impurity states from ΔE_i^0 to ΔE_i [effect (II)]. Note that the effects (I) and (II) correspond, respectively, to the problems (I) and (II) mentioned in the beginning of the present paper. In the single impurity case the impurity density of states ρ_i^0 is obtained from G_i^0 and the Zeeman shift ΔE_i^0 is given as

$$\Delta E_i^0 = \mu H + U\Delta N_i. \tag{7}$$

From Eqs. (5) and (6) we obtain the well-known result of the magnetic susceptibility of an isolated impurity, χ_i^{00} as

$$\chi_i^{00} = \frac{2\mu^2\ \rho_i^0(0)}{1 - U\rho_i^0(0)} \tag{8}$$

Note that the Friedel-Anderson criterion for the occurrence of a localized moment, $1 - U\rho_i^o (0) \leqq 0$, is the condition for the divergence of χ_i^{oo}.

As discussed in detail elsewhere,[2] the impurity density of states is modified from ρ_i^o to ρ_i due to the inter-impurity interaction embodied by the self-energy term, $\Sigma_{i\pm} (\omega)$, in the impurity Green's function, Eq. (2). Formally speaking, this part of the many impurity effect [the effect (I)] comes from the spin independent part of $\Sigma_{i\pm}$, which we denote as Σ_i. The spin dependent part of the self energy is responsible for the change in the Zeeman shift of the impurity state from ΔE_i^o, Eq. (7), to ΔE_i,

$$\Delta E_i = \mu H + U\Delta N_i + \sum_{j(\neq i)} Uc_{ij}\Delta N_j , \qquad (9)$$

where the coupling constant c_{ij} is introduced as

$$U c_{ij} = \frac{\partial}{\partial N_j} \text{Re } \Sigma_i (0) \Big|_{H=0} . \qquad (10)$$

The difference between Eq. (7) and Eq. (9) is clear. In ΔE_i the effect of the molecular field from the surrounding other impurities is included.

From Eqs. (5), (6) and (9), we obtain a coupled equation for χ_i

$$\chi_i = \chi_i^o + \frac{1}{2\mu^2} \chi_i^o \sum_{j(\neq i)} U c_{ij} \chi_j \qquad (11)$$

where we introduced χ_i^o as

$$\chi_i^o = 2\mu^2 \frac{\rho_i (0)}{1 - U\rho_i (0)} \qquad (12)$$

In order to obtain the magnetic susceptibility of the individual impurities we must solve Eq. (11). χ_i^o is introduced purely as a convention, but note that in it the effect (I) part of the impurity-impurity interaction effects is

included. The full interaction effects, (I) and (II), are
included in χ_i. It should be emphasized that for the interacting many impurity system the only physically meaningful
susceptibility is χ_i. Both χ_i^{oo} and χ_i^o do not have any
direct physical meaning.

Equation (11) has a very familiar form. We can obtain
an equation very similar to Eq. (11) starting with a simplified model in which the impurities are represented by
Heisenberg or Ising spins of equal length. In the case of
the simplified model, however, the quantities corresponding
to χ_i^o in Eq. (11) are independent of i, being equal to the
Curie susceptibility of an isolated spin. In our case, χ_i^o
are different for different i's reflecting the difference
in the local environments. Namely, in terms of the Heisenberg type model, the length of the impurity spin is different for different site. Thus, the magnetization at the
individual impurity sites can be different even if the impurities are all the same element. In understanding the
observed different hyperfine fields at different impurity
sites[3] this is the essential point. Note that, in principle, from Eqs. (2) - (4) we can calculate the magnetization,
as well as the susceptibility, of the individual impurities
including the local environment effect.

As an illustration we solve Eq. (11) for the case of a
pair of impurities, whose sites are denoted as 1 and 2:

$$\chi_1 = \chi_2 = 2\mu^2 \frac{\rho(0)}{1 - U(1 + c_{12})\,\rho(0)} \ . \tag{13}$$

If the interaction between the pair of impurities is neglected the susceptibility $\chi_1^{oo} = \chi_2^{oo}$, would be given by Eq. (8).
Due to the interaction, the impurity density of states is
modified from ρ_i^o to ρ_i [effect (I)] and the interaction is
modified from U to $U(1 + c_{12})$.

From the above result we can conceive the possibility
that even if an isolated impurity does not have a local
moment, the paired impurities may have local moments
$[(\chi_i^{oo})^{-1} > 0$, but $(\chi_i)^{-1} \leqq 0]$. The opposite case that
$(\chi_i^{oo})^{-1} \leqq 0$ but $(\chi_i)^{-1} > 0$ is also possible.

Another example is a cluster in which a central impurity (subscript c) is surrounded by Z other impurities [subscript s] in its nearest neighbor sites. For simplicity we assume that the interaction is present only between the central impurity and the surrounding ones. Then Eq. (11) is solved as

$$\chi_c = \frac{\chi_c^o}{C.D.} \; [1 + Zc_{cs} \; \frac{1}{2\mu^2} \; U \; \chi_s^o] \; , \tag{14}$$

$$\chi_s = \frac{\chi_s^o}{C.D.} \; [1 + c_{cs} \; \frac{1}{2\mu^2} \; U \; \chi_c^o], \tag{15}$$

where the common denominator is given by

$$C.D. \equiv 1 - Z \; c_{cs}^2 \; (\frac{1}{2\mu^2} \; U \; \chi_c^o) \; (\frac{1}{2\mu^2} \; U \; \chi_s^o), \tag{16}$$

and the meaning of χ_c^o and χ_s^o would be clear.

The susceptibility of the central impurity, χ_c, is different from that of the surrounding impurities, χ_s. Note, however, that the susceptibilities of the central and surrounding impurities diverge simultaneously when the common denominator, C.D., vanishes. Thus it seems quite natural to associate the vanishing of C.D. with the spin ordering of the impurity cluster. The occurrence of local moments and the spin ordering take place simultaneously in the cluster.

If we consider all the impurities in the system, instead of a cluster, the result would be quite similar to the above example of the cluster. The susceptibilities of the individual impurities, χ_i, are different depending upon their local environments. But the susceptibilities of all the mutually interacting impurities would diverge simultaneously when the common denominator diverges. It would be quite natural to associate the simultaneous divergence of all χ_i's with the appearance of the long range spin ordering of the impurities. Thus, in the present scheme of discussion, which is a direct extension of the Friedel-Anderson theory to the many impurity case, it becomes difficult to visualize the occurrence of the local moments seqarately from the occurrence of the long range spin ordering of the entire impurity system.

Finally, we discuss some interesting aspects of the inter-impurity interaction coupling constants c_{ij} appearing in Eq. (11). As is seen from Eqs. (3) and (10) there are three different components for c_{ij} which are proportional, respectively, to T_{ij}^2, $V^4 F_{ij}(\omega)^2$ and $T_{ij} V^2 F_{ij}(\omega)$. The physical meaning of each term may be clear.[2] In the present paper we study only the first one which is the coupling constant due to the direct transfer interaction. This term, which we denote as $c_{ij}^{(d)}$, seems to be the most important when two impurities are at the nearest neighbor sites. From Eqs. (3) and (10) we obtain

$$
\begin{aligned}
U\, c_{ij}^{(d)} &= T_{ij}^2 \frac{\partial}{\partial N_j} \operatorname{Re}\left[\frac{1}{-(E + UN_j) + i\Delta} \right] \\
&= T_{ij}^2\, U\, \frac{(E + UN_j) - \Delta^2}{[(E + UN_j)^2 + \Delta^2]^2}
\end{aligned}
\tag{17}
$$

Note that $\mid E + UN_j \mid$ is the distance between the center of the Lorentzian density of states of the impurity level and the Fermi level.

The implication of our result, Eq. (17), is very simple: when a pair of impurity states are almost full or empty [$\mid E + UN_j \mid > \Delta$], the interaction is ferromagnetic [$c_{ij}^{(d)} > 0$]. When the impurity states are nearly half filled [$\mid E + UN_j \mid < \Delta$], the interaction is antiferromagnetic [$c_{ij}^{(d)} < 0$]. A conclusion similar to the above was obtained earlier by Moriya[8] for the case in which the impurities are assumed to have local moments.

REFERENCES

1.   D.J. Kim, Phys. Rev. 149, 434 (1966).
2.   D.J. Kim, Phys. Rev. B1, 3725 (1970).
3.   V. Jaccarino and L.R. Walker, Phys. Rev. Letters 15, 258 (1965). Other related references can be found in Ref. 2.
4.   P.W. Anderson, Phys. Rev. 124, 41 (1961).

5.   D.N. Zubarev. Usp. Fiz. Nauk 71, 71 (1960) [Soviet
     Phys.-Usp. 3, 320 (1960)].
6.   D.J. Kim and Y. Nagaoka, Prog. Theoret. Phys. (Kyoto)
     30, 743 (1963).
7.   S. Alexander and P.W. Anderson, Phys. Rev. 133, A
     1594 (1964).
8.   T. Moriya, Prog. Theoret. Phys. (Kyoto) 33, 157 (1965).
9.   B. Caroli, J. Phys. Chem. Solids 28, 1427 (1967).
10. S.H. Liu, Phys. Rev. 163, 472 (1967).

DISCUSSION

R. Tournier: Have you done the calculation for the long
distance interaction?

D.J. Kim: Yes, I have done formal calculations for all
three possible types of interactions, namely the direct
transfer interaction which I have shown, the RKKY type
interaction due to the s-d mixing, the mixed type
interaction of the direct transfer and the s-d mixing
(see Ref. 2). The latter two types of interactions are
of the long range nature and they depend not only on the
partial occupation of the impurity states but also on the
band structure of the host metal and the distance between
the impurities.

ANTIFERROMAGNETIC INSTABILITY IN SUBSTITUTIONAL ALLOYS

Scott Kirkpatrick

IBM T. J. Watson Research Center
P. O. Box 218
Yorktown Heights, New York 10698

We study the Hubbard[1] model of electrons in a disordered narrow band in the limit when the disorder is stronger than the electron-electron repulsion, U. Some new results for the staggered (antiferromagnetic) susceptibility in the small-U limit are obtained, and contrasted with earlier approximate treatments of this model. Disorder appears to enhance the tendency towards antiferromagnetism in the system considered.

The model Hamiltonian to be studied is simple enough to permit detailed calculation and some analytical results, yet contains effects of both disorder and electron correlations. The system, a non-degenerate narrow band, is described in the usual second-quantized form, in a Wannier basis, by

$$H = \sum_{i,\sigma} E_i \hat{n}_i + \sum_{\substack{i, j \text{ neighbors} \\ \sigma}} w\, c_i^\dagger c_j + \sum_i U \hat{n}_{i\uparrow} \hat{n}_{i\downarrow} \tag{1}$$

In (1), i labels the sites and σ the electron spins. Disorder enters through the diagonal elements, E_i, which may vary at random from site to site, while the hopping matrix element, w, is constant. Note that the off-diagonal term of (1) connects sites only if they are nearest neighbors on the lattice.

The model reduces to a commonly studied binary alloy problem if we set U=0 and allow E to take only two values:

$$E_i = 0 \qquad \text{on A sites;}$$

$$E_i = w\delta \qquad \text{on B sites.}$$

The type of spectrum which can result from this binary alloy model have been discussed at length by Velický et. al.[2]. When $\delta > z$, where z is the number of nearest neighbors, the spectrum consists essentially of a single band, and the coherent potential approximation[2,3] (CPA) describes the electronic properties of the model with quantitative accuracy. In the "strong scattering limit" $\delta > 2Z$, the spectrum splits into two separated subbands[2], each consisting mostly of states from one of the two types of atoms. The CPA gives only a qualitative account of the density of states in this regime, and a rather poor description of transport properties[4].

Recently Kirkpatrick and Eggarter[5] have studied the binary alloy model in its strong-scattering regime, emphasizing the limiting case $\delta \to \infty$. Several novel results, not anticipated by the CPA, were obtained. One purpose of the present work is to analyze the magnetic properties of the alloy model (1), in the limit of small U/W, using these new results.

Of the many papers on the effects of correlation in this model without other disorder, the work of Penn[6] on the magnetic phase diagram at small values of U is most relevant to the present discussion. Second-order phase transitions in which various types of spin-density waves are formed were found by noting the appearance of poles in the static spin susceptibility, $\chi(q)$, of the interacting electron system. As calculated in a mean field approximation,

$$\chi(q) = 2\mu_B^2 \, \chi^{(o)}(q) / (1 - U\chi^{(o)}(q)) \tag{2}$$

depends essentially upon the response, $\chi^{(o)}(q)$, of the noninteracting system to an infinitesimal magnetization with wave vector q. This is easily evaluated in the absence of disorder by using the Bloch representation:

$$\chi^{(o)}(q) = -\sum_k (f(\varepsilon_k) - f(\varepsilon_{k+q}))/(\varepsilon_k - \varepsilon_{k+q}) \qquad (3)$$

where f's are Fermi functions. For the simple cubic lattice studied here and in Ref. 6,

$$\varepsilon_k = 2w(\cos k_x a + \cos k_y a + \cos k_z a). \qquad (4)$$

For sufficiently low temperatures and any specific choice of the Fermi energy, E_F, it is possible to find a value of q which maximizes $\chi^{(o)}(q)$. The system is therefore unstable toward the formation of a spin density wave with this wave vector whenever U exceeds $(\chi^{(o)}(q))^{-1}$. When $E_F=0$, (a half-filled band), $\chi^{(o)}(q)$ has its maximum at the reciprocal lattice vector $Q = \frac{\pi}{a}(1, 1, 1)$, which corresponds to a magnetization oscillating with the period of the lattice spacing, a. In addition, $\chi^{(o)}(Q)$ diverges as ln kT at low temperatures, so the paramagnetic state is unstable at sufficiently low temperature for any value of U. The singularity arises because, using (4),

$$\varepsilon_{k+Q} = -\varepsilon_k, \qquad (5)$$

and (3) reduces to

$$\chi^{(o)}(Q) = -\int dE \, \rho(E) \, (f(E)-1/2)/E \qquad (6)$$

with $\rho(E)$ the density of states in the non-interacting system. The stability of this antiferromagnetic phase at values of U above threshold has been discussed recently by Cyrot[7].

The analysis of Ref. 6 has recently been extended to the full disordered system (1) by Fukuyama[8], who calculates $\chi^{(o)}(q)$ in a Bloch representation, treating the effects of disorder as a sort of lifetime broadening. The complex self-energy calculated in CPA provides the necessary lifetime. Although the tendency towards spin ordering remains strong in this approximation, $\chi^{(o)}(Q)$ is found to remain always finite in the presence of disorder, because the singularity in (6) is removed.

However, in the limit of very strong disorder ($\delta \to \infty$), it proves possible to evaluate the staggered susceptibility $\chi^{(o)}(Q)$, and hence $\chi(q)$, without resorting to either the Bloch representation or CPA. As shown in Ref. 5, if we confine our attention to the A-atom subband centered on E=0, the one-electron part of (1) may be replaced by

$$H^A = \sum_{i,j \text{ neighbors}} P^A w \, c_i^{\dagger} \, c_j \, P^A. \tag{7}$$

where P^A is a projection operator selecting only sites containing A atoms.

To derive a mean-field staggered susceptibility in the disordered case, a term describing an infinitesimal external h_{ext} is added to (7):

$$H^{ext} = g \, \mu_B h_{ext} \hat{\sigma}_z T. \tag{8}$$

The operator T in (8) takes advantage of the fact that the simple cubic lattice may be decomposed into two interpenetrating sublattices. We take T to be site-diagonal, with matrix element +1 on one sublattice, -1 on the other. The spatial dependence of an alternating magnetization may be represented in this way on the simple cubic, body-centered cubic, and diamond lattices, among others.

The important property of T is that it anticommutes with H^A,

$$TH^A + H^A T = 0, \tag{9}$$

implying that, if ψ_E is an eigenstate of H^A with energy E,

$$T\psi_E = \psi_{-E} \tag{10}$$

where ψ_{-E} is an eigenstate with energy -E. Thus an alternating magnetization couples only pairs of eigenstates with energies symmetric about the center of the band, and the susceptibility $\chi^{(o)}(Q)$ for the A subband may be calculated using (6), just as in the pure system[9]. As long as $\rho(0) \neq 0$ for the A subband, therefore, the antiferromagnetic instability persists, no matter how small V is.

This result should not surprise[10] readers familiar with the theory of dirty superconductors. T is in fact a time-reversal operator, since

$$T \exp(iH^A t)T^{-1} = \exp(-iH^A t). \tag{11}$$

The two-sublattice antiferromagnetism in this system can thus be viewed as a result of time-reversal pairing, and is a more general effect than the pairing of states on opposite sides of the Fermi surface which the use of the Bloch representation emphasizes. There is no Fermi surface in the limit $\delta \to \infty^5$, but this by itself does not inhibit antiferromagnetism.

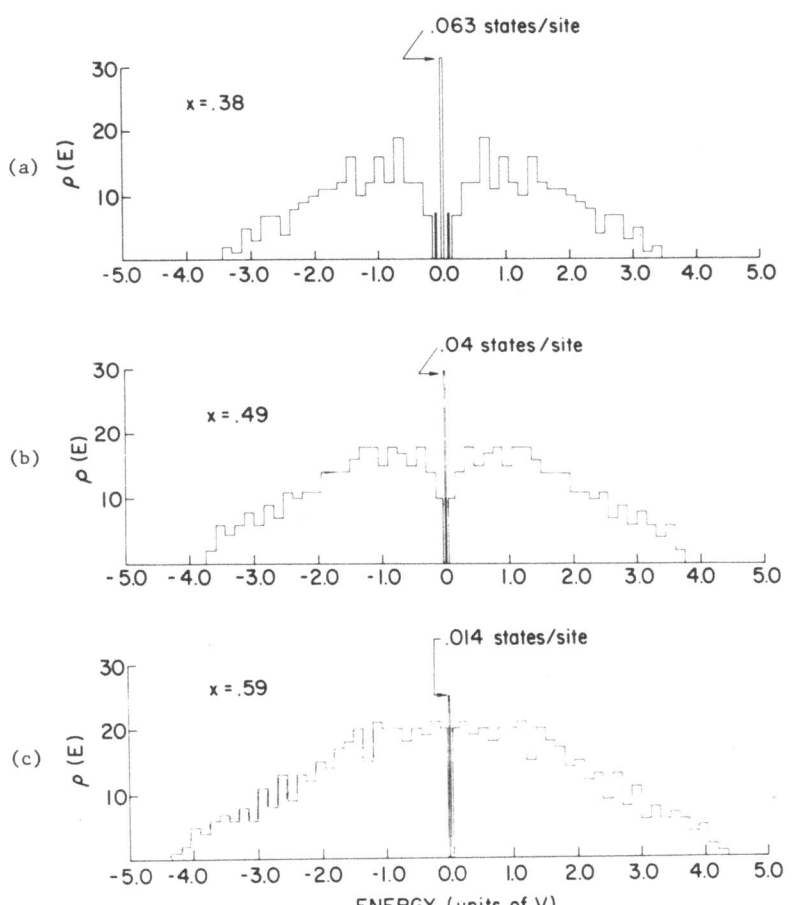

Figure 1. Density of states histograms calculated numerically for 3D Monte Carlo samples of 8x8x20 sites, with periodic boundary conditions. The concentration of A sites, x, is indicated for each case. (From Ref. 5.)

The numerical results of Ref. 5 complicate this picture somewhat. As is evident in the densities of states plotted in Fig. 1, $\rho(E)$ in the strong-scattering limit has a delta-function spike at the origin containing several per cent of the eigenstates of the system. Some of these states consist of electrons on isolated A atoms, surrounded by "forbidden" B sites, and are irrelevant for questions of magnetic order. But most of them (3% in the 50-50 case (b) shown) are not isolated, and have wave functions which vanish identically on one of the sublattices[5]. Because of this characteristic of their wave functions, these states may make a large contribution to alternating magnetization. If their energies are degenerate in with E_F, (6) is incorrect for these states. Instead, they contribute a Curie-like 1/T term to the staggered susceptibility whenever $E_F=0$. Since the spike weight is a few per cent, this occurs over a narrow range of electron concentrations in which the subband is nearly half-filled. The range increases as the concentration of A sites decreases from unity.

In conclusion, we have shown that Fermi surface smearing due to disorder[8] is irrelevant to the antiferromagnetic instability in this model, as the staggered susceptibility is unaffected or even enhanced in the limit $\delta \rightarrow \infty$. However, the uniform effective field theory presented is strictly valid only in the limit of small U. Actual materials of interest, such as $Mn_x Zn_{1-x} F_2$[11], have rather large values of U, as well as band structures in which orbital degeneracy is important. The magnetization in the ordered systems is certainly non-uniform in space, so the present theory does not permit a confident prediction of the Neel temperature for such materials.

REFERENCES

1. J. Hubbard, Proc. Roy. Soc. (London) A276, 238 (1963);
    A277, 237 (1964); A281, 401 (1964).
2. B. Velický, S. Kirkpatrick and H. Ehrenreich, Phys.
    Rev. 175, 747 (1968).
3. P. Soven, Phys. Rev. 156, 809 (1967).
4. B. Velický, Phys. Rev. 184, 614 (1968).
5. S. Kirkpartick and T.P. Eggarter, Phys. Rev. B15, to
    appear in Nov. 1972.
6. D.R. Penn, Phys. Rev. 142, 350 (1966).
7. M. Cyrot, Phil. Mag. 25, 1031 (1972); J. Phys. (Paris)
    33, 125 (1972).
8. H. Fukuyama, Phys. Rev. B5, 2672 (1972).
9. To prove this, one can simply calculate the change in
    total energy, δE, of the electrons in the presence of
    the alternating field, which shifts each eigenvalue
    E_i to a new value, $\text{sgn}(E_i) (E_i^2 + g^2\mu_B^2 h_{ext})^{1/2}$, and
    determine the magnetization from $\delta E = -1/2\ M \cdot h_{ext}$.
10. P.W. Anderson, J. Phys. Chem. Solids 11, 26 (1959).
11. See, e.g., W. Buyers, this conference.

DISCUSSION

D.M. Esterling: In your discussion of the use of the random
phase approximation in the Hubbard model, do you use a real
or effective coulomb potential U?

S. Kirkpatrick: If $\chi^{(0)}(q)$ is infinite it doesn't matter
which one I use.

D.M. Esterling: Suppose U is infinite.

S. Kirkpatrick: In that case, the treatment of Lederer
(this conference) would be appropriate and my results incorrect. But the infinite U limit is not connected to the
results for small U by any sort of extrapolation, and is
irrelevant to the situation I consider. One gets different
results.

D.M. Esterling: Quite different results? In particular, what about your Bloch energies? Are you sure they are the same in the presence of a Coulomb interaction? Are the $\epsilon(k)$'s that you use in your Fermi functions bare Bloch energies, or somehow renormalized?

S. Kirkpatrick: I put the bare ones in, since that is correct to lowest order in U. The only well worked out approximation for going beyond that is the T-matrix approximation, which doesn't affect $\epsilon(k)$.

D.M. Esterling: Doesn't that restrict your result to the low-density limit?

S. Kirkpatrick: Not if U is weak, and perhaps not, for larger U, if the T-matrix theory can be used as an interpolation scheme.

N. Rivier: I am sorry, I probably wasn't very attentive, but could you tell me where the delta function in the density of states comes from in a disordered system? Have you got states with zero restoring force?

S. Kirkpatrick: No, zero energy is just the center of the band by my choice of energies, the energy which an electron sitting on a completely isolated A atom would have. The unusual states in the spike have zero energy because they have high symmetry and are rather localized. They can be viewed as standing waves formed between certain configurations of B atoms, which act as perfectly reflecting hardcore scatterers. There are some pictures of this in the paper (Ref. 5) that I did with Eggarter. The unusual feature of these states is that they occur in the same regions of the material as do the extended states. The spectrum of this three dimensional disordered system is quite unusual, since it contains delta functions that cannot easily be separated from the continuous background. And the nonisolated localized states at zero are just the most common ones. Such states exist at many other energies.

N. Rivier: Does the fact that there are other spike energies than zero mean a broadening of this line?

S. Kirkpatrick: No. The other localized states are too rare, and most of them occur at energies well separated from zero.

E.J. Siegel: If you don't mind, a few comments from the world of real metals. First, does your potential include interactions between parallel spins on any one site? I think you are underestimating the preponderance of magnetic states that exist by not including the Pauli principle in the Hamiltonian.

S. Kirkpatrick: My model has a non-degenerate band. It's really put forward as a counter example to theoretical naivete about the sort of things one will find in disordered systems, not as a description of any particular material.

E.J. Siegel: Well, I just suggest that as a possibility. Secondly, have you tried calculating a free energy and minimizing it with respect to V and band-filling to get a phase diagram in which it's not necessary to appeal to the free electron limit?

S. Kirkpatrick: I restricted myself by considering only second-order transitions. But I'm not sure that the procedure you describe gets around the difficulties of describing the paramagnetic state (as described in Ref. 7).

T.A. Kaplan: You said in an introductory remark that the lack of misfit was going to be very important to your theory. I don't remember you mentioning that again.

S. Kirkpatrick: The existence of a perfect lattice is crucial in proving Eq. 9, which follows from the observation that the Hamiltonian (7) consists only of elements that link one sublattice to the other. Eq. (9) permits me to show that every eigenstate of the disordered system is coupled to its mirror image by the operator, T, from which the singularity in $\chi^{(0)}(Q)$ follows. Eq. (9) would not be true if one had a random network or dislocations present, such that the decomposability of the lattice into two sublattices is lost. So misfit destroys the simple symmetry associated with T.

ELECTRICAL RESISTIVITY ANOMALY IN MICTOMAGNETIC ALLOYS

Paul A. Beck and D. J. Chakrabarti

University of Illinois

Urbana, Illinois 61801

Mictomagnetic alloys[1] are characterized by freezing
of the spin orientations at low temperatures but, unlike in
ferromagnets or in antiferromagnets, <u>without long-range spin
order</u>. The presence of short-range spin order (magnetic
clusters) is indicated by the superparamagnetic behavior,
at least in a certain temperature range.

The most thoroughly studied mictomagnetic alloys are
the f.c.c. solid solutions formed by Cu and Mn. Some typical features of the magnetic behavior of mictomagnetic alloys are briefly reviewed here, using data for Cu-Mn alloys.
The magnetization (σ) vs. field (H) isotherms for such an
alloy, Fig. 1, can be fitted to Eq. 1 by suitable choice of

$$\sigma = \chi' H + \mu cL(\mu H/kT) \qquad (1)$$

the parameters χ' (field-independent susceptibility), μ
(average magnetic cluster moment), and c (magnetic cluster
concentration). L is the Langevin function of $\mu H/kT$. Fig. 1
shows that prolonged aging at a low temperature (100°C),
after quenching the specimen from a high temperature, increases the magnetization. By fitting the data to Eq. 1,
it can be shown that low temperature aging also increases
the average magnetic cluster moment. Fig. 2 shows the dependence on the Mn content of σ_{mc}, the part of the total magnetization due to the clusters, when the cluster moments
are aligned by field cooling, in both the quenched and the
low temperature aged condition. It is clear that the metallurgical condition of the alloys has a profound effect on their

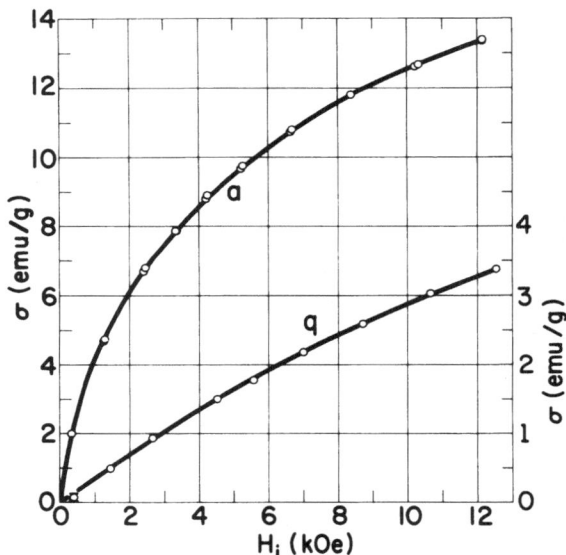

Fig. 1. Magnetization vs. internal field at 78°K for 16.7% Mn alloy quenched, q, and 100°C annealed, a[1a].

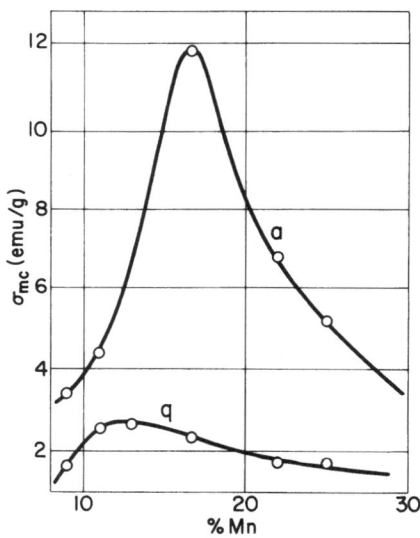

Fig. 2. Total cluster magnetization, σ_{mc}, at 4.2°K vs. Mn content for quenched, q, and for 100°C annealed, a, Cu-Mn alloys[1c].

magnetic properties.

Fig. 3 gives the dependence of the magnetization on the measuring temperature, and it shows that below a certain temperature the measured magnetization depends strongly on the thermomagnetic history of the specimen. In this temperature range, where the spin orientations are at least partially frozen[1], the specimen exhibits remanent magnetization after cooling from a higher temperature in a magnetic field. The highest temperature, T_r, at which remanence can be observed is usually not very different from the temperature at which the magnetization is maximum for a zero-field cooled specimen, measured at increasing temperatures, Fig. 3. In a certain temperature range, below T_r, it is observed that the magnetization changes with time, following a change in the applied field. The "viscous" magnetic effects[2], exemplified by the dependence of the magnetization both on time and on thermomagnetic history, are characteristic of mictomagnets at relatively elevated temperatures, where the spin orientations are not completely frozen.

The typical low temperature magnetic behavior of Cu-Mn alloys, under conditions where the spin orientations are completely frozen, was studied in detail by Kouvel[3]. After cooling to such a low temperature in zero field, the magnetization is proportional to the field. However, if the specimen reaches the low temperature of the measurement by being cooled in a magnetic field, a "hysteresis loop" is observed, which is typically displaced with respect to the zero point of the field in a direction opposite to the field applied during cooling.

Assuming Matthiessen's rule to be valid, and in the absence of magnetic scattering effects, the resistivity of a binary alloy, ρ, is the sum of the residual resistivity, ρ_o, and of the temperature dependent phonon resistivity, $\rho_{ph}(T)$. Nordheim's rule states that ρ_o which, in an alloy, is essentially due to atomic disorder scattering, is proportional to $x(1-x)$, where x is the concentration of one component and 1-x is the concentration of the other component of the binary alloy. Assuming that, in the concentration range of the Cu-Mn alloys to be considered here, ρ_{ph} is sufficiently closely approximated by the phonon resistivity of copper and neglecting ρ_o for copper, ρ_o for the alloy could be obtained by subtracting from ρ, the measured resistivity of the alloy, the phonon resistivity of copper

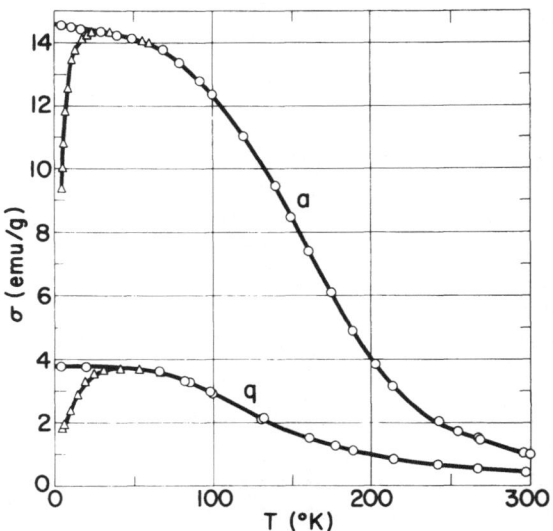

Fig. 3. σ vs. T for quenched, q, and for 100°C annealed, a,
16.7% Mn alloys, field-cooled o and zero-field cooled Δ (1c).

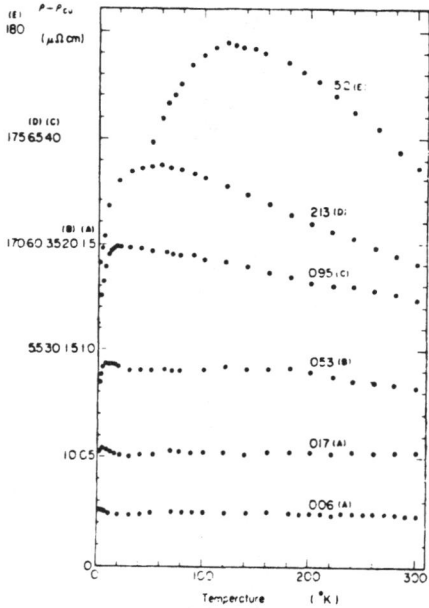

Fig. 4. ρ - ρ_ph vs. T. for Cu-Mn alloys with 0.06% to 5.2%
Mn, slowly cooled(4).

at the same temperature, if magnetic scattering were absent.
The electrical resistivity for Cu-Mn alloys with up to
5.2% Mn, from which the resistivity of copper has been subtracted, $\rho - \rho_{ph}$, is shown in Fig. 4 to be dependent on the
temperature[4]. It is clear that, in addition to the temperature-independent residual resistivity ρ_o , due to atomic
disorder scattering in the alloys, there is present an anomalous temperature-dependent resistivity component, ρ_m, presumably due to magnetic scattering. As seen in Fig. 4, both
the maximum value of ρ_m and the temperature at which this
maximum value is attained, increase with the Mn content.
Extrapolating the curves in Fig. 4 to T=0, the intercepts on
the ordinate axis give approximate values for the temperature-
independent resistivity component. The values obtained for
this component from Fig. 4 and the corresponding values from
Fig. 5 for two alloys with higher Mn contents, fit Nordheim's
rule within the accuracy of the extrapolations used. Accordingly, at least in first approximation, one may consider
these intercepts to represent the ρ_o values due to atomic
disorder scattering. The anomalous resistivity maximum is
at Mn concentrations of 13% and 16.7% so large that it is
clearly observable, even without subtracting the phonon resistivity, as shown in Fig. 5. The maxima of the total resistivity curves here are near room temperature. Assuming
that, even at these higher Mn contents, the phonon scattering
is nearly the same as that of copper, a rough analysis of
the data from Fig. 5, and of those from Fig. 4, gives the
following values (in μ ohm-cm) for the various resistivity
components:

	ρ_o	ρ_m	ρ_{ph}	$T_m (°K)$
0.06% Mn, slowly cooled	0.26	0.01	--	--
0.17% Mn, slowly cooled	0.52	0.02	--	4
0.53% Mn, slowly cooled	1.3	0.14	--	7
0.95% Mn, slowly cooled	3.2	0.3	--	18
2.13% Mn, slowly cooled	5.9	0.48	0.05	50
5.2 % Mn, slowly cooled	16.5	1.35	0.5	120
13 % Mn, as quenched	34.3	4.6	1.2	230
13 % Mn, aged 3 d at 100°C	32.8	5.6	1.3	244
16.7 % Mn, as quenched	43.6	5.9	1.4	265
16.7 % Mn, aged 3 d at 100°C	41.7	7.4	1.6	285

Here ρ_m represents the maximum value of the anomalous temperature-dependent component of the resistivity, which occurs
at the temperature given for each case under T_m; the ρ_{ph}
values listed were determined for the same temperatures.

Comparison of the parameter values listed above for a

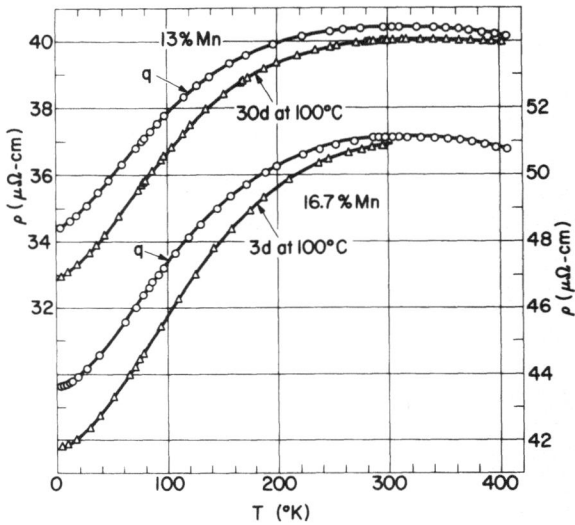

Fig. 5. ρ vs. T for 13% and 16.7% Mn alloys quenched, q , and 100°C annealed (this work).

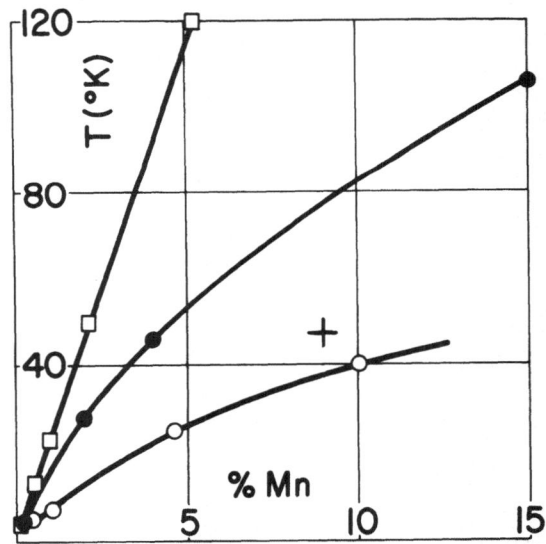

Fig. 6. Temperature of $\rho - \rho_{ph}$ maximum[4] □, of ESR transition ●[5,6], of ME transition +[7], and T_r o[8] for Cu-Mn alloys[1c].

wide range of Mn contents shows the following features: The temperature of the maximum of the magnetic resistivity anomaly rises with the Mn content along a monotonic and smooth curve, which is nearly a straight line up to 5.2% Mn, Fig. 6. The maximum value of the magnetic resistivity component increases proportionally with the Mn content, within the accuracy of the analysis. The results discussed indicate that the low temperature resistivity peak of the dilute alloys should have the same physical origin as the resistivity maximum near room temperature for the two alloys with high Mn contents, Fig. 5.

It is quite unlikely that the resistivity decrease with increasing temperature (at the high temperature side of each peak) is a result of the Kondo effect, as has been suggested[4]. It is much more probable that the temperature-dependent magnetic scattering of the conduction electrons is due mainly to the giant moments of the magnetic clusters[3]. In a mictomagnet this effect may be expected to decrease with decreasing temperature, as the spin orientation of the magnetic clusters is more and more immobilized. The decrease of the magnetic scattering by the giant moments with increasing temperature (above the temperature of the maximum) may be ascribed to more and more of the clusters reaching their intra-cluster Curie temperatures since, as a result, the concentration of the giant moments gradually decreases with increasing temperature. This argument assumes that the many individual Mn spins, into which a magnetic cluster dissociates at its Curie temperature, have a smaller total magnetic scattering effect on the electrical resistivity than has the giant moment of the corresponding magnetic cluster. In the absence of a theory of the scattering of conduction electrons by magnetic clusters, this assumption appears to be reasonable enough. The above analysis of the data for the 13% and 16.7% Mn alloys indicates that an aging treatment at 100°C, which is known to increase the average giant moment (and possibly also the number) of the magnetic clusters at the expense of small clusters and single Mn spins, also increases the magnitude of the resistivity anomaly due to magnetic scattering. This result is consistent with the assumption just mentioned.

The present work was supported by a grant from the National Science Foundation.

REFERENCES

1. a) Paul A. Beck, Met. Trans. $\underline{2}$, 2015 (1971).
   b) Paul A. Beck, J. Less Common Metals, $\underline{28}$, 193 (1972).
   c) Paul A. Beck, in "Magnetism in Alloys," Editors:
       J.T. Waber and Paul A. Beck, TMS, AIME-1972.
2. R. Street, J. Appl. Phys. $\underline{31}$, 310S (1960).
3. J.S. Kouvel, J. Phys. Chem. Solids $\underline{21}$, 57 (1961).
4. A. Nakamura and N. Kinoshita, J. Phys. Soc. Japan $\underline{27}$,
   382 (1969).
5. D. Griffiths, Proc. Phys. Soc. $\underline{90}$, 707 (1967).
6. K. Okuda and M. Date, J. Phys. Soc. Japan $\underline{27}$, 839 (1969).
7. B. Window, J. Phys. C., Solid State Phys. $\underline{2}$, 2380 (1969)
   and B. Window, J. Phys. C., $\underline{3}$, 922 (1970).
8. O.S. Lutes and J.L. Schmit, Phys. Rev. $\underline{125}$, 433 (1961).

DISCUSSION

J.A. Mydosh: The resistivity data shown are somewhat similar to those for $\underline{Au}Fe$ alloys. For the latter we have found a $T^{3/2}$-type initial temperature dependence; did you find similar behavior with $\underline{Cu}Mn$ alloys?

P.A. Beck: We did not look for a $T^{3/2}$-dependence of the resistivity. In Ref. 4 a portion of each ρ vs. T curve for the alloys up to 5.2% Mn was fitted to a logarithmic curve in an attempt to prove that the increase in ρ with decreasing temperature was due to the Kondo effect. Since for the 13% and 16.7% Mn alloys this part of the curve lies above room temperature, this interpretation appears to be incorrect. $\underline{Au}Fe$ alloys up to about 15% Fe are mictomagnetic and above their remanence temperatures they are superparamagnetic, as are the $\underline{Cu}Mn$ alloys. The similar resistivity maxima in the two systems should arise in the same way; the explanation described in the paper (and proposed earlier by Kouvel,[3] see also Houghton, Sarachik and Kouvel, Phys. Rev. Letters $\underline{25}$, 238 (1970)) presumably applies to both systems.

H.C. Siegmann: I do not understand how a displaced hysteresis loop can occur in a disordered material. I think, if the material gives a different response depending on the direction of the field, one must have frozen in some type of spiral, because the magnetic field is an axial vector. Is that correct?

P.A. Beck: In single phase alloys displaced hysteresis
loops have been observed so far only in the atomically disordered, or partially ordered, condition. When a high
degree of long range atomic order is developed, the alloys
change from mictomagnets to either ferromagnets (e.g. Au_4Mn)
or antiferromagnets (e.g. Pd_3Mn). The unidirectional magnetic anisotropy observed in field-cooled mictomagnets
does indeed resemble in many respects a "frozen-in field".[3]

Question: Do the phenomena discussed occur before the onset of crystalline ordering? You were talking about crystalline materials.

P.A. Beck: The single phase alloys in which mictomagnetism
has been studied so far are crystalline. However, it is
quite possible, or even probable, that most of the amorphous
alloys, described until now as ferromagnetic, will prove to
be in fact mictomagnetic on closer examination.

S.M. Shapiro: What is the definition of a mictomagnet?

P.A. Beck: Mictomagnetism may be characterized by the in-
cidence of the various features of complex magnetic behavior
reviewed briefly in the paper. The corresponding magnetic
structure may be described as a dispersion of magnetic
clusters (with giant moments) in a spin-glass matrix.

B.R. Coles: I think that the reversible cluster magnetization, found in CuMn alloys, does not occur in a number of
other magnetic glass systems. It's occurrence may be dependent on the short range ordering effect mentioned by
Prof. Beck. I think that the tendency of the cluster of a
short range ordered region to have a ferromagnetic moment
is known. In Rh-Fe alloys, where no ferromagnetism exists
in the entire solid solution range, one can observe displaced hysteresis loops, but there is no reversible cluster magnetization.

P.A. Beck: It appears that magnetic clusters are in some
single-phase solid solutions (e.g. Au-Fe or Cu-Ni) associated with atomic clusters, while in other alloys (e.g. Cu-Mn
or Pd-Cr) they are connected with regions of short range
atomic order. Magnetic clusters (without mictomagnetism)
also occur in some alloys with a high degree of long range
atomic order (e.g. FeAl or Ni_3Al). The results of Murani
and Coles (J. Phys. C; Metal Phys. Suppl. No. 2, S159 (1970))

strongly suggest that the R̲h̲-Fe solid solutions are mictomagnetic. Thus, one may expect their magnetic behavior to
be very similar in every respect to that of C̲uMn alloys in
which, incidentally, also no ferromagnetism occurs. The
resistivity anomaly in R̲h̲-Fe alloys (B.R. Coles, Phys.
Letters 8̲, 243 (1964)) also is very similar to that in
C̲uMn alloys. The only difference is that in R̲h̲-Fe alloys
the anomaly apparently extends to relatively high temperatures, suggesting that the cluster Curie temperatures are
relatively high in this system.

ANOMALOUS MAGNETIC PROPERTIES OF MICTOMAGNETIC FE-AL ALLOYS

G. P. Huffman

Research Laboratory, U. S. Steel Corporation

Monroeville, Pennsylvania 15146

ABSTRACT

Results of a Mossbauer study of ordered $Fe_{1-x}Al_x$ alloys are presented with emphasis on the anomalous magnetic properties of alloys having $.27 \leq x \leq .33$. A molecular field theory which explains the anomalous behavior and most other magnetic properties of Fe-Al alloys is outlined.

INTRODUCTION

Possibly no other binary alloy has received more intensive magnetic study than has the Fe-Al system. This paper is concerned primarily with the anomalous magnetic properties of ordered alloys having Al concentrations (x) in the range $.27 \leq x \leq .33$. These were first studied by Arrott and Sato,[1] who observed that alloys of both B2 and DO_3 order having $x \sim .30$ exhibited fairly normal ferromagnetic hysteresis loops at room temperature, had zero remanence from 180 down to 60°K, and showed broad hysteresis loops with large coercive forces at 4.2°K. They interpreted this behavior as indicating a transition from ferromagnetism to antiferromagnetism with decreasing temperature. Subsequently, Kouvel observed a shifted hysteresis loops for an alloy with $x = .30$ after field cooling to 1.8°K, indicating the coexistence of ferromagnetic and antiferromagnetic regions.[2] Recently, Beck and his co-workers have made detailed magnetic studies of such alloys over a range of concentrations

and observe superparamagnetic behavior indicative of large
cluster moments in the temperature ranges where remanence
vanishes.[3]   Here, we present the results of a Mössbauer
study of these alloys and outline a molecular field theory
which explains the anomalous behavior and most other magnetic properties of Fe-Al alloys of all concentrations
reasonably well.

EXPERIMENTAL RESULTS

Typical spectra are shown in Fig. 1 for an ordered B2
alloy with x = .298. The solid curves are least squares
fits assuming five six-peak magnetic components; for
$T \leqslant 4C°K$, each component can be tentatively assigned to a
particular nearest neighbor (n.n.) configuration of Fe. At
higher temperatures, however, the small field components
increase in intensity relative to the large field components,
making this identification rather vague. Thus, the temperature dependence of the hyperfine fields of the separate
components is perhaps less meaningful than that of the
average field obtained by weighting the component fields
according to intensity and shown in Fig. 2. Also shown
are the average fields for two DO_3 alloys with x = .298
and .28. We have measured the hyperfine field temperature
dependence for a number of alloys of both crystal structures
having x = .23 to .32, and have previously reported results
for B2 alloys with x = .35 to .50.[4]   The results may be
summarized as follows: (1) for x ≤ .27, the fields show
normal ferromagnetic behavior decreasing as $T^{3/2}$ at low
temperatures; (2) for .27 ≤ x ≤ .33, the field vs. temperature curves show minima such as shown in Fig. 2; the minima
are somewhat deeper for B2 alloys than for DO_3 alloys; (3)
for x ≥ .33 the fields decrease as T^2 at low temperatures
and show no minima.[4]

DISCUSSION OF RESULTS

Previously, we have given a molecular field theory
applicable to Fe-Al and similar alloys.[4,5]   An Fe atom
is assumed to have a spin S = 1 if it has 4 or more Fe n.n.
and S = 0 for < 4 Fe n.n., a reasonable approximation to
the dependence of the Fe moment on n.n. configuration.[6,7]

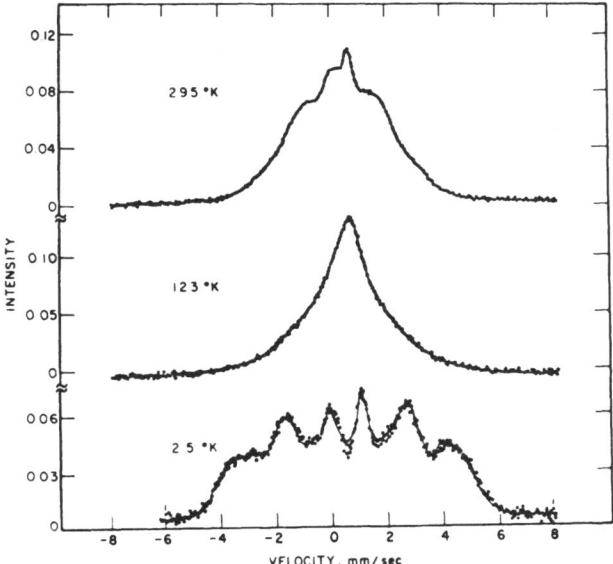

Figure 1. Mössbauer spectra of a B2 alloy with x = .298
at several temperatures.

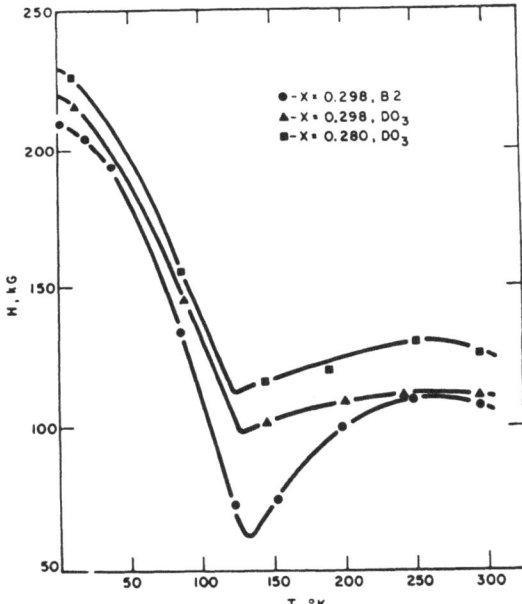

Figure 2. Magnitude of the average hyperfine field vs.
temperature for B2 and DO_3 alloys near x = .30.

For random alloys with $x \leq .15$, essentially all Fe atoms have spins, and the predicted Curie temperature is approximately[5]

$$T_c \cong (1 - x)\ 2S\ (S + 1)\ \{8J_d + \sum_{n=1}^{6} Z_n\ J_n(x)\}/3k_B \quad (1)$$

Here, J_d is the exchange interaction transmitted between n.n. spins by d electrons, Z_n is the number of atoms in the nth n.n. shell, and $J_n(x)$ is the indirect RKKY interaction between nth n.n. spins; its concentration dependence arises because the s electron density of the alloy changes with solute concentration and valence. Equation (1) agrees well with the measured concentration dependences of T_c for $x \leq .15$ to .20.[5] As expected, J_d is positive and about 10 times larger than any of the J_n.

The B2 or CsCl ordered alloys may be viewed as two interpenetrating simple cubic lattices, one of pure Fe and one of composition $Al_{1-c}Fe_c (c = 1 - 2x)$. For $.35 \leq x \leq .5$ most atoms on the Fe sublattice (Fe_F) have S = 0, and the antiferromagnetic behavior of these alloys arises from RKKY interactions between Fe atoms on the $Al_{1-c}Fe_c$ sublattice (Fe_A) which all have S = 1. The predicted Néel temperature is[4]

$$T_N = c\ 2S(S + 1)\ \{12\ J_3(x) - 6\ J_2(x) - 8\ J_5(x)\}/3k_B \quad (2)$$

Thus, 2nd and 5th n.n. spins tend to be aligned antiparallel and 3rd n.n. spins parallel. That the indirect exchange interactions will now have different signs and magnitudes from those we found previously for random Fe-rich alloys[5] is not too surprising since the concentration range and crystal structure are quite different. We feel that the RKKY interaction probably undergoes a discontinuous change when the alloy becomes ordered at around 25 a/o Al. This might arise from a sudden shift of the effective Fermi wavevector to smaller values on contact of the Fermi surface with a Brillouin zone boundary.

For B2 alloys richer in Fe, it is clearly not possible to continue neglecting Fe_F spins; for example, a 30 a/o Al alloy has approximately equal numbers of Fe_F and Fe_A spins. The large d electron interaction aligns all n.n. Fe_A and

Fe_F spins parallel and by random chance, sizable regions in which all spins are parallel are built up. Only if most of these regions are larger than some critical size will such an alloy behave as a normal ferromagnet. Our results and Beck's observations[3] indicate that such normal behavior occurs for $x \leq .27$, and that the parallel spin regions are small enough to behave superparamagnetically for $\geq .27$. Further, the theory predicts DO_3 Curie temperatures which agree with experiment for $x \leq .27$, but lie increasingly above the observed Curie points for $x > .27$.[8] Interpreting the observed values as blocking temperatures,[9] the mean superparamagnetic "particle" radius is estimated to be about 50 to 160°A for $x = .33$ to .27. A model of superparamagnetic "particles" separated by narrow regions of antiferromagnetic (a.f.) order explains all anomalous magnetic properties of these alloys reasonably well. Consider the temperature dependent behavior of an alloy with $x \sim .30$. For $T \geq 200°K$, and below the blocking temperature of most particles,[10] but above the "Néel points" of the a.f. regions separating particles, all spins of a given particle align along some easy axis and sizable hyperfine fields are observed. The remanence and coercive force arise from the ferromagnetic anisotropy energy barrier to rotation of particle moments. For $60 \leq T \leq 200°K$, the spins in the a.f. regions become ordered and exert exchange anisotropy (e.a.) fields on the particles in the manner discussed by Kouvel[11] for Cu–Mn and other alloys. The concept of a "Néel point" for these coupling regions is rather vague, since in many cases they may separate particles by only 2 or 3 lattice spacings, the spin concentration varies from region to region, and there is some n.n. interaction, invalidating Equation (2). There will thus be a range of temperatures below which the spins in various coupling regions order in a basically a.f. fashion. Defining a surface spin as one which has n.n. spins on only one face of its cubic n.n. cell, we can estimate the size of the e.a. field. Consider an Fe_A spin with 2 n.n. Fe_F spins which are 3rd n.n. of each other; 1,6 and 6 of the surface spin's 2nd, 3rd and 5th n.n. sites have no n.n. spins in common with it and will provide an e.a. field if occupied by other Fe_A spins. The field direction will depend in a complicated way on the separation, shape, and relative size of neighboring particles. Its magnitude is approximately

$$H_s \simeq c(2S/g\beta) \; \sigma_A \; \{J_2 - 6J_3 + 6J_5\} \, , \tag{3}$$

where $\mathfrak{I}_A$ is the reduced magnetization of Fe_A spins in the
a.f. region. From Equation (2) and the measured Néel temperatures for $x > .35$,[4] we estimate $(J_2, J_3, J_5) \sim (-8.,$
$+4., -1.5) \times 10^{-4}$ eV, and taking $c = .4$ and $\sigma_A \sim .1$ (slightly below the a.f. ordering temperature), we find $H_s \sim 3 \times 10^4$
gauss. Other possible n.n. configurations for the surface
spin (either Fe_A or Fe_F) give the same order of magnitude
for H_s. The effective e.a. field for the particle is much
smaller, however, since only surface spins have their energies affected significantly by H_s. For a particle 100 Å
in radius having ~1 to 10% of its total surface spins favor-
ing a particular H_s direction over all others, the effective
e.a. field for the whole particle would be $H_{ea} \sim H_s \times (10^{-3}$ to
$10^{-2}) \sim 30$ to 300 gauss. H_{ea} is thus in the right range to
either rotate particle moments directly or to lower the
energy barriers to such rotation sufficiently to allow
thermal excitation to new directions. Moment rotation for
one particle affects the spin orientation in a.f. regions
coupling to other particles which may induce moment rotation
for them and vice versa.[12]

The net effect is to cause a large fraction of the
particles to have spin relaxation times $\leq 10^{-8}$ sec. in this
temperature range leading to the hyperfine field minima shown
in Fig. 2,[13] while nearly all particles will have relaxation times ≤ 10 sec., causing zero remanence and superparamagnetic behavior in magnetic measurements.[1,3] On
further cooling, the reduced magnetization and magnetocrystalline anisotropy energy constant (K_{af}) of the a.f. regions
increase markedly, so that nearly all particles have their
moments locked in some direction, giving Mössbauer spectra
like the bottom spectrum in Fig. 1 for $T \leq 40°K$. From
Kouvel's work,[11] it is clear that if K_{af} is small with
respect to the coupling energy of the e.a. field in the
range $60 \leq T \leq 200°K$, but large at liquid helium temperatures,
field cooling will produce shifted hysteresis loops. At low
temperatures, however, the broad hysteresis loops,[1,2] and
appreciable $\sin 2\theta$ component in torque measurements indicate
that there is a range of K_{af} values associated with various
a.f. regions.

SUMMARY

In summary, the model proposed explains reasonably
well all anomalous properties of ordered Fe-Al alloys in

the range $.27 \leq x \leq .33$. For $x \lesssim .27$, the parallel spin regions are large enough to behave in a normal ferromagnetic fashion, while for $.33 \leq x \leq .50$, there are certainly superparamagnetic regions present,[6,7] but for the most part, these are small enough that their blocking temperatures lie below the Néel temperatures of the a.f. regions, and a.f. behavior dominates.[4] Further detail on the molecular field model and experimental results referred to but not given here will be the subject of a future paper.

The author would like to thank Professor Paul A. Beck of the University of Illinois for communicating his results prior to publication and for a number of very helpful discussions.

REFERENCES

1. A. Arrott and H. Sato, Phys. Rev. 114, 1420 (1956).
2. J.S. Kouvel, J. Appl. Phys. 30, 313S (1959).
3. H. Okamoto, D.J. Chakrabarti and Paul A. Beck, private communication (1971).
4. G.P. Huffman, J. Appl. Phys. 42, 1606 (1971).
5. G.P. Huffman, A.I.P. Conf. Proc. No. 5, Magnetism and Magnetic Materials, 1971, p. 1310 (A.I.P., New York, 1972).
6. H. Okamoto and Paul A. Beck, to be published in Monatsk. Chem. (1971).
7. Paul A. Beck, Metallurg. Trans. 2, 2015 (1971).
8. B2 alloys transform to DO_3 above 300 or 400°C in this concentration range and B2 Curie points have not been measured. However, because of disorder in the DO_3 structure, its magnetic behavior is qualitatively quite similar to that of B2 alloys for $x \geq .27$. Thus, though most of our discussion concerns the simpler B2 structure, it is applicable with but minor changes to DO_3 alloys.
9. I.S. Jacobs and C.P. Bean, Magnetism, v.III, p. 271 (Academic Press, N.Y. 1963).
10. It is worth noting that blocking temperatures for hyperfine field measurements are higher than those of magnetic measurements, since these require spin relaxation times $\geq 10^{-8}$ sec. and 10 sec., respectively.
11. J.S. Kouvel, J. Phys. Chem. Solids 24, 795 (1963).

12. One might say that the tails (a.f. regions) are wagging the dogs (particle moments). However, each "dog" is joined by mutual "tails" to several other dogs and the dogs twist each others' tails, so the effect is a co-operative one.

13. Even in a relatively disordered DO_3 alloy, there would be a fairly large tendency for Fe_A spins to be 3rd n.n. of each other which is not present in B2 alloys. Since J_3 is positive, this could have the effect of making the average particle size slightly larger in DO_3 than in B2 alloys, which would explain the smaller hyperfine field minima observed for the DO_3 structure.

AMORPHOUS MAGNETISM IN F.C.C. VICALLOY II

James P. Cusick, G. Bambakidis, and

Lawrence C. Becker

Lewis Research Center

INTRODUCTION

Although the occurrence of superparamagnetism has been inferred in many binary alloy systems in which one of the alloy components is nonmagnetic (Beck, 1971), relatively few investigations of this phenomenon have been carried out on concentrated ternary alloys. In this report we present the results of magnetization and Mossbauer experiments which give evidence for superparamagnetism in the quenched face-centered-cubic (γ) phase of the ternary system $Fe_{0.34}Co_{0.52}V_{0.14}$ (Vicalloy II). While the magnetization data agree with the limited data of Nesbitt et al. (1967), our more complete study shows clearly that this alloy does not acquire long-range ferromagnetic order, nor does it indicate either antiferromagnetic long-range ordering or mictomagnetic behavior, down to 4.2° K. Our data can best be described as resulting from ferromagnetic clusters in a weakly magnetic lattice.

The Mossbauer data show a slightly broadened single line spectrum at 300° K which broadens with decreasing temperature without development of resolvable hyperfine splitting down to 4.2° K. We attribute the broadening to an increase in the cluster-spin relaxation time resulting from coalescence of the magnetic clusters.

RESULTS AND ANALYSIS

Splat cooled samples of Vicalloy II were obtained from R. Willens of the Bell Telephone Laboratory, Murray Hill, N.J. Magnetometer measurements were performed with a vibrating sample magnetometer mounted in a large iron electromagnet. Calibration was with an iron sample and the effect of sample images in the pole pieces was included in data reduction.[3,4] The data obtained are shown in Fig. 1.

These data are analyzed using the modified Langevin equation:

$$\sigma = \chi \cdot H + N\bar{\mu}\zeta \left(\frac{\bar{\mu} \cdot H}{kT} \right) \tag{1}$$

where σ is sample moment per gram (ergs/gm·Oe), χ is the high field susceptibility (ergs/gm·Oe_e^2), N is cluster number per gram, $\bar{\mu}$ is average magnetic moment per cluster (ergs/Oe), and $\zeta(\bar{\mu}H/kT)$ is the Langevin function. The results of this analysis are shown in Fig. 2 as a plot of $\bar{\mu}$ and N versus temperature. The high field susceptibility was temperature independent with the value $\chi = 0.35 \pm 0.1 \times 10^{-3}$ ergs/gm·Oe_e^2.

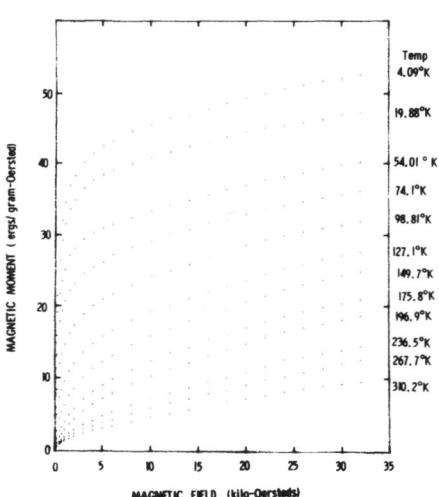

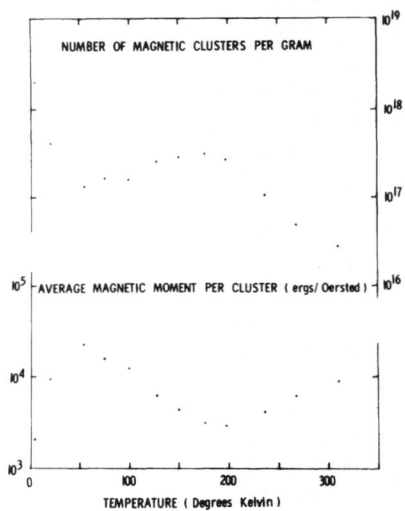

Fig. 1 Magnetic moment per gram vs. magnetic field for splat-cooled Vicalloy II.

Fig. 2 Parameters of Eq. (1) vs. temperature obtained by least squares fit to data of Fig. 1.

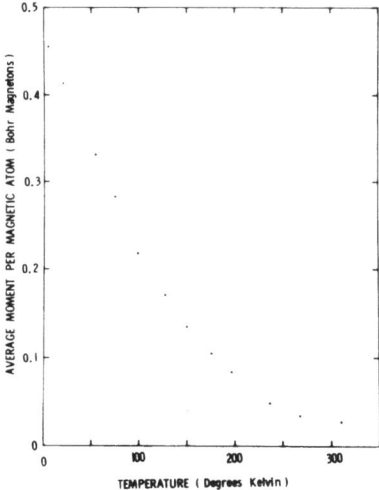

Fig. 3   Average moment per magnetic atom in Bohr magnetons
vs. temperature derived from parameters of Eq. (1).

An estimate of the magnetic moment per magnetic ion can
be made using the parameters of Eq. (1). The result is
shown in Fig. 3 where the average moment in Bohr magnetons
per magnetic ion ($\overline{\mu}_{at}$) is plotted against temperature.

Mossbauer data were taken with a constant acceleration
transmission mode experiment. Absolute velocity calibration was obtained by multiplexing nuclear data with a laser-
interferometer fringe counter. The source was Co^{57} in
palladium. With the exception of room temperature data, all
data were taken through a cryostat with 1 mm of beryllium in
the radiation beam. Parasitic nuclear absorption effects in
the cryostat were corrected. Detection was by CO_2Kr proportional counter. No parasitic absorption was observed in
the beryllium window of the detector. Sample temperature
was controlled within $0.5°$ K or better for all data below
room temperature.

Mossbauer data at room temperature and $4.2°$ K are shown
in Figs. 4 and 5. Error bars indicate range of ± 1 standard
deviation in count away from resonance. Data are analyzed
by least squares with a single Lorentzian line. The parameters of fit are shown in the respective figures. Data at
$4.2°$ K are not fit well by a single Lorentzian.

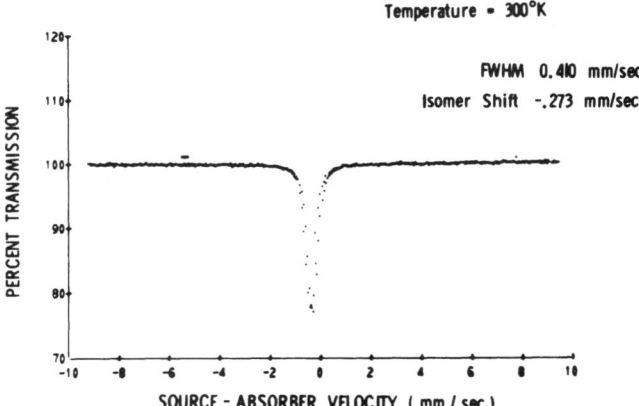

Fig. 4   Mossbauer spectra at 300° K.   Parameters are obtained by least squares fit of data to a single Lorentzian.

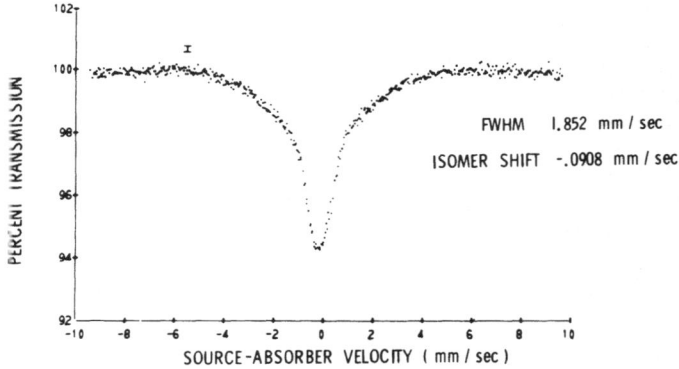

Fig. 5   Mossbauer spectra at 4.2° K.   Single Lorentzian parameters are given, however, data line shape is not
Lorentzian.

The large increase in line broadening at 4.2° K in the
absence of resolvable hyperfine splitting suggests the
existence of a change in the short range magnetic order.
To determine the ordering temperature a thermal analysis
was performed.   The transmitted flux of the 14 keV line was
measured at zero Doppler velocity in the temperature range
of 270° K to 4.2° K.   In order to obtain reproducible measurements at each temperature the observed transmitted flux
at 14 keV was normalized by the observed transmitted flux
at 21 keV.   The results are shown in Fig. 6.

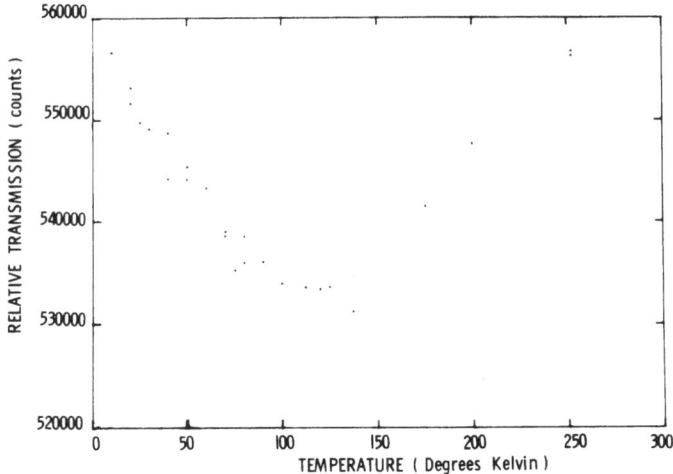

Fig. 6   Relative counts at zero source-absorber velocity vs. temperature.

To identify more clearly the broadening indicated by Fig. 6 additional spectra were taken at 150°, 75°, and 25° K. The results are shown in Fig. 7. The single line

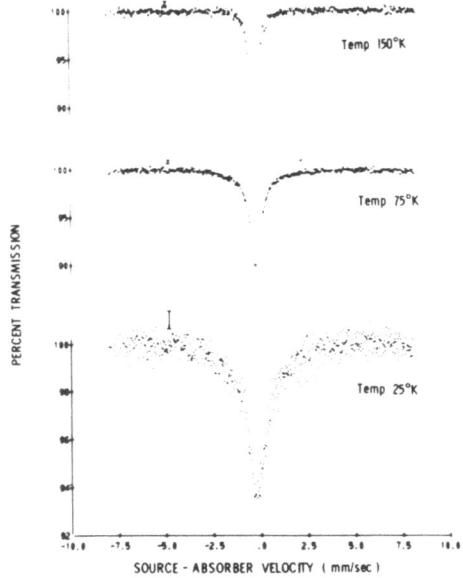

Fig. 7   Mossbauer spectra at 25° K, 75° K, and 150° K.

Temperature, °K	Line width (FWHM), mm/sec	Isomer shift, mm/sec
300	0.410	-0.273
150	0.463	-0.193
75	0.741	-0.153
25	1.390	-0.114
4.2	1.852	-0.091

Table I. - Parameters of least squares fit to Mossbauer spectra shown in Figs. 4, 5, and 7. Isomer shift is relative to Co^{57} in palladium. Line width is full width at half-maximum amplitude.

Lorentzian parameters obtained by least squares fit to the data are tabulated in Table I.

DISCUSSION

The data presented above can be described in terms of a superparamagnetic cluster model. In this model local composition variations produce local regions, which at sufficiently low temperatures, become magnetically ordered. Depending upon local composition, these regions, called clusters, possess a unique magnetic ordering temperature and a unique saturation magnetization at 0° K. An exchange coupling is allowed between clusters in close proximity such that individual clusters may coalesce to form larger clusters. The largest clusters are assumed to be small compared to the size of magnetic domains.

Consider the cluster analysis data of Fig. 2. At 300° K it is assumed that some magnetically ordered clusters exist and are well separated by the remaining paramagnetic material. In the temperature range of 300° to 200° K additional clusters with ordering temperatures in this range contribute to the magnetization and increase the number of clusters per gram. Since these clusters have smaller moments than clusters which order above 300° K, they decrease the average moment per cluster. This process of magnetic cluster formation is thought to exist over the entire range of the measurements.

In the temperature region of 200° to 50° K some clusters which exist are able to coalesce to form larger ordered regions. Sufficient numbers of clusters are lost by this process to more than compensate for the number of new clusters formed in this temperature range. Consequently, the number of clusters per gram decreases and the average moment per cluster increases.

Clusters which become ordered at low temperatures are expected to have the smallest values of magnetic moment. Consequently, the exchange interaction between these clusters and the coalesced clusters may become sufficiently weak at some low temperature to favor the existence of the individual small cluster over continued growth of the coalesced cluster. The data suggest this may occur below 50° K where an increase in cluster number and a decrease in cluster moment begins.

Mossbauer thermal analysis data (**Fig.** 6) and spectra (**Figs.** 4, 5, and 7) suggest the onset of a change in the short range magnetic ordering near 150° K. This transition temperature is slightly below the transition shown in Fig. 2 (i.e., 175° and 200° K) for the onset of cluster coalescence.

The broadening observed in the Mossbauer data is a consequence of a distribution of hyperfine fields among the iron nuclei. In terms of the model given above, the fact that the paramagnetic line at 300° K is only slightly broadened can be accounted for on the basis that a significant number of the clusters at this temperature have spin-relaxation times which are somewhat less than the Larmor period of the iron nucleus. The Mossbauer line width does not change appreciably from 300° K to 150° K since the clusters ordered in this temperature region have smaller moments (and relaxation times) than clusters ordered above 300° K.

Clusters which coalesce at temperatures between 150° K and 200° K do not produce clusters with relaxation times long enough for the cluster moment to be detected by the iron nucleus. Since the coalescence process will produce an abrupt change in the relaxation time of the final cluster, this suggests that coalescence in this temperature range is between clusters having relatively small values of magnetic moment and relaxation time.

In the temperature range 150° K to 50° K the coales-

cence of clusters produces a significant number having relaxation times comparable to or exceeding the nuclear Larmor
period. At temperatures below 50°K these clusters produce
larger hyperfine fields at the iron nucleus due to the increase in cluster spin relaxation time with decreasing temperature. Clusters with ordering temperatures in the range
0°K to 50°K have relaxation times too short to contribute
to the nuclear hyperfine field.

CONCLUSIONS

The magnetometer data clearly show that splat-cooled
FCC Vicalloy II does not exhibit long-range magnetic order.
The magnetometer data together with the Mossbauer data indicate that short-range magnetic ordering occurs which is temperature dependent with coalescence of magnetic clusters
over a limited temperature range. The model presented reasonably accounts for the observed effects and indicates
this ternary system is superparamagnetic down to 4.2°K.
This finding is quite unusual for an alloy with 86 atomic
percent magnetic ions. The reason for this unusual situation is found in the electronic structure and FCC crystal
structure of the individual cluster and will be the subject
of a subsequent report.

REFERENCES

1. Beck, Paul A.: Some Recent Results on Magnetism in
   Alloys. Metallurgical Trans., Vol. 2, no. 8, Aug. 1971,
   pp. 2015-2024.
2. Nesbitt, E.A.; Willens, R.H.; Williams, H.J.; and
   Sherwood, R.C.: Magnetic Properties of Splat-Cooled
   Fe-Co-V Alloys. Jour. Appl. Phys., Vol. 38, no. 3,
   Mar. 1967, pp. 1003-1004.
3. Behrendt, D.R.; and Hegland, D.E.: Saturation Magnetization of Polycrystalline Iron. NASA TM X-2542,
   Apr. 1972.
4. Behrendt, D.R.; and Hegland, D.E.: Image Effects and
   the Vibrating Sample Magnetometer. NASA TN D-6253,
   Mar. 1971.

MAGNETIC PROPERTIES OF AMORPHOUS METALLIC ALLOYS

C. C. Tsuei

W.M. Keck Laboratory of Engineering Materials

California Institute of Technology

Pasadena, California 91109

ABSTRACT

A growing number of amorphous metallic alloys have been obtained by rapid-quenching from the liquid state or by other techniques such as vapor deposition or electrolytic deposition. The purpose of this talk is to review some of the experimental work on the amorphous metallic alloys prepared by the technique of liquid quenching.[1] The emphasis will be on the magnetic properties of these metastable alloys.

TABLE I

Examples of amorphous alloy systems

Alloy Systems	Composition Range	References
$Pd_{100-x}Si_x$	$16 < x < 22$	Duwez et al.[2] (1965)
$(Pd_{50}Ni_{50})_{100-x}P_x$	$16 < x < 27$	Dixmier[3] (1972)
$Pd_{80-x}Fe_xP_{20}$	$13 \leq x \leq 44$	Maitrepierre[4] (1969)
$Fe_{100-x}P_{0.6x}C_{0.4x}$	$18 < x < 26$	Duwez and Lin[5] (1967)
$Au_{80}Si_{20}$	-----------	Klement et al.[6] (1960)

As shown in Table 1, the composition of the metastable amorphous alloys can be expressed by the formula: $TM._{100-x}$ G_x, where TM. stands for transition metal, such as Pd, Pt, Fe, Ni etc., and G refers to glass former, usually a high-valence element such as P, Si, C, or B. The composition range of the amorphous alloys lies in the neighborhood of x = 20 to 25. It is interesting to note that the amorphous alloy $Au_{80}Si_{20}$ is not stable at room temperature. All these amorphous alloys exhibit the characteristics of a metal, for example, metallic luster, certain ductility and relatively high electrical conductivity. These alloys are characterized by a relatively large residual resistivity of about 100 to 300 $\mu\Omega$ cm. In addition, there is a temperature-dependent contribution which increases or decreases with temperature depending on the alloy composition.[7,8] The magnitude of this variation of resistivity with temperature from 4.2°K to 300°K is usually a few per cent of the total resistivity.

A considerable amount of effort has been devoted to the analysis of X-ray diffraction data on the liquid-quenched amorphous alloys. A recent review by Giessen and Wagner gives a summary of the radial distribution functions (RDF) of these amorphous metals.[9] All the RDF are essentially similar if the differences in the atomic sizes of the metals are taken into account. This observation suggests that the short range order might be very similar in all the alloys. Most of the alloys show a splitting of the second peak in RDF which is absent in the RDF of liquid metals. Physical density measurements indicate that there is little or no change (< 2%) between the amorphous and the crystalline state. Several models have been suggested for the structure of the amorphous alloys.[9] Among these models, the one proposed by Polk is particularly attractive.[10] This model is based on a Bernal dense-random packing of the metal atoms[11] with the glass-former atoms occupying the larger holes in such a structure. This structural model not only satisfactorily fits the observed RDF, it also indicates that this kind of atomic arrangement can accommodate only about 20 at.% of glass-former. Furthermore, the glass-former atom is always surrounded by the transition metal atoms as its first neighbors. In this structure, the short range order is similar to that in intermetallic compounds such as Pd_3Si, Ni_3P, or Pd_3P. From the X-ray diffraction analysis, it is also found that the interatomic distance between TM. and G is much less than the sum of their Goldschmidt radii.

For instance, the closest distance between Pd and Si in
amorphous $Pd_{80}Si_{20}$ is about 0.3 A less than the sum of the
Goldschmidt radii. This suggests a strong interaction between TM. and G, and there is probably an electron-transfer
from G to the unfilled d-band of the TM. As a result of
electron-transfer, the size of G atom is reduced so that it
can fit the larger holes in the Bernal structure. The resulting strong bonding between TM. and G presumably stabil-
izes the amorphous structure. This is consistent with the
fact that the amorphous $Au_{80}Si_{20}$ is relatively unstable
even at room temperature. In this case, electron-transfer
becomes energetically unfavorable because the d-band of Au
is filled. This kind of argument also explains why the
amorphous metallic alloys can be obtained only in a certain
restricted composition range (i.e., G is about 20 at.%).

The structural information and the assumption of electron-transfer can also lead to a better understanding of
the electrical and magnetic properties of these amorphous
alloys. For instance, the negative temperature coefficient
of the electrical resistivity of the amorphous alloys
$(Pd_{50}Ni_{50})_{100-x}P_x$ for x > 23 can be explained in terms of
localized electron states contributed by the glass-former
atoms (in this case, P atoms) which occupied the metal sites
in the Bernal structure.[12] Results of magnetic susceptibility measurement indicate that the amorphous $Pd_{80}Si_{20}$ is
diamagnetic and the amorphous Pd-Ni-P alloys exhibit weak
Pauli-type paramagnetism down to the helium temperature
range. These experimental findings again are in accordance
with the idea that the d-band of Pd or Ni atoms is filled
by electrons transferred from the glass-former G (Si or P).
On the other hand, measurements of the magnetization and
Mössbauer spectrum[13,14] show that some of the amorphous
metallic alloys are ferromagnetic. For example, amorphous
$Fe_{75}P_{15}C_{10}$ is ferromagnetic below ∿580°K. The magnetic
moment per iron atom is 2.1 μ_B. The small reduction in
magnetic moment compared to pure iron suggests there is a
small amount of electron transfer from phosphorus and carbon atoms to the 3d band in iron.

The magnetic ordering in amorphous alloys of composition $Fe_xPd_{80-x}P_{20}$ (13 ≤ x ≤ 44) is of particular interest.
The variation of magnetic transition temperature with Fe
concentration bears a close resemblance to that of AuFe
system. The long-range magnetic order which prevails in

the higher-Fe-concentration alloys breaks down for alloys
with x < 26, giving rise to a spin-glass type of more local
ordering. In general, it is found that atoms of Fe, Co, Mn,
and Cr in the amorphous alloys carry localized magnetic moments while Pd, Pt and Ni do not. Also, the value of the
localized moment decreases with increasing content of G.

The Mössbauer study of a typical amorphous ferromagnet
(Pd$_{36}$Fe$_{44}$P$_{20}$) is discussed in some detail. This is because
Mössbauer spectroscopy has proven to be a unique tool for
studying the magnetic properties of amorphous alloys. It
permits examination of the properties of single atoms, rather than complicated assemblages of atoms as in conventional magnetization measurements. Furthermore, it yields valuable information about true zero-field magnetization. A-
bove the Curie temperature, typical Mössbauer spectra for
amorphous ferromagnets such as Pd$_{36}$Fe$_{44}$P$_{20}$ and Fe$_{75}$P$_{15}$C$_{10}$
show two peaks which are characteristic of a quadrupole
splitting.[13,14] Below the Curie temperature, the spectra
consist of six broad peaks symmetrically located about
their center of gravity. The outer peaks are much broader
than inner ones. The apparent absense of quadrupole interaction below the Curie temperature is attributed to a dir-
ectional range of electric field gradients, with respect to
the hyperfine field which gives rise to a broadening rather
than a line shift. Similar phenomenon has been observed
and discussed in other alloy systems.[15,16] Since the quadrupole interaction produces only a small perturbation compared to the hyperfine interaction, the Mössbauer spectrum
can therefore be fitted in terms of a distribution of hyperfine field P(H), and average isomer shift and a line
width[14] Γ. The best fit of experimental data has been obtained with the following functional form of P(H):

$$P(H) \propto \begin{cases} 1/[(H-H_o)^2 + \frac{1}{4}\Delta^2_o], & o \le H \le H_o \\ e^{-(H-H_o)^2/2\Delta_1^2}, & H > H_o \end{cases}$$

and $\int_o^\infty P(H)\ dH = 1.$

Since the hyperfine field H at the nucleus of Fe is
proportional to its magnetic moment, the temperature vari-
ation of the average hyperfine field $\overline{H} = \int_o^\infty P(H)H\ dH$ should
follow that of zero-field magnetization of the amorphous

ferromagnet. The experimental results so obtained from the fitting of Mössbauer spectra at various temperatures are compared with that predicted by existing theories.

At the end of the talk, some important problems of future interest in the field of amorphous ferromagnetism are discussed.

REFERENCES

*Work supported by U.S. Atomic Energy Commission.

1. Pol Duwez and R.H. Willens, Trans. AIME, 227, 362 (1962). P. Pietrokowsky, J. Sci. Instrs. 34, 445 (1962);
2. Pol Duwez, R.H. Willens and R.C. Crewdson, J. Appl. Phys. 36, 2267 (1965).
3. J. Dixmier and Pol Duwez (to be published in J. Appl. Phys. 1972).
4. P.L. Maitrepierre, J. Appl. Phys. 40, 4826 (1969).
5. Pol Duwez and S.C.H. Lin, J. Appl. Phys. 38, 4096 (1967).
6. W. Klement Jr., R.H. Willens and Pol Duwez, Nature 187, 869 (1960). This amorphous alloy is not stable at room temperature.
7. Ashok K. Sinha, Phys. Rev. B 1, 4541 (1970).
8. B. Boucher, J. Non-Cryst. Solids, 7, 277 (1972).
9. B.C. Giessen and C.N.J. Wagner, "Structure and Properties of Non-Crystalline Metallic Alloys Produced by Rapid Quenching of Liquid Alloys", in "Physics and Chemistry of Liquid Metals", S. Beer, Ed., Marcel Dekker, New York (1972).
10. D.E. Polk, Scripta Metallurgica, 4, 117 (1970).
11. J.D. Bernal, Nature, 185, 68 (1960).
12. C.C. Tsuei and R.N.Y. Chan (to be published).
13. C.C. Tsuei, G. Longworth, and S.C.H. Lin, Phys. Rev. 170, 603 (1968).
14. T.E. Sharon and C.C. Tsuei, Phys. Rev. B 5, 1047 (1972).
15. G.K. Wertheim, V. Jaccarino, J.H. Wernick and D.N.E. Buchanan, Phys. Rev. Letters 12, 24 (1964).
16. B. Window, Phys. Rev. B 6, 2013 (1972).

DISCUSSION

B.R. Cooper: As you go up in temperature, you go into the
liquid metals regime for the same alloy. Would you anti-
cipate a continuous behavior?

C.C. Tsuei: Concerning that, we have only some results on
the electrical resistivity of the alloy $Pd_{80}Si_{20}$. According to the work of Crewdson, the portions of the resistivity
versus temperature plot relating to the amorphous and liquid
states are found indeed to be a smooth curve. As far as
the electrical resistivity is concerned, the liquid state
is therefore sort of a continuous extension of the amorphous
state. No data on magnetic properties in this respect are
available.

J. Wong: What is the structural significance of the doublet in the second peak of the radial distribution function?

C.C. Tsuei: It indicates definitely some short-range order
in the structure of these alloys. I would like to mention
that recent experimental results of Dixmier and Duwez show
that if the glass-former concentration is much larger than
20 at.%, then the double-peak becomes a broad one. This
could mean that the short-range order is reduced by the G
atoms which occupy the metal sites.

P.A. Beck: How can one be sure that these alloys are indeed ferromagnetic? Surely, the fact that the alloys are
attracted by the magnet is not sufficient.

C.C. Tsuei: In addition to the magnetization and the
Mössbauer effect measurement, the B-H loops for some of
the amorphous ferromagnets have been measured. These alloys
definitely have remanence and coercive force.

P.A. Beck: These alloys still could be mictomagnetic.

C.C. Tsuei: For alloys such as $Fe_{75}P_{15}C_{10}$, no thermomagnetic effect which is characteristic of mictomagnetic alloys
has been observed in magnetization measurements.

R. Tahir-Kheli: Could you confirm or deny the existence of
the $T^{3/2}$ law of magnetization at low temperatures for the
amorphous ferromagnet?

C.C. Tsuei: For one of the amorphous Fe-P-C alloys, a $T^{3/2}$ dependence of magnetization has been observed at low temperatures.

D.J. Sellmyer: Is it true that in your glassy alloys the paramagnetic Curie temperature is zero as suggested by Kok and Anderson?

C.C. Tsuei: No. It is non-zero. For example, the Kondo alloy $Pd_{80}Si_{20}$ containing a small amount of chromium gives rise to a negative paramagnetic Curie temperature.

J.A. Mydosh: Would you be able to reduce the composition of your glass-former if you were able to quench at a lower temperature?

C.C. Tsuei: Probably yes. If you could increase the cooling rate fast enough, you would obtain an amorphous material. It is not clear it will be stable.

M.B. Stearns: Can you prepare these alloys in large quantities?

C.C. Tsuei: There are techniques with which you can do this.

M.B. Stearns: Why do you see a double peak in the Mössbauer spectrum above the transition temperature?

C.C. Tsuei: This is closely related to the amorphous nature of the alloys. The electric field gradient in these amorphous materials is primarily determined by only the charge distribution of the neighbors within a few interatomic distances. The metallic nature of these alloys also prevents the existance of longer range electrostatic effects. On the other hand, the cooperative magnetic ordering of the Fe atoms extends over a range considerably larger than an interatomic distance. This is the reason for the apparent disappearance of quadrupole interaction below the Curie temperature.

H. Alperin: In the paper on $TbFe_2$ (these proceedings), I think we can answer some of the questions that you raise. As far as domains go we have evidence that they do exist. We do a neutron diffraction experiment above and below the

Curie temperature, and if you substract these two diffraction patterns, only the magnetic scattering is left. And we see evidence for small angle scattering in the magnetic pattern which does not exist in the pattern above the Curie temperature. In other words, the chemical structure does not have any domains, but the magnetic structure does. And the size of these domains would be much smaller than those in crystalline ferromagnets.
Just another comment to Professor Beck's question: Of course, a neutron diffraction experiment can be done in zero magnetic field. So one can confirm the existence of ferromagnetism without applying a magnetic field.

C.C. Tsuei: I would like to have some one work on these amorphous alloys with neutron diffraction, so we can have more evidence for the ferromagnetism.

E.J. Siegel: Can carbon and nitrogen act to form glasses?

C.C. Tsuei: In certain alloys, I think they probably can.

G.S. Cargill: You mentioned that in the Pd-Fe-P alloys it is difficult to reach the saturation of magnetization. In what fields do you think it would saturate and why do you think it's so difficult?

C.C. Tsuei: I don't know exactly. I think probably the amorphous nature of the alloys makes it difficult to saturate. This is expecially true for the low-Fe-concentration alloys in which the Fe atoms are not close enough for alignment of their moments. On the other hand, if there are enough iron atoms in an alloy like $Fe_{80}P_{15}C_{10}$, there is no problem of reaching the saturation.

E.J. Siegel: Have you ever observed, say a transition from paramagnetism to ferromagnetism or ferromagnetism to diamagnetism by quenching at different rates?

C.C. Tsuei: No. The cooling rate is extremely difficult to control using the present quenching technique.

H.C. Siegmann: If you make two successive samples, are they the same or are they different?

C.C. Tsuei: Their properties are almost the same (say within a few per cent).

S.C. Moss: Are there large distributions of nearest neighbor distances in the amorphous alloys? Is it the same as you would get for thermal broadening?

C.C. Tsuei: Yes, and it is more than a thermal broadening.

S.C. Moss: So that would put it in a distribution of exchange interactions, and would that smear the magnetic transition?

C.C. Tsuei: No. In higher-Fe-concentration alloys, the magnetic transition is as sharp as crystalline alloys. Broad transitions have been observed in low concentration alloys.

ANOMALIES IN AMORPHOUS KONDO ALLOYS

Ryusuke Hasegawa

W.M. Keck Laboratory of Engineering Materials

California Institute of Technology

Pasadena, California 91109

Recently an anomaly in the low temperature electrical resistivity has been found in amorphous Ni-Pd-P alloys containing iron[1]. The anomalous resistivity varies as $T^{-1/2}$ and was attributed to the interference between scattering by magnetic impurities and that by non-magnetic impurities [2]. A comparison of the present results with the previous ones for other amorphous alloys [3] has lead to the statement that the non-magnetic impurities must have d-character and the electron mean-free path should be of the order of several Angstrom for the anomaly to be observed. For the alloys $Ni_{40}Pd_{40}P_{20}$ in which Ni is replaced by Fe, the $T^{-1/2}$ term dominates over the usual Kondo term for small concentrations of Fe and persists for Fe concentrations up to about 25 at.%. For the amorphous alloys containing more than 25 at.% Fe, the resistivity obeys a logarithmic law below the resistivity minimum temperature and above 5 ~ 10°K. This ln T dependence occurs below the ferromagnetic Curie temperature of the alloys, and therefore may be due to an electron-magnon interaction discussed elsewhere [4]. To understand the new $T^{-1/2}$ resistivity term, temperature dependence of the magnetoresistivity and magnetic susceptibility were studied for the present alloy system.

A conventional analysis of the isothermal magnetization curves does not give a well-defined ferromagnetic Curie temperature for alloys containing less than 10 at.% Fe. In addition, the magnetization versus temperature curves for these alloys show a maximum around 10°K. These characteristics are typical of mictomagnetic or spin-glass alloys. The spin value S per Fe atom is about 5/2 for the Fe concentrations between 2 and 15 at.%. Below 2 at.% Fe, the value of S decreases rapidly with decreasing Fe concentration. For example, the S value for the 0.4 at.% Fe case is about 3.5/2 and the paramagnetic Neel temperature is about 5°K. To summarize these results, we may say that the present alloys tend to be antiferromagnetic for Fe concentrations less than 0.8 at.% and ferromagnetic at higher Fe concentrations. In other words, the magnetic interactions seem to be of the RKKY type in the amorphous alloys studied here.

The results of the magnetoresistivity measurements of the present alloys seem to be consistent with the above findings. An analysis of the magnetoresistivity data for the alloys containing 0.25 and 2.4 at.% gives $2S \approx 1.5$ and 4.5 respectively. These results were obtained by taking $E_F = 3.6$ eV and J_{sd} (s-d exchange integral) $= -0.28$ eV. A minimum in the negative magnetoresistivity versus temperature occurs as expected, around the temperature at which the magnetization as a function of temperature becomes a maximum.

Although the role of the non-magnetic states is not quite clear, the present study indicates that the mechanism leading to the $T^{-1/2}$ resistivity term could be compatible with the RKKY type spin polarization around magnetic impurities.

REFERENCES

(1) R. Hasegawa, Phys. Rev. Lett. <u>28</u>, 1376 (1972).
(2) H.U. Everts and J. Keller, Z. Phys. <u>240</u>, 281 (1970).
(3) R. Hasegawa and C.C. Tsuei, Phys. Rev. B <u>2</u>, 1631 (1970) and B <u>3</u>, 214 (1971); R. Hasegawa, J. Phys. Chem. Solids <u>32</u>, 2487 (1971).
(4) R. Hasegawa, Phys. Lett. <u>36A</u>, 207 (1971).

DISCUSSION

B.R. Coles: Can I make a comment and ask a question. My comment is that in Rh-Fe, which Professor Beck also called mictomagnetic, again one can get a resistance minimum associated with the magnetic ordering, as your system seems to show. The other thing is: What is the field dependence of the magnetization in the vicinity of your magnetization anomaly?

R. Hasegawa: The magnetization versus field curve looks like that of a typical superparamagnet in the vicinity of the magnetization anomaly.

P. Lederer: If I understand that you have this $T^{-1/2}$ term also for the iron concentration of 10 at.%, it is very hard to believe that it is due to a simple impurity.

R. Hasegawa: Yes, it is difficult to believe, when we are dealing with crystalline alloys. In amorphous alloys, d-d exchange interactions are considered to be small. Therefore we can introduce more impurities in amorphous hosts before the single impurity assumption breaks down. However, the real problem here does not seem to be whether or not the effect is due to a single impurity, considering the fact that the effect also persists to a dilute case.

SOME STRUCTURAL AND MAGNETIC PROPERTIES OF AMORPHOUS COBALT-PHOSPHORUS ALLOYS*

G. S. Cargill III

Department of Engineering and Applied Science
Yale University, New Haven, Connecticut 06520

and

R. W. Cochrane

Eaton Electronics Research Laboratory
McGill University, Montreal 101, Quebec

INTRODUCTION

The first observations of ferromagnetism in amorphous alloys appear to have been those of Brenner, et al.[1], in 1950 and of Fisher and Koopman[2] in 1964 for electrodeposited and chemically deposited Co-P alloys respectively. Although several workers[3-6] have subsequently reported magnetic measurements on amorphous alloys in the Co-P system, no careful structural analyses of these alloys have been published.

In this paper we report some structural and magnetic properties of amorphous Co-P alloys. Our study employed x-ray diffraction, physical density measurements, and low field magnetization measurements at several temperatures between 77 and 525°K.

SAMPLE PREPARATION

Ring shaped foils, having inner and outer diameters 1.6 cm and 3.3 cm, 0.025 cm thick were electrodeposited from

* This work was supported in part by the National Science Foundation.

baths based on those described by Brenner et al.[1] which have
been summarized in Table I. As well, a rectangular foil,
2.0 cm x 2.5 cm, 0.01 cm thick, was prepared from bath B.
All foils were plated onto copper substrates from fresh,
agitated baths at 75°C with 100 ma/cm^2 current density using
cobalt anodes. Results of chemical analyses and density
measurements are also given in Table I. The latter were
several percent less than the densities expected for the
equilibrium mixtures of hcp cobalt and orthorhombic Co_2P
phases.[7-9]

Table I. Plating baths, foil compositions and densities

Sample	H_3PO_3 g/ℓ	$CoCO_3$ g/ℓ	$CoCl_2 \cdot 6H_2O$ g/ℓ	H_3PO_4 g/ℓ	Foil Comp. at.%P	Density g/cm^3
A	65	39.4	181	50	15.6	7.90±.04
B	60	37.5	182	50	14.5	7.89
C	55	35.6	184	50	13.5	7.94
D	50	33.7	185	50	12.6	7.97

STRUCTURAL CHARACTERIZATION

All samples were examined by x-ray scattering using a
General Electric diffractometer with a LiF diffracted-beam
monochromator and MoK$_\alpha$ radiation. No crystalline diffraction
peaks were found. All samples produced nearly identical
liquid-like patterns similar to those reported for electrodeposited Ni-P alloys and for most "splat-cooled" amorphous
metallic alloys.[10]

Quantitative, step-scan scattering measurements were
performed on the 14.5 at.% P rectangular sample. These data
were used to calculate the radial distribution function (RDF)
shown in Fig. 1. Experimental details and methods of data
analysis were similar to those described in reference 10.
The RDF has a well resolved first neighbor peak and a split
second peak; these features are seen in RDF's of many amorphous metallic alloys.[10] The first neighbor peak, which is
dominated by Co-Co pairs, occurs at 2.58±.05A, which is 3%
larger than the nearest neighbor spacing in hcp cobalt.[8]
The first peak area is approximately 13, compared with 12
near neighbors in hcp cobalt.

The histogram in Fig. 1 is a RDF calculated for a Bernal dense random packing of equal sized hard spheres.[11-13] The only adjustable parameter in this comparison is the value taken for the hard sphere diameter. The agreement in relative peak positions, first peak areas, and second peak splitting indicates that the arrangement of atoms in the Co-P alloys is similar to the dense random packing of hard spheres. Cargill[13] previously pointed out the applicability of dense random packing as a structural model for many amorphous metallic alloys. Polk[14-16] has suggested that the semimetal components of these alloys occupy "holes" within the Bernal packing. However, experimental determination of the location of the phosphorus atoms in the amorphous Co-P structure is difficult because cobalt has a much greater x-ray scattering factor than phosphorus.

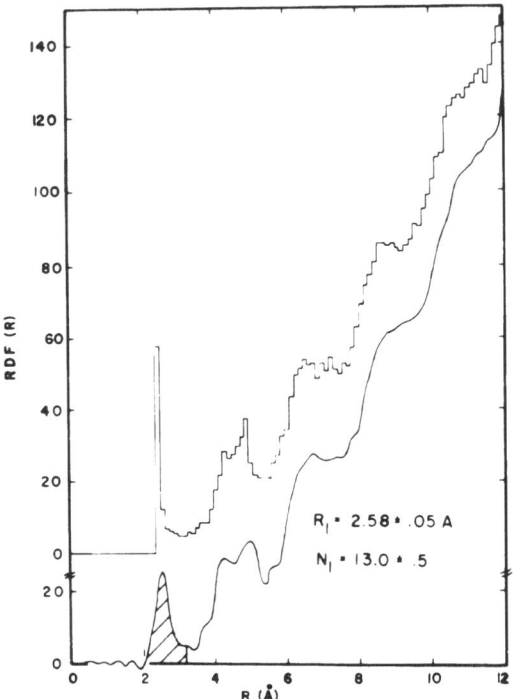

Figure 1: Upper: RDF for Bernal random packing of hard spheres.
           Lower: RDF for 14.5 at.% P alloy.

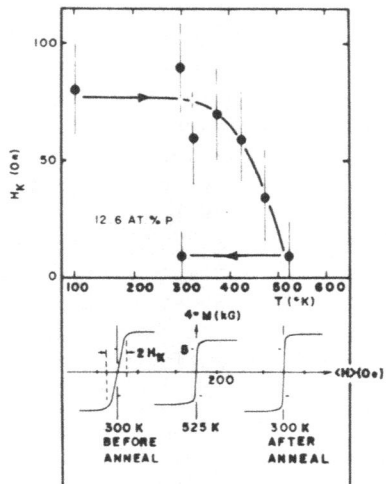

Figure 2: Upper: Effect of temperature on saturating
                   field H_K for 12.6 at.% P.

           Lower: Typical M-H curves for amorphous Co-P
                   alloys, shown for 12.6 at.% P.

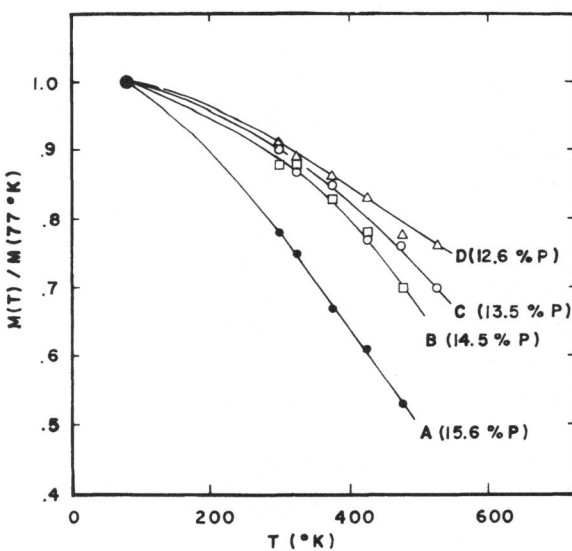

Figure 3: Temperature dependence of saturation magnetiza-
           tion for the four amorphous Co-P alloys.

MAGNETIC PROPERTIES

Magnetization curves were taken for the four ring shaped foils at several temperatures between 77 and 525°K by means of a conventional hysteresis loop tracer[17] incorporating an operational amplifier as integrator and oscilloscope display. The samples were mounted in phenolic holders and wound with primary coils; M-H curves were obtained directly by including an empty holder with identical coils in the circuit. Sinusoidal drive fields at 60 and 400 Hz with amplitudes up to 400 Oe were used.

The magnetization data for all four samples were similar to those shown in Fig. 2 for the 12.6 at.% P alloy and clearly indicate *ferromagnetic order*. The distinctive shape of these curves--a linear rise to saturation with negligible hysteresis or remanence--was identical at both 60 and 400 Hz and has been characterized by the saturating field H_K. On heating to 525°K, H_K decreases continuously and irreversibly as shown in the upper portion of Fig. 2. Nevertheless, thermal cycling has produced no change in the saturation magnetizations or in the x-ray diffraction patterns. The temperature dependence of the saturation magnetization evaluated from the M-H curves is shown in Fig. 3 normalized to M(77°K) to reduce the uncertainties in the absolute calibration. In all cases the Curie temperatures T_c are greater than 525°K and decrease with increasing phosphorus content. The saturation values extrapolated to 0°K are listed in Table II and illustrated in Fig. 4 together with measurements of Simpson and Brambley[5] for a chemically deposited Co-P alloy in both amorphous and crystalline forms, and measurements of Kanbe and Kanematsu[18] for electrodeposited solid solutions of phosphorus in a mixture of hcp and fcc cobalt. The solid line in Fig. 4 corresponds to a rigid band model calculation in which each phosphorus atom contributes five electrons to filling the 3d band of cobalt.[5] Except for our most phosphorus rich sample there is good agreement among all these data.

The most striking effect of structural disorder in the Co-P alloys is their magnetic softness, which has also been noted by Bondar, et al.[3], de Lau[4], and Simpson and Brambley[5], although no previous workers have published M-H curves for their specimens. Generally, the magnetic softness of the amorphous alloys has been attributed to the absence of magnetic and crystallographic anisotropy.

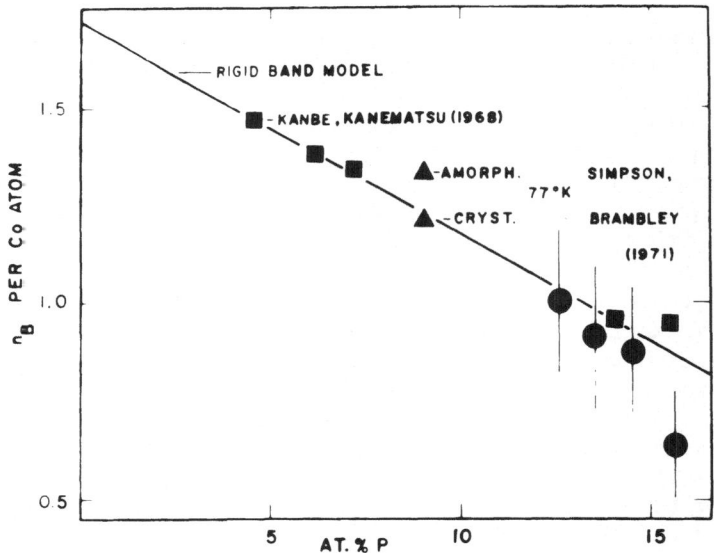

Figure 4:   Magnetic moment per cobalt atom in amorphous Co-P
            alloys from data extrapolated to 0°K (circles),
            values from Simpson and Brambley[5] (triangles),
            values from Kanbe and Kanematsu[18] (squares), and
            rigid band model (solid line).

Table II.   Saturation magnetizations for amorphous Co-P
            alloys, rigid band model values, and saturating
            fields before and after heating samples to 525°K.

Sample	Comp. (at.%P)	$n_B(0°K)$	Rigid Band Model n_B	Initial $H_K(300°K)$	Final $H_K(300°K)$
A	15.6	.64±.13	.87	70 Oe	25 Oe
B	14.5	.88±.18	.93	70	25
C	13.5	.92±.18	.98	60	20
D	12.6	1.01±.20	1.03	90	10

The linear closed shape of the magnetization curve is
quite similar to that found in uniaxial crystals with an
easy axis perpendicular to the magnetic field.[19] However,
it seems unlikely that a unidirectional anisotropy has
developed in our alloys in view of their amorphous structure
and also the temperature dependence of the saturating field
H_K. The radial direction in the foil plane cannot be
uniquely characterized to develop a magnetic anisotropy.
On the other hand, the foil normal is the growth direction
and is the most likely axis for strains, composition gradients or microstructure to induce magnetic anisotropy.
Such terms must then be just large enough to over balance
the considerable shape anisotropy in this direction in
order to give the relatively small observed values of H_K.
Moreover, this balance must be maintained at all temperatures even though the magnetization and hence the shape
anisotropy changes significantly. We would conclude that
the magnetization curves are representative of a random
state of strains, microstructure, etc. in the samples
rather than any specific uniaxial property. This random
structure responds to thermal annealing as observed in both
de Lau's and our own samples. It is possible that with
proper treatment these amorphous alloys might have considerable technological potential.

We hope that the structural simplicity of these materials
and their availability over a range of compositions as metastable amorphous bulk alloys will facilitate further experimental and theoretical progress in understanding ferromagnetism in structurally and compositionally disordered
solids.

REFERENCES

1. A. Brenner, D.E. Couch and E.K. Williams, J. Res. Natl.
    Bur. Std. 44, 109 (1950).
2. R.D. Fisher and D.E. Koopman, J. Electrochem. Soc. 111,
    263 (1964).
3. V.V. Bondar, K.M. Gorbunova and Y.M. Polukarov, Phys.
    Met. Metallog 26, 193 (1968).
4. J.G.M. de Lau, J. Appl. Phys. 41, 5355 (1970).
5. A.W. Simpson and D.R. Brambley, Phys. Stat. Sol. (b)
    43, 291 (1971).

6. B.G. Bagley and D. Turnbull, Bull. Am. Phys. Soc. 10, 1101 (1965).

7. M. Hansen, Constitution of Binary Alloys, Second Edition (McGraw-Hill, New York, 1958), p. 488.

8. R.P. Elliot, Constitution of Binary Alloys, First Supplement (McGraw-Hill, New York, 1965), p. 870.

9. W.B. Pearson, Handbook of Lattice Spacings and Structures of Metals, Vol. I (Pergamon, New York, 1967), p. 518.

10. G.S. Cargill III, J. Appl. Phys. 41, 12 (1970).

11. J.D. Bernal, Proc. Roy. Soc. (London) A280, 299 (1964).

12. J.L. Finney, Proc. Roy. Soc. (London) A319, 479 (1970).

13. G.S. Cargill III, J. Appl. Phys. 41, 2249 (1970).

14. D.E. Polk, Scripta Met. 4, 117 (1970).

15. C.H. Bennett, D.E. Polk and D. Turnbull, Acta. Met. 19, 1295 (1971).

16. D.E. Polk, Acta Met. 20, 485 (1972).

17. S. Chikazumi, Physics of Magnetism (John Wiley & Sons, New York, 1964), p. 32.

18. T. Kanbe and K. Kanematsu, J. Phys. Soc. Japan 24, 1396 (1968).

19. J.B. Goodenough and D.O. Smith, in Magnetic Properties of Metals and Alloys, (American Society for Metals, Cleveland, 1959), p. 112.

LOW TEMPERATURE HEAT CAPACITY OF DISORDERED ZrZn$_2$[†]

R. Viswanathan, L. Kammerdiner and H. L. Luo

Dept. of Applied Physics & Information Science
University of California, San Diego
La Jolla, California   92037

It was pointed out recently[1] that the Curie temperature of the weak ferromagnet ZrZn$_2$ is quite sensitive to lattice order and that it decreases with increased disorder. A well-disordered (or amorphous) ZrZn$_2$ can be prepared effectively by quick condensation; we report here our results of low temperature heat capacity on such samples.

The samples were prepared by RF co-sputtering of the elements in appropriate proportions[2-4] onto molybdenum substrates. The temperature of the substrates was kept below 100°C. The rate of sputtering was estimated at ∼ 1 μ/hr. A sample large enough for heat capacity measurements could usually be collected in 10 hours of continuous sputtering. The sputtered material was peeled off the substrate and pressed into a pellet in a tungsten carbide die. Part of this pellet was used directly for heat capacity measurements, while the rest was annealed at ∼ 850°C in a sealed fused quartz tube under He-atmosphere. Droplets of Zn were added to these tubes to compensate for Zn losses in the annealing.

The actual compositions of the sputtered material were measured with an electron microprobe, which also showed the uniformity of composition throughout the sample. The crystal structures were investigated with both a Debye-Scherrer

† This work was supported by the U.S. Atomic Energy Commission under contract AEC AT(04-3)-34.

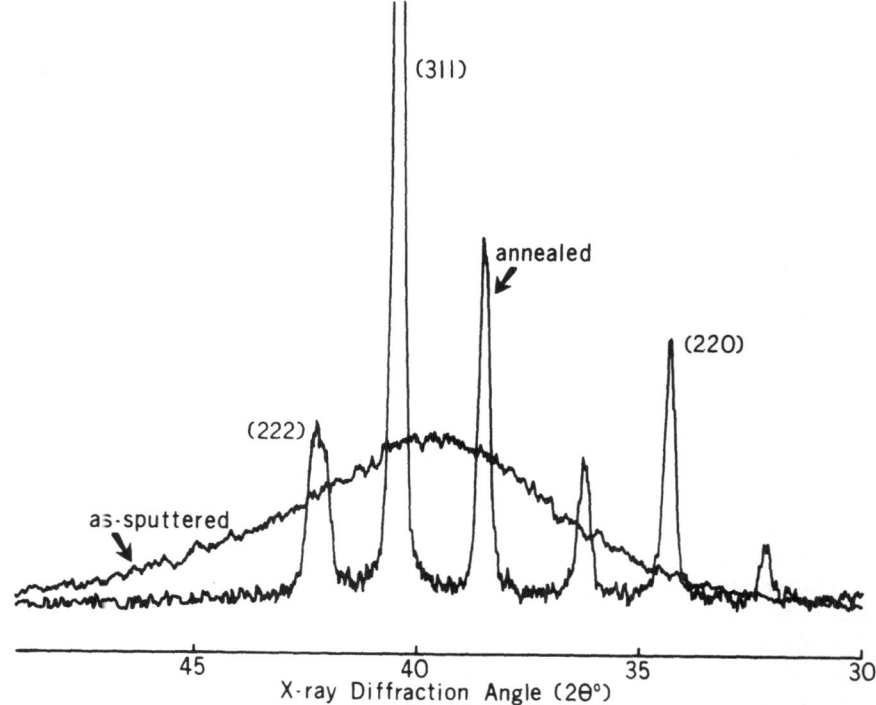

Figure 1. X-ray diffraction traces of as-sputtered (amorphous) and annealed (crystalline) $ZrZn_2$. The indexed peaks correspond to those of $ZrZn_2$ and non-indexed ones to the second phase.

camera and a diffractometer. The traces from the latter, before and after anneal, are shown in Figure 1, indicating that appreciable disorder existed in samples in the as-sputtered state, whereas the annealed samples were well crystallized.

The low temperature heat capacity between 1.8 and 25°K was measured by a.c. calorimetry technique using a laser beam as heat source[5] The results are shown in Figure 2 in the conventional C/T vs. T^2 plot. The pertinent parameters are given in Table 1.

The amorphous samples show very similar rapid increase in their heat capacity above 5°K compared to the crystalline material[1](#43). They also have a much lower electronic heat capacity coefficient γ and Debye temperature θ_D. It is known

Table 1

Electronic Heat Capacity Coefficient γ
and Debye Temperature θ_D of ZrZn$_2$ Samples

Sample		γ in mJ/gm-mole°K^2	β in mJ/gm-mole°K^4	θ_D in °K
Amorphous	#44	6.0	0.750	198
Amorphous	#47	15.0	0.630	210
Amorphous	#49	7.0	1.000	180
Crystalline	#43	40.0	0.190	313

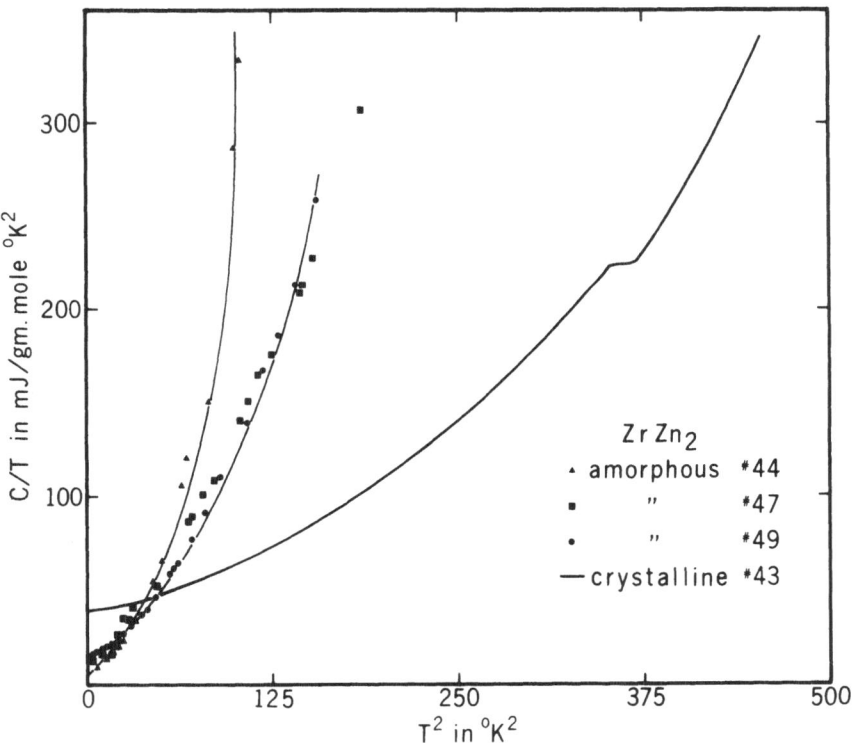

Figure 2. Heat capacity of as-sputtered and annealed ZrZn$_2$.

that $ZrZn_2$ is a ferromagnet by virtue of its large density of states peak at its Fermi level[6] reflected in its large γ-values[1] Appreciable disorder in its lattice will undoubtedly affect its band structure and hence destroy its ferromagnetism. In fact, our amorphous samples showed low γ-values and were all non-magnetic down to 1.5°K. However, ferromagnetism with Curie temperatures $\sim$ 25°K together with the $ZrZn_2$ structure were always detected[7] in the annealed samples, even though a second phase corresponding to $ZrZn_3$ often coexisted.

The variations in the γ-values (Table 1) indicate that the disordered samples are not exactly identical in the degree of disorder. The large decrease in θ_D, resulting in huge lattice contributions to the heat capacity for the amorphous samples, is quite similar to (though much larger than) that found in disordered NbZr alloys[7] and glassy PdSiCu[8] This could be due to either characteristic softening of transverse phonons[8] or to low frequency vibrations of loosely bound clusters of atoms.

ACKNOWLEDGMENT

The authors wish to thank Dr. J. C. Clinton for magnetic measurements for T_c on annealed samples.

REFERENCES

[1] R. Viswanathan, H.L. Luo and D.O. Massetti, AIP Conf. Proc. on Magnetism & Magnetic Materials, #5, 1290 (1971).

[2] D.O. Massetti, M.S. thesis, University of California at San Diego (1971).

[3] L. Kammerdiner, M.S. thesis, University of California at San Diego (1971).

[4] L. Kammerdiner and H.L. Luo, J.Appl.Phys. (October 1972).

[5] R. Viswanathan and H.L. Luo, Solid State Comm. 9,1733 (1971).

[6] B. Veal, F.M. Mueller, R. Afshar and J. Shaffer, AIP Conf. Proc.on Magnetism & Magnetic Materials, #5, 1285 (1971).

[7] J.R. Clinton, private communication.

[8] R. Viswanathan and L. Kammerdiner, Phys.Letters (submitted for publication).

[9] B. Golding, B.G. Bagley and F.S.L. Hsu, Phys.Rev.Lett. 29, 68 (1972).

FERROMAGNETISM OF AMORPHOUS IRON ALLOYS

T. Mizoguchi, K. Yamauchi and H. Miyajima

Department of Physics, Gakushuin University

1-5-1 Mejiro, Toshimaku, Tokyo, Japan

INTRODUCTION

It is well known that ferromagnetism appears in 3d metals (Fe, Co and Ni) and alloys whose average outer electron concentration, N, ranges from about 6.5(Fe-Cr) to 10.6 (Ni-Cu). The saturation magnetization of these metals and alloys is expressed by the so-called Slater Pauling curve. The crystal structure of these alloys in ordinary phases is generally body centered cubic or face centered cubic for N < 8.5 or N > 8.7 respectively. There is some dincontinuity between the saturation moment of bcc alloys and that of fcc alloys. The Curie temperature is much more sensitive to the crystal structure. There are interesting topics for Invar alloys in a region close to the phase boundary between fcc and bcc phase.[1] The magnetic properties of metals and alloys are necessarily affected by their crystal structures.

It would be interesting to see how 3d electrons behave in a random lattice, that is, in an amorphous phase. In this case we can investigate the dependence of magnetic properties of alloys on electron concentration alone, disregarding the effect of the crystal structure. Some information about fundamental properties of 3d electrons would be obtained from a study of magnetism in amorphous alloys.

The amorphous phase can be obtained by evaporation onto a cooled substrate[2] or by rapid quenching from the liquid state.[3] The amorphous metal films obtained by the former

technique generally crystallize during heating to the room
temperature. By the latter technique it has been scarcely
possible so far to get amorphous phase of pure metals or alloys containing no glass former atoms such as B, C, Si or
P. In this experimental study quasibinary amorphous alloys
having the composition of $(Fe_{1-x}M_x)_{0.8}B_{0.1}P_{0.1}$, where M =

Ni, Co, Mn, Cr and V, were prepared by rapid quenching technique, and fundamental magnetic properties of these amorphous alloys were measured.

EXPERIMENTAL PROCEDURE

Mother alloys were prepared beforehand by powder metal-
lurgy technique in order to prevent volatilization of phosphorus. Powdered raw materials of intended proportion were
mixed well and agglomerated in a steel die under the pressure of 2 ~ 7 ton/cm^2. These specimens were slowly heated
up to 400°C, kept at that temperature for a day and then at
800°C for several hours in evacuated silica tubes, during
which reactive sintering takes place. These sintered alloys
were melted once at 1250°C in the silica tubes and quenched
to the room temperature. A small piece of these mother alloys was melted on a water cooled copper hearth in a plasma
jet furnace, and then it was clapped suddenly by a flat copper plate to be quenched rapidly.[4] The thickness of a foil
produced by this method is 60 ~ 100 μ. X-ray diffraction
of these quenched alloy specimens shows a broad halo around
$sin\theta/\lambda = 0.24A^{-1}$, indicating an amorphous structure. The
static measurements of magnetic properties of these amorphous alloys, that is, the field and temperature dependence
of the magnetization were carried out using a magnetic balance. As the amorphous phase is metastable, these alloys
crystallize at about 400°C during measurement with increasing temperatures. The Curie temperature higher than the
crystallizing temperature was determined by extrapolation
of the magnetization of amorphous phase to high temperatures.

RESULTS AND DISCUSSION

The behavior of 3d electrons in these amorphous alloys
is considered to be affected by two factors. The first is

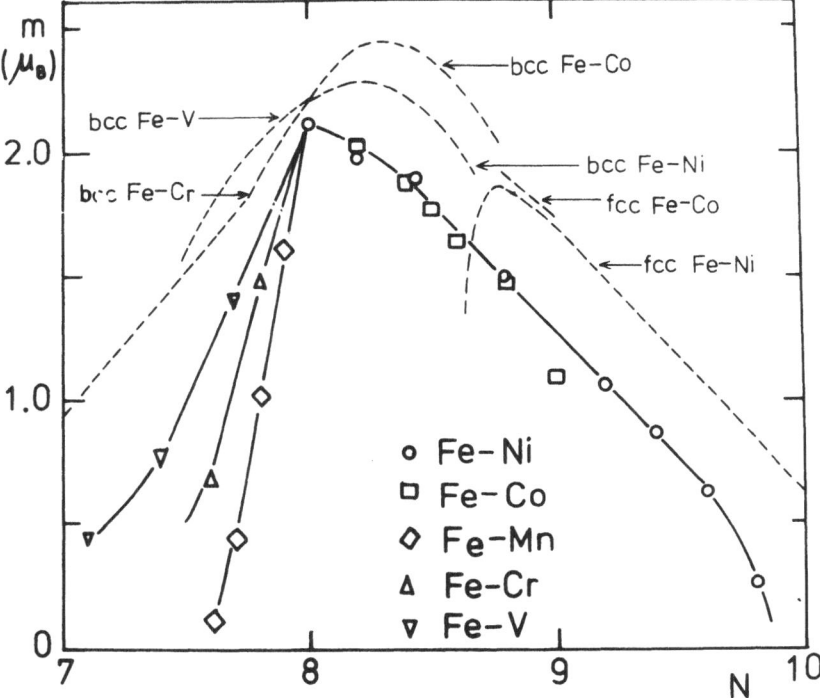

Fig. 1   The average saturation moment, m, per metallic atom,
         (Fe$_{1-x}$M$_x$), plotted as a function of the average
         outer electron concentration, N, of a metallic atom

the randomness of the structure and the second is effects
by B and P atoms. In order to fix the latter effect, the
concentration of B and P is kept constant at 10% respectively and a part of Fe atoms is replaced by other transition
metals.

    The experimental results are summarized in Fig. 1 and
Fig. 2, where the average saturation moment, m, per metallic atom, (Fe$_{1-x}$M$_x$), or the Curie temperature, T$_c$, is plotted as a function of the average outer electron concentration, N, of a metallic atom.

    The remarkable feature of these results is that for
Fe-Ni and Fe-Co quasibinary amorphous alloys the saturation
moment decreases linearly with increasing N, tracing a
universal line parallel to the so-called Slater-Pauling

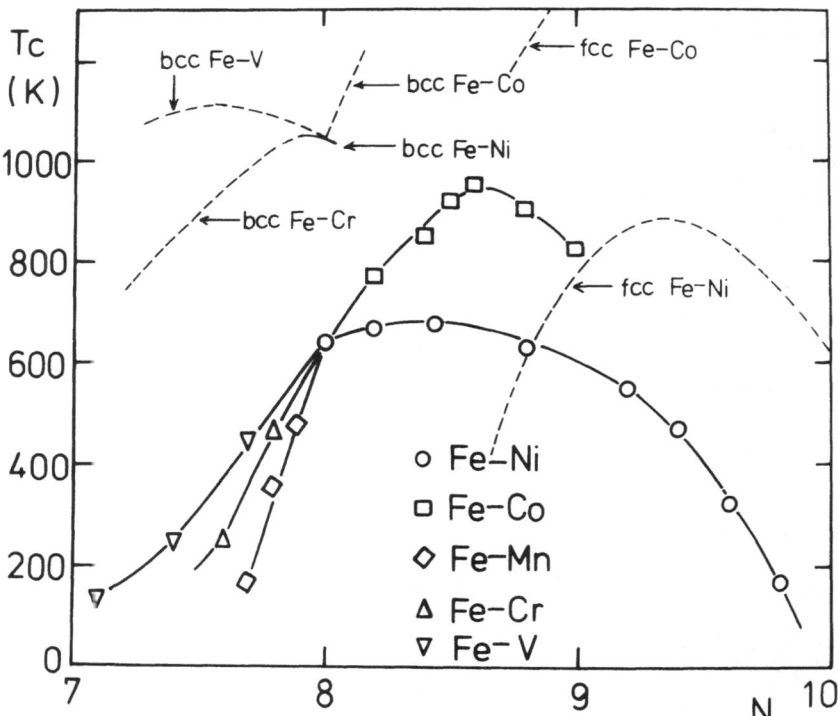

Fig. 2   The Curie temperature of amorphous Fe alloys plotted
         as a function of the average outer electron concen-
         tration of a metallic atom, $(Fe_{1-x}M_x)$.

curve for ordinary crystalline alloys. This experimental
result does not conflict with the simple rigid band model.
The derivative of this line, dm/dN, is equal to -1, which
means that up-spin band is already full, and increasing
electrons enter into the down-spin band. The above simple
rigid band model concerns only   the energy spectrum of
electrons and says nothing about the spacial distribution
of the magnetic moment. Taking account of the screening
effect, we have the complemental spacial aspect.[5] Roughly
speaking, Fe or Co seems to have 2 or 1 μ_B per atom respectively and Ni has no magnetic moment in these amorphous alloys. The composition dependence of magnetization can be
also interpreted with this picture. It should be emphasized
that these two pictures, that is, the rigid band model and
the screening consideration are not contradictory, but com-

plementary.   They are the energetic and the spacial aspect
of the same first order approximation.

The transition metals such as Mn, Cr and V which have
less 3d electrons than Fe reduce greatly the net moment of
the amorphous alloys when they replace a part of Fe atoms.
They seem to have antiparallel moments to the host matrix
of Fe moments.   If we extrapolate the reduction of the magnetization with initial slope to X = 1, we find that Mn, Cr
or V has roughly -3, -4 or -5 μ_B per atom, respectively. It
is interesting that the atomic moment of transition elements
increases linearly from 0 for Ni to 5 μ_B for V with decreasing atomic number in these amorphous alloys.   This would be
interpreted if a certain extent of intra-atomic exchange interaction is assumed.

The Curie temperature of ordinary crystalline alloys
depends not only on their composition but also on their
crystal structure.   That of amorphous alloys shows smooth
dependence only on their composition.   For Fe-Ni and Fe-Co
amorphous alloys it shows maximum at a certain intermediate
composition, which may be interpreted by a simple pair interaction model, assuming that Fe-Ni or Fe-Co pair interaction is greater than that between same species of atoms.
On the other hand, Mn, Cr and V reduce the Curie temperature
of amorphous Fe alloys, being consistent with the assumption that they prefer negative coupling to the Fe moments.

The electronic structure in random lattice must be very
complicated and difficult to solve.   However, as long as we are
concerned with a problem which can be treated with an electron gas model as the first approximation, the situation
may be rather favorable.   The state density of electrons in a
random lattice should have no fine structure resulting from
the periodicity of a crystal lattice, which would permit a
crude but simple interpretation for the magnetism of amorphous alloys.

REFERENCES

1.   S. Chikazumi, T. Mizoguchi, N. Yamaguchi and P. Beck-
      with: J. Appl. Phys. 39 (1968) 939. T. Mizoguchi: J.
      Phys. Soc. Japan 25 (1968) 904.
2.   S. Fujime: Japan J. Appl. Phys. 6 (1967) 305. S.
      Mader and A.S. Norwick: Appl. Phys. Letter 7 (1965) 57.

3.   P. Pietrokowsky: Rev. Sci. Inst. 34 (1963) 445. P.
     Duwez: Prog. Solid State Chem. 3 (1966) 377.
4.   K. Yamauchi and Y. Nakagawa: Japan J. Appl. Phys. 10
     (1971) 1730.
5.   J. Friedel: Nuovo Cimento Suppl. 2 (1958) 287.

SEMI-CLASSICAL THEORY OF SPIN WAVE EXCITATIONS IN

AMORPHOUS FERROMAGNETS

R.G. Henderson and A.M. de Graaf[*]

Department of Physics, Wayne State University

Detroit, Michigan 48202

ABSTRACT

The Landau-Lifshitz theory of ferromagnetic spin waves is generalized to amorphous ferromagnetic materials. It is shown that this theory predicts the existence of both extended and localized spin-wave states. The localized states are due to the randomness of the ferromagnet. It is postulated that the localized spin waves can be resonantly excited by means of a time-varying electromagnetic field. The anomalously large widths of the resonance lines observed in recent spin wave resonance experiments performed on amorphous thin nickel films can be explained in terms of the localized spin-wave states.

I. INTRODUCTION

The importance of spin waves in the study of the properties of crystalline magnetic materials at low temperatures has been well established.[1] In the last 40 years several microscopic quantum mechanical theories of spin waves have been constructed, demonstrating over and over again the existence of this type of excitation in crystalline magnetic materials. Most of these theories make use of the Heisenberg hamiltonian and are primarily applicable to magnetic insulators, while others[2] are designed to describe spin wave excitations in itinerant electron systems. However, one of the first theories of spin waves was macroscopic and was introduced by Landau and Lifshitz[3] and was later elaborated upon by Herring and Kittel.[4] The virtue of this theory, besides its simplicity, is its wide applicability to both insulators and metals. In particular the theory of spin-wave resonance is almost entirely based on it.[5] The natural drawback of a macroscopic theory is the appearance of empirical constants (in this case the exchange constants) which have to be determined by comparing the predictions of the theory with experiment.

The purpose of the present paper is to describe an attempt to extend the Landau-Lifshitz theory to amorphous magnetic materials. We will restrict the discussion exclusively to ferromagnets. We shall begin (Section II) by writing down the increase of the energy density associated with a local deviation from perfect ferromagnetic order. It proves convenient to divide the total magnetization density in two parts, $\vec{M}(\vec{r})$ and $\vec{m}(\vec{r})$. The part $\vec{M}(\vec{r})$ is due to position dependent deviations from the ferromagnetic ground state, while $\vec{m}(\vec{r})$ results from the randomness of the amorphous ferromagnet. The average of $\vec{m}(\vec{r})$ over the volume of the sample is zero, and the magnitude of $\vec{M}(\vec{r})$ is equal to M_0, the average saturation magnetization of the ferromagnetic sample. The expression for the energy density is then used to derive an equation of motion for $\vec{M}(\vec{r})$, which turns out to contain a term involving the first derivative of $\vec{M}(\vec{r})$. This term would vanish in a crystalline ferromagnet. We shall show that this equation of motion predicts the existence of both extended and local spin-wave states. The local spin-wave states are a consequence of the first-derivative term in the equation of motion.

In Section III we postulate that the local states can participate in the spin-wave resonance in amorphous ferromagnetic thin films and we estimate their contribution to the resonance line width for the case of amorphous nickel. Reasonable agreement with the results of recent ferromagnetic resonance experiments in amorphous nickel films has been obtained.[6]

II. EQUATION OF MOTION

We shall characterize the groundstate of an amorphous ferromagnet by a magnetization density $\vec{M}_0(\vec{r})$, which is a randomly but smoothly varying function of the position vector

$\vec{r}$. The vector $\vec{M}_0(\vec{r})$ points everywhere in the same direction, which is taken as the z-direction. The volume average of the magnitude of $\vec{M}_0(\vec{r})$ will be denoted by M_0. As soon as the ferromagnet is excited slightly above the groundstate, the magnetization density acquires x- and y-directed components. The magnetization density in the excited state consists of a slowly varying part $\vec{M}(\vec{r})$ and a randomly varying part $\vec{m}(\vec{r})$. The contribution $\vec{m}(\vec{r})$ is entirely due to the randomness of the ferromagnet, while $\vec{M}(\vec{r})$ results from position dependent deviations from perfect ferromagnetic order. The average of $\vec{m}(\vec{r})$ over the volume of the sample is zero, and the magnitude of $\vec{M}(\vec{r})$ is constant and equal to M_0. Here, we have assumed (as is usually done) that the magnitude of the magnetization density at every point remains constant when exciting the ferromagnet. The increase in energy density, $\Delta E(\vec{r})$, associated with such an excitation can now be expressed in terms of $\vec{M}(\vec{r})$ alone, and is given by[7]

$$\Delta E(\vec{r}) = \frac{1}{2} C_{ij}(\vec{r}) \frac{\partial M_\alpha}{\partial x_i} \frac{\partial M_\alpha}{\partial x_j} \quad . \tag{1}$$

Equation (1) is compatible with an isotropic exchange interaction energy, and also satisfies the requirement that $\Delta E(\vec{r})$ does not change sign when $M_x(\vec{r})$, $M_y(\vec{r})$, and $M_z(\vec{r})$ change sign. The coefficients $C_{ij}(\vec{r})$ are related to the exchange coupling constants. Since these coefficients at a given point $\vec{r}$ depend in general on the structure of the immediate environment of that point, they will also be randomly but smoothly varying functions of position. In order to make this argument more transparent, we may for example think of the amorphous ferromagnet as consisting of a collection of randomly distributed magnetic atoms. Now, making use of an Heisenberg model we can obtain microscopic expressions for the coefficients C_{ij} and we will find that at a given atom they depend on the relative positions of the neighboring atoms. Since the relative positions of the neighboring atoms vary randomly from atom to atom for the amorphous material, the coefficients C_{ij} will be random functions of position. The main objective of the present paper is to explore the consequences of this spatial dependence of the coefficients C_{ij}.

It is worth emphasizing that the quantity $\vec{m}(\vec{r})$ does not appear in Eq. (1). The reason for this is that $\vec{m}(\vec{r})$ is independent of whether or not $M_0(\vec{r})$ is rotated through the same angle at every $\vec{r}$. It is a property of the isotropic exchange interaction energy that $\Delta E(\vec{r})$ becomes different from zero only when $M_0(\vec{r})$ is rotated through different angles at different positions. If $M_0(\vec{r})$ is rotated through the same angle at every $\vec{r}$, so that M_x, M_y, and M_z would be independent of position, then $\Delta E(\vec{r})$ would be zero, despite the fact that $\vec{m}(\vec{r})$ is position dependent.

In principle $\Delta E(\vec{r})$ could also contain a term of the form

$$\frac{1}{2} C_i(\vec{r}) \; M_\alpha \; \frac{\partial M_\alpha}{\partial x_i} \; , \tag{2}$$

which is also compatible with an isotropic exchange interaction energy, and is also invariant with respect to a change of sign of $M_x(\vec{r})$, $M_y(\vec{r})$, and $M_z(\vec{r})$. However, it can easily be shown that this term may be disregarded. To this end we simply integrate Eq. (2) over the volume of the sample, obtaining

$$-\frac{1}{4} \int (\partial C_i / \partial x_i) \; [M_x^2(\vec{r}) + M_y^2(\vec{r}) + M_z^2(\vec{r})] d^3 r$$

$$+ \text{ Surface terms,} \tag{3}$$

which is equal to

$$-\frac{1}{4} M_0^2 \int (\partial C_i / \partial x_i) d^3 r + \text{ Surface terms.} \tag{4}$$

Since we are only interested in bulk effects, the surface terms will be neglected. Furthermore, the first term does not depend on the derivatives of $M_x(\vec{r})$, $M_y(\vec{r})$, and $M_z(\vec{r})$, and can therefore also be discarded.

We now wish to find the equation of motion for $\vec{M}(\vec{r})$. Integrating $\Delta E(\vec{r})$, Eq. (1), over all volume, we find

$$\int \Delta E(\vec{r}) d^3 r = - \frac{1}{2} \int \left[\frac{\partial C_{ij}}{\partial x_i} \frac{\partial M_\alpha}{\partial x_j} + C_{ij} \frac{\partial^2 M_\alpha}{\partial x_i \partial x_j} \right] M_\alpha \, d^3 r. \quad (5)$$

This equation can be interpreted to mean that an effective field $\vec{H}_{eff.}(\vec{r})$ is acting on $\vec{M}(\vec{r})$. Inspection of Eq. (5) reveals that $\vec{H}_{eff.}(\vec{r})$ is given by

$$\vec{H}_{eff.}(\vec{r}) = \vec{B}(\vec{r}) \cdot \vec{\nabla} \, \vec{M}(\vec{r}) + \overleftrightarrow{C}(\vec{r}) \cdot \vec{\nabla} \, \vec{\nabla} \, \vec{M}(\vec{r}) \quad , \quad (6)$$

where the components of the vector $\vec{B}(\vec{r})$ are defined as

$$B_j(\vec{r}) = (\partial C_{ij} / \partial x_i) \quad . \quad (7)$$

The components of the tensor $\overleftrightarrow{C}(\vec{r})$ are given by C_{ij}. This effective field will give rise to a torque acting on $\vec{M}(\vec{r})$, whose equation of motion is therefore

$$d\vec{M}(\vec{r})/dt = -\gamma \vec{M}(\vec{r}) \times \vec{H}_{eff.}(\vec{r}) \quad . \quad (8)$$

Equation (8) is very similar to the original Landau–Lifshitz equation.[3,4] However, the two equations are not identical. First of all, Eq. (8) is the equation of motion for $\vec{M}(\vec{r})$, which is not the total magnetization density but just its slowly varying part. The random contribution to the total magnetization density, $\vec{m}(\vec{r})$, is missing from Eq. (8). Further, the effective field $\vec{H}_{eff.}(\vec{r})$ contains the first derivatives of $\vec{H}(\vec{r})$, a feature not occurring in the Landau–Lifshitz equation. It should be pointed out that these first derivatives are due to the fact that the coefficients C_{ij} are position dependent. We recover of course the Landau–Lifshitz equation if we apply our formalism to a crystalline

ferromagnet. In that case both $\vec{B}(\vec{r})$ and $\vec{m}(\vec{r})$ would be zero,
so that $\overleftrightarrow{M}(\vec{r})$ becomes the true magnetization density.

Our next task is to explore the novel properties of
Eq. (8). Of course, Eq. (8) cannot be solved directly due
to the random (and unknown) spatial dependence of $\vec{B}(\vec{r})$ and
$\overleftrightarrow{C}(\vec{r})$. However, we can obtain information about the solutions
of Eq. (8) by dividing the volume V of the sample in small
volumes Ω. We require Ω to be much smaller than V, but
large enough to contain several atoms comprising the amorphous ferromagnet.[8] We then take the average of Eq. (8)
over Ω. It should be recalled at this point that $\vec{M}(\vec{r})$ and
its derivatives are slowly varying functions, so that these
quantities do not vary appreciably within one small volume
Ω. On the other hand, $\vec{B}(\vec{r})$ and $\overleftrightarrow{C}(\vec{r})$ vary randomly and
rapidly within Ω. Therefore, after averaging Eq. (8) over
Ω, we can replace $\vec{B}(\vec{r})$ and $\overleftrightarrow{C}(\vec{r})$ by their averages over Ω.
These averages will be indicated by the symbol < >. If Ω
is large enough (say larger than a certain volume Ω_0, where
Ω_0 contains a sufficiently large number of atoms), then
$<\vec{B}(\vec{r})>$ and the off-diagonal elements of $<\overleftrightarrow{C}(\vec{r})>$ will be
negligibly small, and Eq. (8) reduces to

$$d\vec{M}(\vec{r})/dt = -\gamma\vec{M}(\vec{r}) \times [<\vec{C}(\vec{r})> \cdot \nabla^2\vec{M}(\vec{r})], \qquad (9)$$

where $<\vec{C}(\vec{r})>$ represents the diagonal elements of $<\overleftrightarrow{C}(\vec{r})>$.
This equation has the same form as the Landau-Lifshitz
equation, and can be solved with standard methods. The
solutions are extended spin wave states whose dispersion
relation, for propagation in the z-direction, is given by

$$\omega = \gamma M_o C_e k^2 , \qquad (10)$$

where the constant $C_e = <C_z(\vec{r})>$.

An entirely new situation arises when Ω is chosen
smaller than Ω_0, so that $<\vec{B}(\vec{r})>$ can not be neglected. It
should be clear from the above discussion that we have de-
fined Ω_0 such that $<B(\vec{r})>$ can be neglected when $\Omega > \Omega_0$, and
must be taken into account when $\Omega < \Omega_0$. The equation of

motion for $\vec{M}(\vec{r})$ is now given by

$$d\vec{M}(\vec{r})/dt = -\gamma\vec{M}(\vec{r}) \times [<\vec{B}(\vec{r})>\cdot\vec{\nabla}\,\vec{M}(\vec{r})$$

$$+ <\overleftrightarrow{C}(\vec{r})>\cdot\vec{\nabla}\,\vec{\nabla}\,\vec{M}(\vec{r})]. \tag{11}$$

It can easily be shown that the solutions of Eq. (11), after linearizing, are of the form

$$M_+(\vec{r},t) = [M_x(\vec{r},t)+iM_y(\vec{r},t)] \propto e^{-|\vec{\kappa}\cdot\vec{r}|}e^{-i(\omega t-\vec{k}\cdot\vec{r})}, \tag{12}$$

where $\kappa_z = B/2C_\ell$, with $B = <B_z(\vec{r})>$. The frequency ω and the wavevector $\vec{k}$ are now related by

$$\omega = \gamma M_o[C_\ell k^2 + (B^2/2C_\ell)] . \tag{13}$$

It should be noted that a different constant, namely C_ℓ appears in Eq. (13). This is so because the averaging volume used to calculate C_ℓ is smaller than Ω_o, whereas the averaging volume for C_e appearing in Eq. (10) is greater than Ω_o. The two constants C_e and C_ℓ are not necessarily equal. The spin wave states described by Eq. (12) are localized due to the appearance of the factor $\exp(-|\vec{\kappa}\cdot\vec{r}|)$. It can be seen very clearly that these localized states are a consequence of the randomness of the amorphous ferromagnet. In the case of a crystalline ferromagnet B would be zero and the localized states would become extended. Equation (13) would reduce to Eq. (10).

III. SPIN WAVE RESONANCE

Recently Ajiro, Tamura, and Endo reported the results of a ferromagnetic resonance experiment performed on amorphous thin films of nickel.[6] This experiment showed that

it is possible to excite long wave length spin waves in
amorphous ferromagnets. It was found that the widths of
the resonance lines were broader than is usually obtained
in thin film resonance experiments. In permalloy films,
for instance, the line width is on the order of one to two
hundred gauss,[9] while the results of Ref. 6 show a line
width of four to five hundred gauss. The authors of Ref. 6
attributed the extra width to the scattering of the spin
waves caused by the random distribution of the atoms in
the amorphous ferromagnet. However, it can be shown that
this is likely not the case. Ziman has demonstrated that
the mean free path of a wave propagating trough a random
medium is given by[10]

$$
\Lambda = \frac{1}{4\pi^{7/2}} \frac{v^2 \lambda}{(\delta v)^2 L} \, , \quad \lambda >> L. \tag{14}
$$

Here, v is the average group velocity of the wave, and λ is
the wavelength. In deriving Eq. (14), Ziman assumed that
the group velocity varies randomly from point to point in
the material. The root mean square deviation is denoted
by δv. Further, L is a correlation length, i.e., the
distance over which the group velocity itself does not
change appreciably. In order to obtain a lower bound for Λ
we assume that $\lambda = d$, where d is the thickness of the film.
Since the films used in the experiment were approximately
1000 Å thick, and it seems probable that L is much less
than that, we set L = d. We assume further that $(\delta v/v) =$
0.1, which is almost certainly too large. With these
values we find $\Lambda \gtrsim d$, which implies that there is hardly any
scattering of the long wave length spin waves in these films
due to the randomness alone.

The anomalous width of the resonance lines can be
accounted for if it is assumed that the localized spin waves
can be excited resonantly by means of an electromagnetic
field. In that case they should also contribute non-
negligibly, to the intensity of the resonance lines. In
the presence of a uniform magnetic field H, the frequencies
of the extended and localized spin waves are given by

$$\omega = \gamma M_o C_e k^2 + \gamma H , \tag{15}$$

and

$$\omega = \gamma M_o [C_\ell k^2 + B^2/2C_\ell] + \gamma H , \tag{16}$$

respectively. Since the wavelengths of the waves that can be excited in a film of thickness d, when the magnetic field is perpindicular to the film surface are equal to $\lambda = (2d/n)$, where n = 1, 3, 5,...., Eqs. (15) and (16) give rise to the following resonance fields[11]

$$H_{res.} = \frac{\omega}{\gamma} - \frac{\pi^2 M_o n^2 C_e}{d^2} , \tag{17}$$

and

$$H_{res.} = \frac{\omega}{\gamma} - \frac{\pi^2 M_o n^2 C_\ell}{d^2} - \frac{M_o B^2}{2C_\ell} . \tag{18}$$

Let us concentrate for the moment on the resonance line for width n=3. It should now be realized that B and C_ℓ assume a range of values due to the fact that we are at liberty to vary the magnitude of Ω providing Ω remains smaller than Ω_0. If we change the volume Ω such that it contains on the average one atom more or less, then B and C_ℓ take on different values. We see therefore that Eq. (18) gives rise to a range, ΔH, of resonance fields. Hence, the resonance line will appear to be broadened by an amount ΔH as compared to the corresponding resonance line for the crystalline ferromagnet.

In order to estimate ΔH, we assume that B takes on values between $-B_0$ and B_0, where B_0 is greater than 0, such that the average of B is zero. Correspondingly, C_ℓ will

take on values between $C_e - C_o$ and $C_e + C_o$, the average
being C_e. Further we define a quantity which is a measure
of the randomness of the amorphous ferromagnet, by setting
$B_o = \epsilon C_e/a$, where a is the lattice parameter of the crystalline ferromagnet (a = 3.5 × 10^{-8} cm for nickel). We
find then that ΔH for the n=3 resonance is given by

$$\Delta H = M_o C_e \left[\frac{9\pi^2}{d^2} \left(1 - \frac{C_\ell}{C_e}\right) + \frac{\epsilon^2}{a^2} \right] \tag{19}$$

It is reasonable to assume that $[1 - (C_\ell/ C_e)]$ is of the
order of ϵ, and since the value of ϵ that we obtain below
is on the order of 10^{-3} while a is three orders of magnitude
smaller than d, the second term in Eq. (19) will dominate.
Hence to a good approximation

$$\Delta H = M_o \frac{\epsilon^2 C_e}{2a^2} \tag{20}$$

Using the values of Ref. 6, viz. $M_o = (2.7/4\pi)$ kG, $C_e =$
$(1.2\times10^{-7}/M_o)$erg/cm, and $\Delta H = 400$ gauss, we obtain $\epsilon = .003$.
Using a Heisenberg model one would find that $C_e \sim Ja^2$,
where J is the exchange coupling constant between nearest
neighbor atoms, while $B \sim \delta(Ja)$ which is the root mean
square deviation of the quantity Ja about zero. From the
defining equation for ϵ we see that $[\delta(Ja)/(Ja)]\sim\epsilon$. Thus
an average random fluctuation in the quantity Ja of only
.3% can account for the increased broadening of the resonance lines.

In the above discussion we have assumed that the usual
mechanisms of relaxation, such as magnon-magnon, magnon-
phonon, and magnon-conduction electron scattering are not
significantly altered. This is a question which requires
further investigation.

In conclusion we may say that the Landau-Lifshitz
theory of ferromagnetic spin waves, when extended to amorphous materials, predicts the existence of both extended and
localized spin-wave states. We have shown that, if these

localized spin waves can be resonantly excited, they contribute significantly to the width of the ferromagnetic resonance lines. We have obtained reasonable agreement with the recent experimantal data of Ajiro, et.al.

REFERENCES

*   Supported in part by the National Science Foundation

1. F. Keffer in ENCYCLOPEDIA OF PHYSICS, ed. S. Flügge, Springer-Verlag, Berlin, 1966.

2. For a comprehensive review see C. Herring in MAGNETISM, Vol. IV, eds. G. Rado and H. Suhl, Academic Press, New York, 1966.

3. L.D. Landau and E.F. Lifshitz, Physik. Z. Sowjetunion, 8, 153 (1935).

4. C. Herring and C. Kittel, Phys. Rev. 81, 869 (1951).

5. S.V. Vonsovskii ed., FERROMAGNETIC RESONANCE, Pergamon Press, New York, 1966.

6. Y. Ajiro, K. Tamura and H. Endo, Phys. Letters, 35A, 275 (1971).

7. L.D. Landau and E.F. Lifshitz, ELECTRODYNAMICS OF CONTINUOUS MEDIA, Addison Wesley, Reading, 1960, p. 159.

8. Of course we must also require that $\Omega^{1/3} << \lambda$, where λ is the wavelength of a spin wave.

9. M. Nisenoff and R.W. Terhune, J. Appl. Phys. 36, 734, (1965).

10. J.M. Ziman, ELECTRONS AND PHONONS, Oxford University Press, London, 1960, p. 254.

11. This resonance condition is also reasonable for the localized spin waves. It will turn out that κ^{-1} is on the order of 2500 Å, hence the magnitude of the localized spin waves will remain almost constant across the film.

12. R.G. Henderson and A.M. de Graaf, unpublished.

MAGNETIC PROPERTIES, ANOMALOUS ELECTRICAL RESISTIVITIES,
AND SPECIFIC HEATS OF ORDERED AND DISORDERED NiPt AND
NEARBY COMPOSITIONS

D. J. Gillespie, C. A. Mackliet, and
A. I. Schindler

Naval Research Laboratory

Washington, D. C. 20390

INTRODUCTION

Alloys which are subject to an order-disorder transformation are in some respects a link between amorphous
solids on the one hand and crystalline elemental solids on
the other hand. For example, alloys in the disordered
state and amorphous solids both have non-periodic structures (although of somewhat different natures), whereas
alloys in the ordered state and crystalline elemental solids
both have periodic structures (again of somewhat different
natures).

The Pt-Ni system is subject to order-disorder trans-
formations in two composition regions and has some particularly interesting additional properties as well. The elements Pt and Ni are miscible in all proportions, and the
alloys are f.c.c. except for ordered alloys in the vicinity
of Ni_3Pt and NiPt. The latter alloy[1] has an ordering temperature of $\sim645°C$ and a tetragonal (Ll_0) structure with
$c/a = 0.939$. This alloy and those having nearby compositions are of special interest for the following reasons:
(1) the critical composition for the occurrence of ferromagnetism in disordered Pt-Ni alloys lies within the region
in which ordering (to the tetragonal structure) is possible
and it is therefore possible to study the connection between
magnetic order and atomic order; (2) the Pt-Ni system is
generally considered to be exchange enhanced, with the like-
lihood that the properties will be substantially modified,

especially in the neighborhood of the critical composition; it should therefore be possible to study the interrelationship between exchange enhancement and the presence or lack of atomic order. We have accordingly carried out low temperature measurements of the magnetization, electrical resistivity, and specific heat of a series of Pt-Ni alloys in both the ordered and disordered states.

EXPERIMENTAL DETAILS

All alloys were prepared by induction melting high purity (99.999% pure) Pt powder and Ni rod in fused silica crucibles under an argon atmosphere. The magnetization samples consisted of cylinders (0.6" long and 0.155" dia.) or spheres (0.125" dia.). The resistivity samples consisted of wires (0.010" to 0.015" dia.). The specific heat samples consisted of irregular ingots with 2 flat faces (.27 to .34 moles). All alloys were given homogenizing anneals of 20 hrs. at 1200°C under an argon atmosphere, were slowly cooled to 800°C, and were quenched to room temperature in ~1/2 minute or less. The samples were given ordering anneals as required (see the table below). The experimental procedures have been described elsewhere[2,3,4].

FINAL ANNEALS FOR ORDERING

Magnetization Specimens			Resistivity Specimens			Specific Heat Specimens		
Ni (at%)	Temp (°C)	Time (days)	Ni (at%)	Temp (°C)	Time (days)	Ni (at%)	Temp (°C)	Time (days)
38	400	66	38	400	66	-	-	-
40	400	66	40	400	66	-	-	-
$43\frac{1}{2}$	400	57	$43\frac{1}{2}$	425	35	$43\frac{1}{2}$	400	35
$44\frac{1}{2}$	400	57	-	-	-	-	-	-
46	580	22	46	500	139	-	-	-
50	600	37	50	600	68	50	600	64

RESULTS

Magnetization

The magnetization of a series of both ordered and disordered Pt-Ni alloys has been measured as a function of magnetic field (up to 87 kG) at temperatures of 4.2 and 1.5 K. The results at 4.2 K for all disordered specimens with Ni concentrations of 50% or less and for the ordered 50% Ni specimen are shown in Fig. 1 on an Arrott plot. The results for the other ordered specimens differed little from those for the 50% Ni specimen and could not conveniently be plotted. The results for 1.5 K showed little or no difference from those at 4.2 K except for the disordered 43½ and 44½% Ni specimens, in which cases some relatively small increases in magnetization were observed as the temperature was lowered. The Arrott plot was used in the usual manner to determine the extrapolated zero-field susceptibilities, which are shown in Fig. 2. It is evident from Fig. 1 that the critical composition in these alloys is ~44% Ni. It is evident from Fig. 2 that the susceptibility of the disordered alloys diverges at the critical composition (as expected) while the susceptibility of the ordered alloys is markedly smaller and essentially independent of composition.

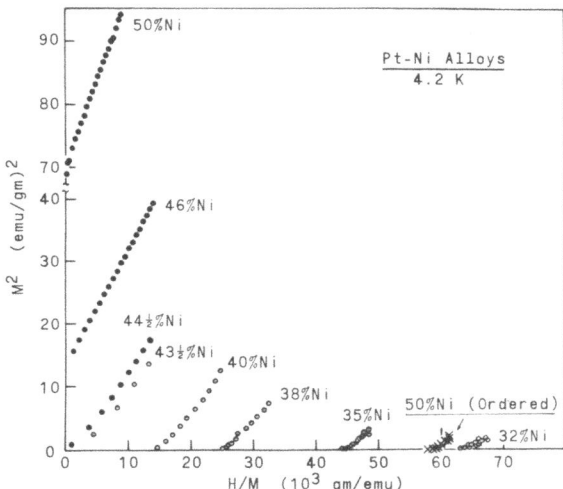

Fig. 1 - Arrott plot for Pt–Ni alloys at 4.2 K. (M is the magnetization; H is the magnetic field intensity.)

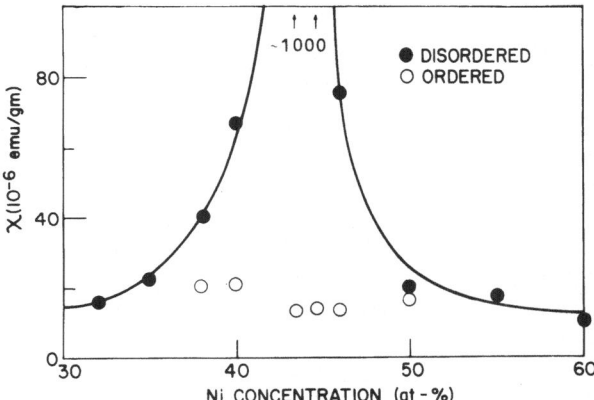

Fig. 2 - Extrapolated zero-field susceptibilities for
Pt-Ni alloys.

Resistivity

The electrical resistivity of a series of ordered and
disordered Pt-Ni alloys has been measured at temperatures
ranging usually from 1.3 to 4.2 K for the ordered alloys and
from 1.3 to 300 K for the disordered alloys. The temperature
dependence of the resistivity of all specimens (both ordered
and disordered) could be generally represented by

$$\rho = \rho_o + AT^2 - C \ln T,$$

where ρ_o is the residual resistivity, AT^2 is an electron-
electron interaction term, and $C \ln T$ is an anomalous term
whose presence yields a resistance minimum (since A and C
were both positive constants). For the **disordered** specimens
with Ni concentrations of $43\frac{1}{2}\%$ or more, the temperature at
the minimum ($\sim$2 to 3 K) was within the range of measurement
and a minimum in the resistivity was clearly observed, as
may readily be seen for example in Fig. 3 for the case of
the 50% Ni specimen. For the **disordered** specimens having
Ni concentrations from 40% down to as little as 20%, the
temperature at the minimum was below the range of measurement, so that while a ln T term was definitely found, no
resistance minimum was actually observed in these cases.
For all **ordered** specimens except those containing 38 and
$43\frac{1}{2}\%$ Ni, the ln T term was not statistically significant.
For the "ordered" 38% Ni specimen a significant ln T term

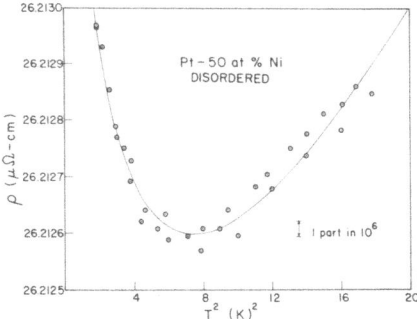

Fig. 3 - Electrical resistivity vs temperature-squared.

was observed, because a specimen with a composition so far from stoichiometry can develop only a very limited amount of order at the given annealing temperature. For the ordered 43½% Ni specimen, the observed ln T term was very small (~6% of the value for the disordered state), probably as a result of insufficient annealing time and, consequently, a limited amount of ordered phase. The resistivity results are summarized in Fig. 4, in which the coefficients of both

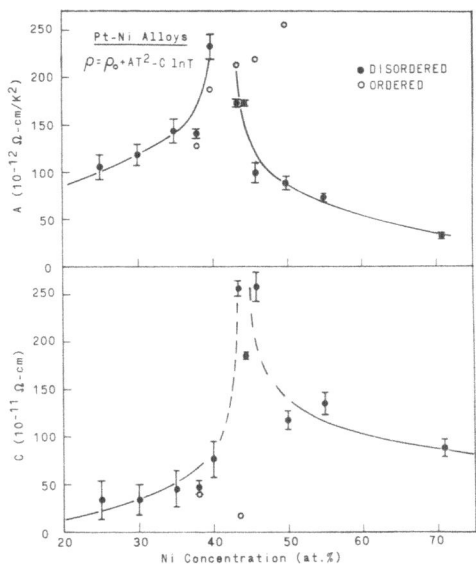

Fig. 4 - Resistivity coefficients[12] A and C for Pt-Ni alloys.

the T^2 and the $\ln T$ terms are plotted as functions of composition. As may be seen in Fig. 4, the values of A and C for the disordered alloys both peak strongly in the vicinity of the critical composition for the occurrence of ferromagnetism, while for the ordered alloys A increases monotonically with increasing Ni concentration and C is essentially zero (with the exceptions noted above).

Specific Heat

The specific heat of Pt-Ni specimens containing $43\frac{1}{2}$ and 50% Ni has been measured in the 1.1 to 4.2 K range for both the ordered[5] and disordered states. The results appear in Fig. 5 in the usual C/T vs T^2 format. It may be noted that the data plotted in this form are well fitted by straight lines and, consequently, that no unusual specific heat terms occur. Values of the electronic specific heat coefficient γ (the zero-temperature intercept in Fig. 5) and of the slope (β) are given below. It is evident that ordering significantly reduces the value of γ, while the value of β is essentially unchanged for the 50% Ni specimen and substantially increased for the $43\frac{1}{2}$% Ni specimen.

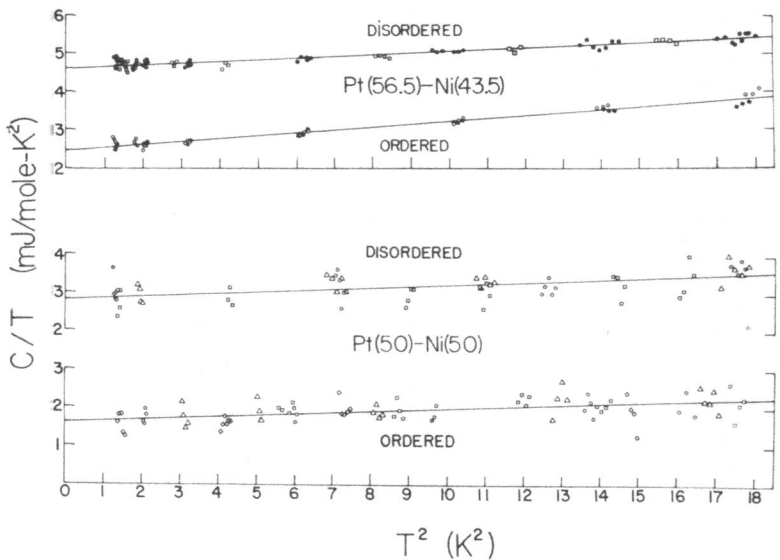

Fig. 5 - Specific heat of Pt-Ni alloys.

Specific Heat Parameters[12]

	Pt-43½at.%Ni		Pt-50at.%Ni	
	γ	β	γ	β
Disord	$14.61 \pm .03$	$.045 \pm .004$	$12.81 \pm .15$	$.041 \pm .014$
Ord[5]	$12.47 \pm .04$	$.075 \pm .005$	$11.61 \pm .13$	$.035 \pm .012$

(γ in mJ/mole-K^2 and β in mJ/mole-K^4)

SUMMARY AND DISCUSSION

We have found that the magnetic properties, the electrical resistivity, and the specific heat of Pt-Ni alloys with concentrations near the critical concentration for the occurrence of ferromagnetism are all strongly dependent upon the degree of atomic order. The ordering process in these alloys changes the structure[1] from f.c.c. to a layered one in which (for the 50-50 composition) there are planes containing only Ni atoms, alternating with planes containing only Pt atoms. Several important features of this change in structure may be noted: (1) the immediate environments of both Ni atoms and Pt atoms have been substantially changed, since in the disordered alloy a given atom of Ni, for example, can have from zero to 12 (with an average of 6) nearest-neighbor Ni atoms, whereas in the ordered alloy (again for the 50-50 composition), a given Ni atom will have precisely 4 nearest-neighbor Ni atoms and 8 nearest-neighbor Pt atoms. (2) The nearest-neighbor distances between Ni-Ni, Pt-Pt, and Pt-Ni atom pairs have all been changed, the first 2 increasing slightly (~2%) and the 3rd decreasing slightly (~1%).

It appears that some of the most significant changes observed in the properties of Pt-Ni alloys upon ordering may be qualitatively interpreted in terms of a reduced exchange interaction brought about by the above-mentioned features of the ordering process. In particular, (1) ordering causes the susceptibility of the _paramagnetic disordered_ alloys to decrease markedly and it causes the _ferromagnetic disordered_ alloys to become paramagnetic, the results in both cases being expected from a reduced exchange interaction; (2) ordering of the specific heat specimens[5] causes γ of the disordered alloys to decrease significantly, as would be expected according to recent theories[6] relating the enhancement in γ to the exchange enhancement;

(3) ordering causes the value of A for the paragmagnetic disordered alloys to decrease[7] substantially, as would be expected from both uniform[8] and local[9] enhancement theories.

     Although a substantially reduced enhancement upon ordering appears to account for a number of the observed properties of Pt-Ni alloys, one of the most unusual features of these alloys, namely, the occurrence of an anomalous ln T term in the electrical resistivity, does not seem to be related primarily to the enhancement. Resistivity minima are typically associated with the occurrence of local moments (Kondo effect), but this effect does not seem likely in this system[3]. Resistivity minima have also been found to be associated with clustering, as for example in the Cu-Ni system[10]. The extent of clustering in Pt-Ni alloys could not be very great, for several reasons : (1) the specific heat data show no sign of the constant term appropriate for magnetic clusters; (2) the magnetization data show only a very modest change from 1.5 to 4.2 K, which precludes the possibility of large cluster effects; (3) with the exception of showing a minimum at low temperatures, the resistivity data for these alloys in the 1.3 to 300 K range show none of the anomalous behavior reported for Cu-Ni alloys[11]. Yet even these limitations on the maximum possible amount of clustering do not completely rule it out as the source of the ln T term, since the observed minima are actually rather small.

     It is appropriate, finally, to mention several other features associated with the ln T term: (1) the value of C peaks strongly about the critical composition and is larger for the ferromagnetic specimens; (2) the value of C goes to zero upon ordering. The former behavior contrasts strongly with that found in Cu-Ni alloys, in which case C does increase as the critical composition is approached from the paramagnetic side but is not observed in the ferromagnetic alloys. The second feature (the disappearance of the ln T term upon ordering) is probably consistent with a possible clustering origin of the ln T term since ordering in these alloys markedly reduces the maximum possible number of nearest-neighbor Ni atoms about a given Ni atom and this in turn might eliminate clustering entirely.

REFERENCES AND FOOTNOTES

1. M. Hansen, Constitution of Binary Alloys (New York: McGraw Hill, 1958).

2. D. J. Gillespie and A. I. Schindler, Proc. of Conf. on Magnetism and Magnetic Materials, 1971, p. 461.

3. A. I. Schindler and D. J. Gillespie, Proc. of 12th Int. Conf. on Low Temp. Physics (Kyoto, Japan, Academic Press of Japan, 1971, p. 277).

4. C. A. Mackliet and A. I. Schindler, J. Phys. Chem. Solids 24, 1639 (1963).

5. The 43½% Ni specific heat specimen showed only slight long-range order. More recent work (currently in progress) on the same specimen after much more extensive ordering apparently yields the same simple $\gamma T + \beta T^3$ dependence and a still further reduction in γ.

6. See, for example, J. R. Schrieffer, J. Appl. Physics 39, 642 (1968).

7. For this particular resistivity specimen, A(ord) was larger than A(disord), as was found for the ferromagnetic specimens. It is possible that this particular resistivity specimen was in fact ferromagnetic, since the properties of these alloys depend sensitively upon thermal treatment; for example, doubling the length of the homogenizing anneal for samples near the critical composition significantly reduced the magnetization. This sensitive dependence of the properties upon thermal treatment could account for the differences between the present results and those of Herr and Besnus (Physics Letters 39A, 83 (72)) who found, for example, that specimens with as little as 41 at.% Ni were ferromagnetic.

8. A. I. Schindler and M. J. Rice, Phys. Rev. 164, 759 (1967).

9. P. Lederer and D. L. Mills, Phys. Rev. 165, 837 (1968).

10. R. W. Houghton, M. P. Sarachik, and J. S. Kouvel, Solid State Comm. 8, 943 (1970).

11. R. W. Houghton, M. P. Sarachik, and J. S. Kouvel, Phys. Rev. Letters 25, 238 (1970).

12. Indicated errors are for a 95% confidence level; for the resistivity data, an additional uncertainty of ~3% arises in connection with the determination of the form factor.

DISCUSSION

P. Lederer: Is the ordered 50% alloy fully paramagnetic?

A.I. Schindler: As far as we can tell, yes.

P. Lederer: Down to what temperature?

A.I. Schindler: We measured the susceptibility down to 1.5 K. Furthermore, the susceptibility was roughly temperature independent, so it is unlikely to become ferromagnetic at lower temperatures.

P. Lederer: What is your estimate for the exchange enhancement?

A.I. Schindler: We haven't made any estimate for the exchange enhancement of these alloys. Since we do not know the variation with composition of the density of states or of the unenhanced susceptibilities, we have no way of unambiguously determining the exchange enhancement.

P. Lederer: Do you have γ?

A.I. Schindler: Yes, but we nontheless have not calculated values of the exchange enhancement, since we do not know what enhancement model to use.

B.R. Coles: It would be rather interesting to compare your 50-50 alloy with Ni_3Al, because this seems to be a case in which the ordering, by keeping nickel atoms further apart from one another than in some of the random clusters in the disordered state, just stops the system from being ferromagnetic. That seems to be the case at least on the slightly aluminum rich side of Ni_3Al.

A.I. Schindler: That is true, but as I recall Ni_3Al is ferromagnetic on stoichiometry, while NiPt is paramagnetic when ordered.

B.R. Coles: If you go to the slightly Al rich side then it would be comparable.

A.I. Schindler: Perhaps.

J.A. Mydosh: Would you have any speculations regarding the
origin of the resistivity minimum that you observed in the
disordered system?

A.I. Schindler: We felt it might have originated with the
scattering of s-electrons by magnetic clusters, similar to
the ideas proposed by Sarachik for the resistivity anomalies
observed in some Cu-Ni alloys.   However there is one distinct
difference between the Cu-Ni alloys displaying anomalous resistivities and the disordered Ni-Pt.   In disordered Ni-Pt
the minimum occurs when the sample is ferromagnetic, but in
Cu-Ni the low temperature anomalous properties are observed
in paramagnetic samples.   In fact, we really don't know the
origin of the resistance minimum in disordered Ni-Pt.

R. Tournier: You are seeing the same variation in the resistivity as Tholence has observed in the CuFe system for
very dilute alloys.   And he sees for the giant moments,
which are <u>pairs</u> - in CuFe, a T^2-term which varies at T_c^2
and then he observed nearly magnetic pairs.   He observes
also a Log T term for the magnetic pairs which give magnetic ordering at very low temperatures.   I think by analogy
that you are observing two types of clusters, magnetic clusters which are probably giant moments as Kouvel and Comly
has observed in the CuNi system, and secondly you are observing a nearly magnetic cloud with a T^2 behavior in the
susceptibility.   My question is have you studied the susceptibility at high temperatures?

A.I. Schindler: As I mentioned earlier, I feel that there
are some distinct differences between Cu-Ni and Ni-Pt near
the critical composition for the onset of ferromagnetism.
But to answer your specific question, no, we haven't examined the susceptibility at high temperatures.   When these measurements were made we really didn't have the ability to
measure straightforward Curie-Weiss type normal susceptibility temperature dependencies.   We just measured the susceptibility at 4°K and below.   We now can make measurements of
the variation of susceptibility with temperature and may
make such measurements in the near future.

R. Tournier: I do not understand how you have evaluated
your susceptibility.   Your susceptibility is infinite at
the critical concentration.

A.I. Schindler: The susceptibility is taken right off of
the Arrott-plot in the standard fashion; from the intercept
of the Arrott-plots.   It diverges at the critical composition because we are taking extrapolated zero field values.

EXCHANGE INTERACTIONS IN STRONGLY PARAMAGNETIC ALLOYS

R. Harris and M.J. Zuckermann

McGill University, P.O. Box 6070, Montreal 101,

Quebec, Canada

In this contribution we show that the properties of strongly paramagnetic disordered alloys can be described in terms of a concentration dependent exchange interaction. We will apply our theory to experimental results for the magnetic susceptibility and spin disorder resistivity of Pd-Ni alloys.

Experiment shows that the low temperature spin susceptibility χ of pure Pd is 9.7 times larger than the value obtained from band structure calculations. When a small concentration c of Ni impurities is added to the Pd, χ increases linearly with c[1,2], but, as c is further increased, χ continues increasing non-linearly until the alloy becomes ferromagnetic at c = 2.3 at.%. The low temperature spin disorder resistivity ρ_{sd} of pure Pd shows a quadratic behavior[3] with temperature T, i.e. $\rho_{sd} \sim AT^2$ and the coefficient A also increases in a non-linear fashion as Ni is added to Pd. The Sommerfeld constant of the electronic specific heat also increases with Ni concentration, but more weakly than χ or A.[2]

These experimental results have been successfully accounted for to lowest order in Ni concentration by the theory of localized spin fluctuations (LSF) in strongly paramagnetic alloys due to Lederer and Mills,[4] Englesberg et al.[5] and Kaiser and Doniach.[6] In order to account for the non-

355

linear behavior in χ at higher Ni concentrations we have
extended the theory of Lederer and Mills[4] (which describes
LSF in the random phase approximation) by use of an analogy
to the coherent potential approximation (CPA). The dynamic
susceptibility $\chi(q, \omega)$ can then be written in terms of an
effective frequency dependent exchange interaction[7], i.e.,

$$\chi(q, \omega) = \chi_0(q, \omega)/[1 - I_{eff}(\omega)\chi_0(q, \omega)]. \tag{1}$$

$\chi_0(q, \omega)$ is the dymanic susceptibility of the conduction
electrons in the absence of interactions and $I_{eff}(\omega)$ is
given in terms of the exchange constant of pure Ni and pure
Pd by the following self-consistent equation:

$$I_{eff}(\omega) = I_{Pd} + c(I_{Ni} - I_{Pd})$$
$$\tag{2}$$
$$- (I_{eff}(\omega) - I_{Pd}) \chi_{loc}(\omega)(I_{eff}(\omega) - I_{Ni}).$$

Here $\chi_{loc}(\omega) = \sum_q \chi(q, \omega)$ and $\chi(q, \omega)$ is given by (1). Since
it was difficult to fit the data for the magnetic suscep-
tibility of Pd-Ni as a function of c using the full self-
consistency of equation (2), equations (1) to (2) were
iterated in a consistent manner and a fit to the data of
Williams was made (see Fig. 1) using values of I_{Pd} and
I_{Ni} obtained by other workers (see Ref. 7). In particular
we found that $\chi_{loc}^{(0)}/\chi = 0.4$ for pure Pd, and predicted a
critical concentration of 2.3 at.% Ni for the occurence of
ferromagnetism which agrees with the experimental data of
Tari and Coles.[3] The fit was made under the assumption
that the band structure of dilute paramagnetic Pd-Ni is
identical to that of pure Pd, a reasonable assumption in
view of the low critical concentration.

A calculation of the spin disorder resistivity of Pd-Ni
requires a detailed knowledge of the full dymanic susceptibility. This is difficult to obtain from the CPA because
of the self-consistency of equations (1) and (2). However,
since the sub-band of localized spin fluctuations due to
the Ni impurity lies inside the band of spin fluctuations
of pure Pd, the average t-matrix approximation (ATA) may
be used in place of the CPA.[8] Since the ATA is a first
iteration of the CPA equations, the expression for $\chi(q, \omega)$

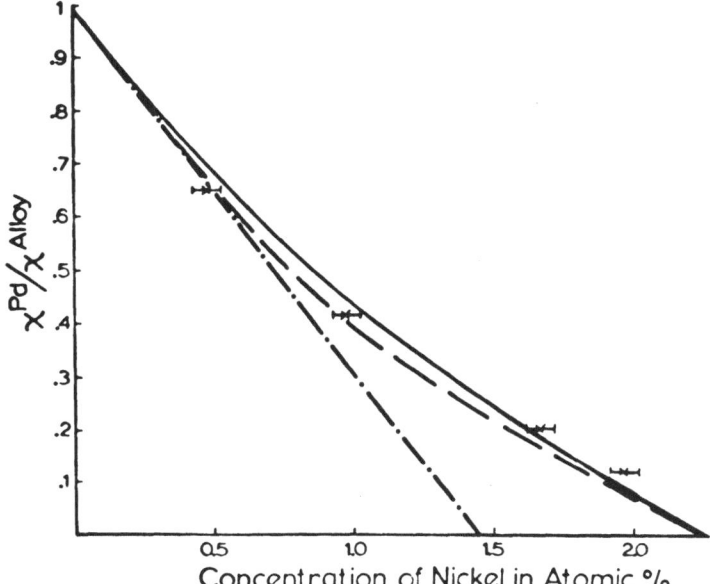

Figure 1: Theoretical fit to the data of Williams for the spin susceptibility of Pd-Ni alloys (Schindler et al.[1]). Error bars for the concentrations (A.I. Schindler, private communication) are shown. The full curve uses $N(0)I_{Pd} = 0.88$ and $N(0)I_{Ni} = 1.16$. The dashed curve uses $N(0)I_{Pd} = 0.8$ and $N(0)I_{Ni} = 1.2$. $N(0)$ is the density of states of conduction electrons at the Fermi level of pure Pd. The dashed dotted line corresponds to the theory of Ref. 5.

(see Ref. 9) is non self-consistent and explicit.

A modified version of the Lederer and Mills model[4] for the dynamic frequency dependent susceptibility $\chi_0(q, \omega)$ of pure Pd is used. The two parameters α and γ which define this model[9] can be determined by a best fit to the suscep-

tibility data for the alloys, and good agreement with
Diamond's theoretical calculation of $\chi(q, \omega)$ for the alloy
can then be calculated without much difficulty and the
analysis of Kaiser and Doniach[6] is used to calculate the
spin disorder resistivity due to localized spin fluctuations (LSF). For temperature $T \ll T_{sf}(c)$ (where $T_{sf}(c)$ is
the degeneracy temperature for spin fluctuations as a
function of Ni concentration) ρ_{sd} is given by:

$$\rho_{sd} = A(c)T^2 \tag{3}$$

where $A(c)$ is proportional to c and to the bulk susceptibility of the alloy. This dependence comes directly from the
2-dimensional nature of the Lederer and Mills model[4] for the
susceptibility. Figure 2 shows $A(c)$ as a function of c for
$\alpha = 1$ and different values of γ near $\gamma = 0.37$. These values
lie close to those giving the best fit to the susceptibility
data. For $\alpha = 1$ and $\gamma = 0.37$ the theoretical curve of $A(c)$
lies above the experimental points of Tari and Coles.[3]
However Tari and Coles find the law:

$$\rho_{sd} = A'(c)T^n \tag{4}$$

where $1 < n \leq 2$. Coles has suggested that this occurs because $T_{sf}(c)$ goes to zero as c approaches the critical concentration c_{crit} and hence the experimental temperatures are
not low enough to see a T^2 power law. Use of Kaiser and
Doniach's analysis shows that $A(c) > A'(c)$ and hence our
calculations are qualitatively correct. Further experimental work is required at lower temperatures to obtain
$A(c)$ experimentally.

Since the appearance of our first article, there have
been several reports of further research work on Pd-Ni.
Kato and Shimizu,[11] using the CPA approach outlined above,
calculate the static susceptibility self-consistently using
calculated values for $\chi_0(q, 0)$ based on Diamond's work.[10]
They find that $\chi_{loc}(0)/\chi = 0.063$ for Pd (in contrast to our
value of 0.4) and $c_{crit} = 1.5$ at.%. This value for
$\chi_{loc}(0)/\chi$ agrees with the value obtained from neutron
scattering data for Pd-Fe alloys, but is only correct if
the range of the exchange interaction I_{Pd} in (1) is assumed
to be zero. If we use the expression of Englesberg et al.[5]
for $\chi_{loc}(0)$ with their calculated value for the range of

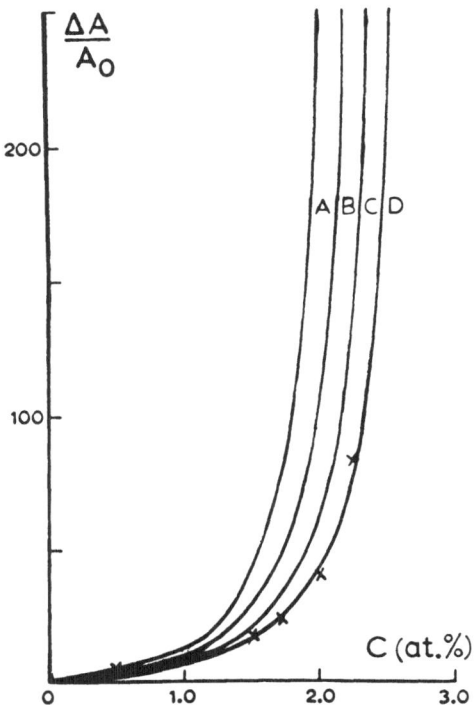

Figure 2: Fractional change in the coefficient A(c) of the
T^2 term in the spin disorder resistivity with
impurity concentration. The parameter α has the
value 1.0 for all curves. γ has the values 0.37,
0.372, 0.374 and 0.376 for curves A, B, C, D respectively. The crosses represent the experimental results of reference 3 for A'(c).

I_{Pd}, then we can show that Kato and Shimizu underestimated
$\chi_{loc}(0)/\chi$ in Pd by a factor of nearly 3. Such a factor
would bring their value of $\chi_{loc}(0)/\chi$ for pure Pd closer in
line with our value. However, the lack of knowledge of the
actual spatial variation of the exchange interaction implies
that $\chi_{loc}(0)/\chi$ is not accessible to calculation. Levin
et al.[12] and Hasegawa and Kamamori[13] have developed a CPA

method which combines the effect of potential scattering
with the effect of exchange interaction in the Hartree-
Fock approximation. Such an approach is correct for alloys
where potential scattering effects dominate but cannot
describe the properties of Pd-Ni since dynamic spin fluctuation effects are not indluded. Fukuyama[14] obtains
general self-consistent equations for $\chi(q, \omega)$ in the Random Phase Approximation for all strengths of potential
scattering. He also shows that our CPA approach and that
of Levin et al.[12] correspond to limiting cases of his
general formalism.

Both the CPA and the ATA are effective crystal approximations. In consequence local properties such as the
spatial variation of the magnetization $M(r)$ near an impurity site cannot be correctly obtained from either approximation. Recently Edwards et al.[15] were able to obtain
an expression for $M(r)$ near a Ni-impurity in dilute Pd-Ni
alloys using an extension of the Landau theory of phase
transitions. For more concentrated alloys the Ni impurities were assumed to form a regular lattice with the
lattice constant determined by the concentration, and a
fit was obtained to Chouteau's susceptibility data.[2]
While their assumption of a solute "super-lattice" is
clearly a gross simplification, this approach is in many
ways complementary to the CPA and ATA effective crystal
models. It is likely that further progress will be made
by some combinations of these two techniques.

We wish to thank Professor Coles for his hospitality
and assistance.

REFERENCES

1.   A. Schindler and C. Mackliet, Phys. Rev. Letters 20,
      15 (1968); E. Fawcett, E. Bucher, W.F. Brinkman and
      J.P. Maita, Phys. Rev. Letters 21, 1183 (1968).
2.   C. Chouteau, R. Fourneaux, K. Gobrecht and R. Tournier,
      Phys. Rev. Letters 20, 193 (1968).
3.   A. Tari and B.R. Coles, J. Phys. F 1, L69 (1971).
4.   P. Lederer and D.L. Mills, Phys. Rev. 165, 837 (1968).
5.   S. Englesberg, W.F. Brinkman and S. Doniach, Phys. Rev.
      Letters 20, 1040 (1968).
6.   A.B. Kaiser and S. Doniach, Int. J. Mag. 1, 11 (1970).

7. R. Harris and M.J. Zuckermann, Phys. Rev. B5, 101 (1972).
8. L. Schwartz, F. Brouers, A.V. Vedyayev and H. Ehrenreich,
    Phys. Rev. B4, 3383 (1971).
9. R. Harris and M.J. Zuckermann, J. Phys. F. to be pub-
    lished.
10. J.B. Diamond, Int. J. Mag. to be published.
11. T. Kato and M. Shimizu, J. Phys. Soc. Japan 33, (1972)
    to be published.
12. K. Levin, R. Bass and K.H. Benneman, Phys. Rev. Letters
    27, 589 (1971).
13. H. Hasegawa and J. Kamamori, J. Phys. Soc. Japan 31,
    382 (1972).
14. H. Fukuyama, Phys. Rev. B to be published.
15. D.M. Edwards, J. Mathon and E.P. Wohlfarth, J. Phys.
    F, to be published.

DISCUSSION

[Editors Note: This paper was read by Professor B.R. Coles]

D.M. Esterling: I want to make some comments about some of
the approximations which go into theories based upon an RPA
or Hartree-Fock approximation. I think it is probably pre-
mature even at this conference to talk about itinerant magnetism in disordered systems since we are quite a way from
understanding itinerant magnetism in ordered systems. But
there are some things that we do know about itinerant magnetism in strictly ordered systems and that is that the use
of the Hartree-Fock and RPA approximations is not necessarily very reliable. These approximations tend to overesti-
mate the tendency towards magnetism, since they overesti-
mate the correlation energy of the antiparallel spin electrons. There are many effects which are lost by use of
this sort of approximation. In particular, I believe the
Coulomb interaction in Ni and Pd has been estimated pretty
clearly and my feeling is that the estimation is probably
far too low, precisely because the Random Phase Approximation is used.

B.R. Coles: Perhaps Dr. Lederer would like to reply to
this comment?

P. Lederer: I think it is clear that the Hartree-Fock approximation is wrong in the sense that it predicts a local
divergence in the response function for the alloy when there
are two or three degrees of freedom, but my general feeling
about this is that an exact theory would yield the same formal type of structure as theories for which the dynamic susceptibility goes linearly with frequency, but with an exchange enhancement which has to be computed self-consistently.

D.M. Esterling: We don't have an exact theory.

P. Lederer: I agree. I would like to comment on the use
of the CPA approximation. In fact, there is no critical
concentration for this alloy or other nearly magnetic alloys
which become magnetic, as a whole body of experiments made
by the Grenoble group show. In fact, above 1% or 1.5%, I
think that saturation magnetization measurements in Pd-Ni
show that you already have pairs which are magnetic, and
practically the same thing occurs in CuNi and other such
systems. Using a CPA approximation, which amounts to finding a susceptibility which is translationally invariant,
amounts to neglecting cluster effects. One has neglected
the possibility of having one impurity becoming magnetic or
having very low T_K because of the proximity of another one.
I think in general that the CPA approximation has drawbacks
which are known for non-magnetic alloys, but it is certainly a very valid approximation for any almost magnetic or
magnetic alloy.

B.R. Coles: I think for this particular system to be fair
to the theorists concerned, there is a well defined transition from paramagnetism to ferromagnetism. Experimentally
there is a critical transition, a critical composition for
that transition to the right of which there is ferromagnetism and to the left of which, as you go down no matter how
low in temperature, there is no ordering effect, there is
therefore no well defined moment which is ordering. There
is no magnetic glass here. There is a clear cut transition
from paramagnetism to ferromagnetism.

P. Lederer: I don't agree with that. The experiments show
clearly that you have a subtransition from a state where
the susceptibility is linear in field, and above 1.2 or
1.3%, you have clusters of magnetic impurities. You do

have a spin glass regime in Pd-Ni. Just look at the data
by Chouteau and others for the saturation magnetization.

B.R. Coles: No, I am sorry you do not. I have my own data
and I have seen Chouteau's data and I have seen no evidence
in them of a spin glass. There may be cluster moments at
finite temperatures for $c < c_{crit.}$; there is no evidence
yet that they order.

COMMENTS ON THE HUBBARD MODEL OF MAGNETISM IN DISORDERED

TRANSITION METAL ALLOYS*

D. M. Esterling and R. A. Tahir-Kheli[†]

Department of Physics, Indiana University

Bloomington, Indiana 47401

INTRODUCTION

The Hubbard model[1] has been used extensively in the discussion of magnetism in metals[2]. It has also been employed as a model of disordered magnetic alloys[3]. Here we present a straightforward extension of Hubbard's original approximation for the model. The disorder is introduced in the hopping elements and in the interaction term. We shall argue that in many cases the former type of disorder is more important than the latter.

The solution follows from a rather natural extension of the work of Blackman, Esterling and Berk[4] for off-diagonal randomness within the coherent potential approximation[5] (CPA). The form of the general solution will be presented. We will then apply our results to a simple cubic lattice with nearest neighbor hopping in the extreme split band limit where energy separation between A and B atomic levels is infinite - so B sites are "excluded". We then test for ferromagnetic stability within this <u>approximation</u> (keeping in mind the fact that in the ordered lattice this approximation is overly restrictive towards[6] magnetic order.) As expected, ferromagnetic stability is not found for any electron density or concentration of A atoms (c_A). In an indirect way, however, we can get a feel for the effect of spatial disorder on the tendency toward ferromagnetism by analyzing the difference between

the ground state energy in the ferromagnetic and paramagnetic configurations for various concentrations c_A. Our
conclusion is that increase in spatial disorder reduces the
tendency toward ferromagnetism.

MODEL HAMILTONIAN

The model Hamiltonian in an A-B alloy assumes the form:

$$H = \sum_{i\sigma} \varepsilon_i c_{i\sigma}^+ c_{i\sigma} + \sum_{ij\sigma} t_{ij} c_{i\sigma}^+ c_{j\sigma} + \sum_{i\sigma} I_i n_{i\sigma} n_{i-\sigma} \quad (1)$$

Here ε_i is the atomic level for Wannier state associated
with site i ($\varepsilon_i = \varepsilon_A$ or ε_B), I_i is the intra-atomic coulomb
interaction and t_{ij} are the hopping elements (t_{ij}^{AA}, $t_{ij}^{AB} =$
t_{ij}^{BA}, t_{ij}^{BB}). Here we take $I >> \Delta$ where $\Delta = \max \{t_{ij}\}$ and
$I = \min (I_i)$ (except for the excluded volume case indicated
below). This corresponds to an alloy of two constituents
each of which has a narrow energy band. We feel that in
this limit the disorder in the hopping elements plays a
more important role than disorder in the interaction
strengths (I_A and I_B). For suppose Δ is of the order of
one electron volt and I_A and I_B are much larger (on the
order of ten electron volts) -- as may very well be the
case in narrow band materials -- then the physical effect
of I_A and I_B is to disallow double occupancy (spin up and
spin down) simultaneously on the same site. This effect of
will be essentially the same if say $I_A \simeq 8$ ev and $I_B \simeq 14$
ev. Hence we may as well consider $I_A \simeq I_B$ and focus our
attention on the contribution of disorder in the hopping
elements.

This assumption would not be appropriate if we had an
alloy such that one constituent had a wide band and the
other had a narrow band (e.g. CuFe). Here we would expect
the difference between I_A and I_B to be important. We will
retain the general case ($I_A \neq I_B$, $t^{AA} \neq t^{BB} \neq t^{AB}$ in our
formalism).

We will, in addition, consider the excluded volume case
($\varepsilon_B \to \infty$) where the B sites are effectively forbidden. This
corresponds to an alloy where the constituents are quite
different so that the separation between atomic levels
(ε_A and ε_B) is much greater than the bandwidth.

This situation should apply to the d-electron contribution in alloys of transition metals with aluminum where the aluminum atoms are essentially forbidden sites.[7]

SOLUTION

Our problem is an interacting electron system in a disordered lattice. We will treat the disorder using the widely used coherent potential approximation (CPA). We include the possibility of off-diagonal randomness and so employ an appropriate extension[4] of the CPA. This solution may be summarized as follows: The single particle propagator $G_k(\omega)$ is given by the matrix expression

$$G_k(\omega) = \begin{pmatrix} G^{AA} & G^{AB} \\ G^{BA} & G^{BB} \end{pmatrix} = \begin{pmatrix} B - \beta_k & \zeta_k - U_3 \\ \zeta_k - U_3 & A - \alpha_k \end{pmatrix} / D_k \qquad (2)$$

where

$$D_k = (AB - U_3^2) - (A\beta_k + B\alpha_k - 2U_3\zeta_k) + (\alpha_k\beta_k - \zeta_k^2) \qquad (3)$$

$$A(\omega) = (\omega - \epsilon_A) + c_B/\gamma^A(\omega) \qquad (4)$$

$$B(\omega) = (\omega - \epsilon_B) + c_A/\gamma^B(\omega) \qquad (5)$$

$(\alpha_k, \beta_k, \zeta_k)$ are the Fourier transforms of

$$(t_{ij}^{AA}, t_{ij}^{BB}, t_{ij}^{AB}) \qquad (6)$$

and where γ^A, γ^B, and U_3 (all functions of energy) are given self-consistently by the following three equations:

$$\gamma^A = (1/N) \sum_k (B - \beta_k)/D_k \qquad (7)$$

$$\gamma^B = (1/N) \sum_k (A - \alpha_k)/D_k \qquad (8)$$

$$0 = (1/N) \sum_k (\zeta_k - U_3)/D_k \qquad (9)$$

The total average propagator $<G>$ is then given by

$$<G_k(\omega)> = G_k^{AA}(\omega) + G_k^{AB}(\omega) + G_k^{BA}(\omega) + G_k^{BB}(\omega) \qquad (10)$$

The question of how to treat the interaction is a much more difficult one. As a first step in treating the disordered Hubbard model in the narrow energy band limit ($\Delta \ll I$), we will work within the spirit of Hubbard's initial approximation.[1] This approximation in the pure metal was equivalent to replacing the exact mass operator $\Sigma_{k\sigma}(\omega)$ by its expression in the atomic limit. Here the mass operator in the pure metal is defined in the usual way:

$$G_{k\sigma}(\omega)^{-1} = \omega - \Sigma_{k\sigma}(\omega) - \varepsilon_k \qquad (11)$$

and the Hubbard approximation is then

$$G_{k\sigma}(\omega)^{-1} = \omega - \Sigma_\sigma^{at}(\omega) - \varepsilon_k \qquad (12)$$

where

$$\Sigma_\sigma^{at}(\omega) = n_{-\sigma} I\omega/[\omega - I(1 - n_{-\sigma})] \qquad (13)$$

This is a local approximation in that the mass operator is k-independent. Observe that the approximate Green's function is identical in form to the free-particle Green's function with the energy, ω, replaced by a renormalized energy

$$\Omega_\sigma(\omega) = \omega[1 - n_{-\sigma} I/(\omega - I(1 - n_{-\sigma}))] \qquad (14)$$

Now Now we indicate how to combine the two techniques. One may exploit the fact that both the Hubbard and the extended[4] CPA are single-site approximations. Therefore, within the spirit of the Hubbard approximation indicated above, one replaces the atomic limit propagators,[4] i.e.,

$$g_A(\omega)^{-1} = \omega - \varepsilon_A; \quad g_B(\omega)^{-1} = \omega - \varepsilon_B \qquad (15)$$

by the following which are appropriate to the interacting system

$$g_A(\omega)^{-1} = \Omega_\sigma^A(\omega) - \varepsilon_A; \quad g_B(\omega)^{-1} = \Omega_\sigma^B(\omega) - \varepsilon_B . \qquad (16)$$

Here

$$\Omega_\sigma^{A,B}(\omega) = n_{-\sigma}^{A,B} I_{A,B}\omega/[\omega - I_{A,B}(1 - n_{-\sigma}^{A,B})] \qquad (17)$$

and $n_\sigma^{A,B}$ is the average number of electrons of spin on site A or B. We may now do the diagrammatic expansion as in Ref. 4. We then recover exactly the same expressions as given above in Eqs. (2)-(9), except that $A(\omega)$ and $B(\omega)$ become

$$A(\omega) = (\Omega_\sigma^A(\omega) - \varepsilon_A) + c_B/\gamma_\sigma^A(\omega); \; B(\omega) = (\Omega_\sigma^B(\omega) - \varepsilon_B) +$$

$$c_A/\gamma_\sigma^B(\omega) \tag{18}$$

where

$$\gamma_\sigma^A(\omega) = (1/N) \sum_k G_{k\sigma}^{AA}(\omega); \; \gamma_\sigma^B(\omega) = (1/N) \sum_k G_{k\sigma}^{BB}(\omega) \tag{19}$$

RESULTS AND DISCUSSION

We now apply the technique of the previous section to the excluded volume case ($\varepsilon_B \to \infty$). In this case $G^{AB} = G^{BB} = 0$. We also let $I_A \to \infty$. The averaged propagator becomes

$$G_{k\sigma}(\omega)^{-1} = (\Omega_\sigma(\omega) - \varepsilon_A) + c_B/\gamma_\sigma(\omega) - \varepsilon_k \tag{20}$$

where

$$\Omega_\sigma(\omega) = \omega/(1 - n_{-\sigma}); \; \gamma_\sigma(\omega) = (1/N) \sum_k G_{k\sigma}(\omega) \tag{21}$$

The ground state energy E_{GS} was calculated using an exact expression derived by Bari and Lange[8], which for our purposes has the form:

$$E_{GS} = \sum_{k,\sigma} \alpha_k n_{k\sigma} \tag{22}$$

where $n_{k\sigma}$ is the momentum distribution, calculated in the usual way. The results for nearest neighbor hopping in a simple cubic lattice are given in the Table. Within a given decoupling approximation ferromagnetism is not obtained for any density (n) or concentration (c_A). This is to be contrasted with the conclusion of one of us[2] in the ordered lattice case ($c_A = 1$) using a more reliable technique which generates the spectral weight function from its calculated moments. Moreover, the trend away from magnetic order as a function of increasing density for a given concentration is opposite to what might be expected at least for high atomic concentration on the

TABLE

n \ c_A	.1	.5	.9
$.25 * c_A$	0.001 (0.20)	0.008 (0.15)	0.02 (0.17)
$.50 * c_A$	0.003 (0.54)	0.027 (0.44)	0.05 (0.35)
$.75 * c_A$	0.005 (1.16)	0.048 (1.02)	0.07 (0.64)

Table Caption: Difference in ground state energy for ferromagnetic and paramagnetic configurations $(E_F - E_p)$ for $\alpha_k = (1/3) [\cos k_x + \cos k_y + \cos k_z]$ and $\varepsilon_B \to \infty$. The density n is total electron density $(n = n_\uparrow + n_\downarrow)$. The quantity in paranthesis is $(E_F - E_p)/E_F$.

basis of the following argument: Consider the expression for E_{GS} given in Eq. (22). Suppose c_A is close to unity. In the ferromagnetic state $n_{k\sigma}$ will then have a fairly sharp cut-off. In the paramagnetic state the exact $n_{k\sigma}$ will not be a sharp distribution due to the electron-electron scattering. At low densities the electrons will not scatter much and the paramagnetic distribution will approach a sharp free-electron shape. This latter distribution will then correspond to the lower energy since lower energies are sampled in the k-summation in Eq. (22) -- just the usual argument. At higher densities, there will be more scattering. The tail on the paramagnetic distribution will extend further out, sampling higher energy states, while the ferromagnetic distribution retains its cut-off (recall $c_A \simeq 1$). Eventually it is possible that the paramagnetic state may become energetically unfavorable, but this can only occur at the high density end. This is exactly the behavior found by one of us in the ordered $(c_A = 1)$ case. The situation is not so clear when the ferromagnetic and paramagnetic configurations involve scattering $(c_A \neq 1)$. However the trend in the above table as a function of the electron density is in the wrong direction if we are to accept the above argument. As to the question of the effect of disorder

on magnetic stability, we note the following. Our dis-
order parameter ($\varepsilon_B - \varepsilon_A$) is infinite. However varying
c_A corresponds to varying the disorder in a certain sense.
In comparing different c_A columns, one must recognize that
both the ferromagnetic and the paramagnetic energies
become small as $c_A \to 0$. In order to scale this effect,
it may be reasonable to compare the quantities given in
parentheses in the table. One concludes that increasing
c_A (i.e. increasing the spatial order) enhances the
tendency towards magnetism.

References

*Supported by the U. S. Army Research Office, Durham, N.C.
†Permanent address: Department of Physics, Temple University, Philadelphia, Pa. 19122

1. J. Hubbard, Proc. Roy. Soc. A276, 238 (1963).
2. A short review of the current literature is contained
   in the work of D. M. Esterling, Phys. Rev. B (Dec.
   1972).
3. R. Harris and M. J. Zuckermann, Phys. Rev. B5, 101(1972)
   and other articles in this symposium.
4. J. Blackman, D. Esterling, and N. Berk, Phys. Rev.
   B4, 2412 (1971).
5. P. Soven, Phys. Rev. 156, 809 (1967).
6. A. B. Harris and R. Lange, Phys. Rev. 157, 295 (1967).
7. E. Stern, Physics (USA) 1, 255 (1965).
8. R. Bari and R. V. Lange, Phys. Lett. 30A, 418 (1969).

AMORPHOUS SPIN POLARIZATION IN A Tb-Fe COMPOUND*

J. J. Rhyne, S.J. Pickart, and H. A. Alperin

Naval Ordnance Laboratory

White Oak, Maryland   20910

ABSTRACT

Neutron diffraction and magnetization measurements on a sample of 33.3 at.% Tb, 66.7 at.% Fe prepared by sputtering show clear evidence for an amorphous atomic arrangement, and in addition exhibit an amorphous spin polarization distribution. Data presented both above and below the 388 K Curie temperature and a difference pattern show explicitly the diffuse scattering of both nuclear and spin origin. A low angle "tail" of purely magnetic origin is observed which may be related to the occurrence of longer range inhomogeneities or domain-like regions in these materials.

INTRODUCTION

The binary compound $TbFe_2$ prepared by sputtering[1] has been shown to have an amorphous structure and to exhibit magnetic ordering in which the spin distribution is also of an amorphous nature. The samples were in bulk from approximately 25 mm in diameter and one mm thick and were investigated by elastic neutron scattering at the National Bureau of Standards Reactor.

TbFe$_2$ normally crystallizes in the Laves phase (MgCu$_2$) structure and exhibits a ferrimagnetic alignment of Tb (9 μ_B) spin and Fe (approximately 2 μ_B) giving rise to an experimentally determined spontaneous magnetization of 4.7 μ_B per molecule[2] at 0 K. The Curie temperature is 710 K. Magnetization curves of the polycrystalline phase show a high field susceptibility arising from anisotropy effects.

II. MAGNETIZATION MEASUREMENTS

Magnetic moment measurements[3] shown in Figure 1 made on the sputtered material with composition 33.3 at.% Tb,

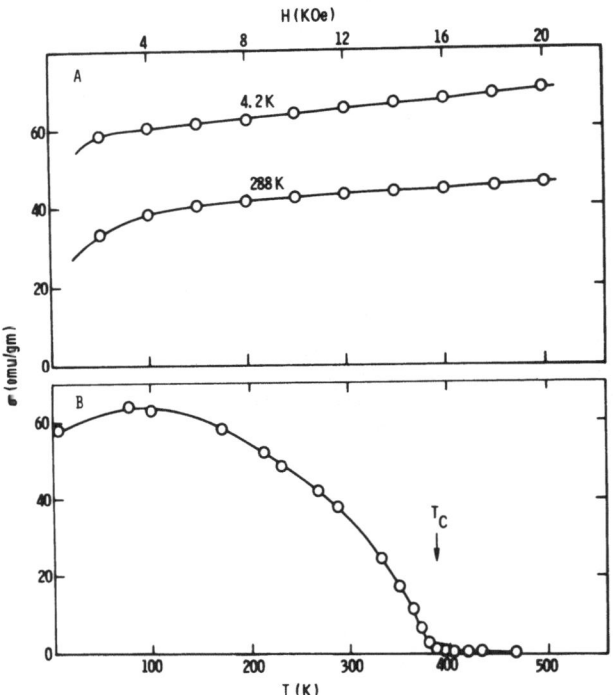

Figure 1.   a) The magnetic moment of sputtered TbFe$_2$ versus field at 288 K and at 4.2 K.   b) The spontaneous magnetization (σ extrapolated to H = 0) versus temperature. The measurements above room temperature are relative values taken in a constant 1 kOe field.

66.7 at.% Fe showed clearly that the samples were indeed macroscopically ordered yet exhibited significant differences from their crystalline counterpart. In particular, the Curie temperature was much lower, 388 K, and the zero Kelvin spontaneous magnetization was reduced to 2.8 μ_B per formula unit. Similar reductions in Curie temperature and moment were observed in ternary Fe-Pd-P compounds[4] shown to be amorphous. The isotherms of Figure la show effective saturation in fields of 2-3 kOe followed by a high field susceptibility which reflects either an increase in the ferromagnetic spin alignment as discussed later or a remnant magnetic anisotropy in the amorphous state. This may arise from the highly non-spherical Tb charge distribution interacting with neighboring positively charged atom cores although not on conventional lattice sites. Prolonged annealing of the sputtered samples at 650 K in the magneto-meter showed an increase in magnetization with time reflecting the partial recrystallization of the TbFe$_2$ Laves phase at this temperature. No evidence of recrystalli-zation was observed at 425 K and below, which was the range explored in the following neutron measurements.

III. NEUTRON DIFFRACTION RESULTS

Neutron data were taken with an incident wavelength of 1.35Å with collimation of 20' in-pile, zero between monochrometer and sample, and 40' before the counter. Reducing the counter collimation to 20' did not alter the diffraction maxima and thus instrumental resolution corrections have been neglected. Scattering data taken as a function of wavenumber $\kappa = 4\pi \sin \theta / \lambda$ at 423 K (above T_C) and at 4 K (below T_C) are shown in Figures 2a and b with the instrumental background subtracted. The positions of the Bragg diffraction lines in the crystalline counterpart TbFe$_2$ are shown by the arrows.

The high temperature pattern is characteristic of atomic scattering from an amorphous material with the addition of a superimposed form factor from the diffuse paramagnetic scattering which alters the normal angular dependence. The broad maximum is observed to lie near the positions of the (113) and (222) Laves phase lines which are the strongest nuclear reflections in the crystalline compound. The amorphous pattern is, however, not the result of broadened Laves phase peaks since strong peak intensity would then

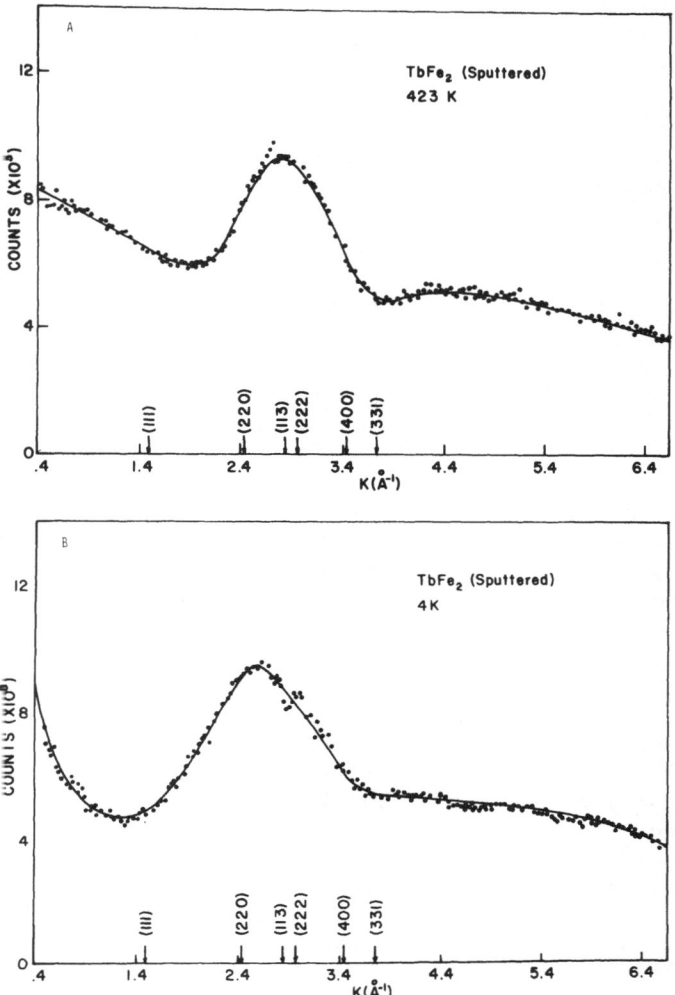

Figure 2. Neutron diffraction intensity from sputtered TbFe$_2$
a) above the magnetic ordering temperature and b) at 4.2 K.
The small "pip" near the top of the broad peaks is the remnant
of an Al (111) line arising from the radiation shield.

occur both at higher and lower angles than we observe in
Figure 2, assuming that the relative structure factors are
unaltered. In addition, microcrystallinity on this scale
would give rise to significant increased low angle scattering
which is absent from our results. An application of the
Scherrer formula indicates a range of atomic correlations of

6-8Å which is comparable to the TbFe$_2$ lattice parameter of
7.34Å. We therefore conclude that the pattern results from
truly amorphous atomic scattering.

The low temperature data of Figure 2b show several
pronounced differences from the high temperature results.
The diffuse peak is further broadened and shifted to
lower angle (lower κ) and also an additional low angle
scattering has appeared which was absent in the nuclear
scattering shown in Figure 2a. These features are attributed
to the added scattering from the macroscopic spin polarization.
This scattering of entirely magnetic origin can be explicitly
recovered by subtracting the high temperature from the low
temperature patterns numerically as shown in Figure 3.

This procedure removes the nuclear (atomic) scattering
contribution and also eliminates the effects of absorption,
nuclear incoherent scattering and inelastic phonon scattering
to the extent that they are temperature independent. The
resulting intensity is again characteristic of an amorphous
structure and displays directly the amorphous nature of the
magnetic spin polarization. The scattering results of
Figure 2a and Figure 3 thus show separately the diffuse
nature of both the atomic and the spin systems in this
material.

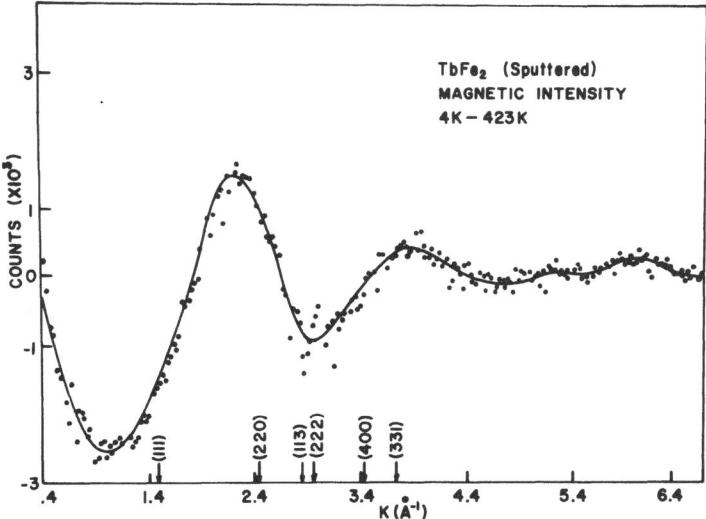

Figure 3. Difference pattern (low temperature data minus
high) showing the scattered intensity of purely magnetic
origin in sputtered TbFe$_2$.

IV. DISCUSSION

The magnetic intensity of Figure 3 is of the form
necessary to obtain a spin correlation function of the spin
system by Fourier transform since the intensity at high
temperatures contains the total magnetic scattering
distributed into a paramagnetic "form factor". The
subtraction procedure then gives the magnetic intensity
less the square of the form factor (Fig. 3) which is a
function which oscillates about a fixed value at large κ
as required for the transform. A straightforward Fourier
transform applied to the magnetic data (Fig. 3) produced
the results of Figure 4 which show a strong positive spin
correlation maximum near 3.6Å, a large negative extremum
near 4.8Å, and other subsidiary peaks which may in part
arise from statistical errors and truncation effects.
Inasmuch as there are two dissimilar atom spins involved,
the distribution of Figure 4 involves a superposition of
three pair distribution functions in both space and spin
coordinates, and thus an unambiguous interpretation is not
possible. The rare earth-rare earth (RE-RE) pair distribution
function would be expected to dominate the magnetic scattering
due to its larger moment and it is noted that the positive

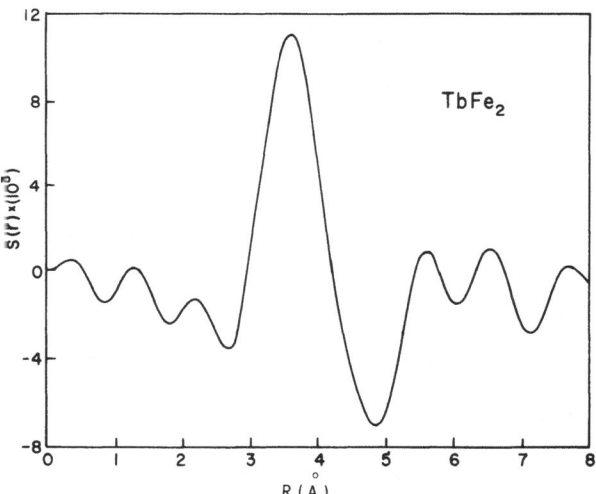

Figure 4. Magnetic spin correlation function found by
Fourier transform of the data of Figure 3.

peak in the radial distribution function occurs close to the
nearest neighbor RE-RE approach distance in the crystalline
cell of 3.2Å and the negative peak at one of the Tb-Fe
distances. The dominance of the rare earth also partically
accounts for the shift in the peak intensity in Figure 2
to lower angles in the ordered temperature range as opposed
to the high temperature (atomic scattering) case in which
all atom correlations contribute essentially equally.

It is of interest to note that if the immediate
environment of a rare earth site in the amorphous structure
looked like the crystalline ferrimagnetic arrangement,
then magnetic intensity would be expected near the (111)
angular position (the strongest crystalline magnetic line)
which is not observed. This fact plus the lack of good
agreement between the atom separations and the extrema of
the spin correlation transform lead one to speculate that
the magnetic arrangement may be altered in the amorphous
state. It is also necessary to assume that some of the rare
earth spins are either anti-aligned or disordered (a plausible
case assuming a RKKY-type exchange coupling and the random
spacing of spins) in order to account for the greatly reduced
bulk magnetization. Either of these can also account for
the large high field susceptibility observed in the magnetic
results. In the absence of   further information including
the magnitude of the Fe atomic moment and the degree of
possible delocalization, an explicit model of the spin
distribution is difficult to construct.

The small angle scattering "tail" observed in Figure 3
and 2b and absent in the atomic scattering of Figure 2a is
of magnetic origin and may reflect regions of inhomogeneity
which are of longer range than the average near-neighbor
spin correlations in the magnetic structure. These regions
may be analogous to magnetic domains or regions of spin
order transitions which give rise to the observed additional
scattering, and are perhaps intrinsic to the magnetic
amorphous state in general. The strength of the low angle
scattering is sensitive to magnetic fields and to temperature
cycling. Application of a 3 kOe field along the scattering
vector produced a significant increase in the low angle
magnetic scattering and a small decrease in the diffuse
peak intensity. The latter indicates that only a small
fraction of the spins were rotated by the field. The first
effect was sharply reduced by field cooling the specimens
in a 3 kOe field at right angles to the scattering vector.

These effects are being investigated further as well as the intriguing question of the nature of the critical scattering near the Curie temperature and also the existence of localized spin-wave type excitations in these materials.

We are indebted to A. E. Clark for making available the sample, which is one of a series prepared for magnetostriction studies,[5] and to T. R. McGuire for expediting the magnetic measurements.

V.   REFERENCES

*   This work was supported in part by the Naval Ordnance Laboratory Independent Research Fund and the Office of Naval Research.
1. The material was prepared by Battelle Institute Northwest under contract with the Advanced Research Projects Agency.
2. K. H. J. Buschow and R. P. VanStapele, J. Appl. Phys. 41, 4066 (1970).
3. We are indebted to Mr. H. Lilienthal of the IBM Research Center for making these measurements.
4. T. E. Sharon and C. C. Tsuei, Phys. Rev. B 5, 1047 (1972).
5. A. E. Clark and H. S. Belson, Phys. Rev. B 5, 1498 (1972).

DISCUSSION

B.R. Coles: Can you actually, at low temperatures, rotate the magnetization with an external field and look at the changes in the magnetic scattering that way?

J.J. Rhyne: Of course, that's the ideal way to do it. Unfortunately these materials still have a very high anisotropy, and would require a very large magnetic field to rotate the spins into the scattering vector and eliminate the magnetic intensity. Ideally that's an excellent way to do it.

C.D. Graham: You say there is a high anisotropy in an amorphous material?

J.J. Rhyne: The only evidence we have for that is that the magnetostriction is large. It's about 1/3 of that of the polycrystalline material.

Question: You said the susceptibility was done. What does it look like?

J.J. Rhyne: The susceptibility has not been done. A moment study in one kilogauss was done to get the Curie temperature.

NEUTRON SCATTERING OBSERVATION AND THEORY OF SPIN-WAVES IN DISORDERED MAGNETIC INSULATORS[*]

W. J. L. Buyers[†]

Oak Ridge National Laboratory

Oak Ridge, Tennessee   37830

Neutron scattering provides a direct spectroscopic probe of the spin-wave response in substitutional disordered magnetic materials. The spin-wave frequencies, damping and detailed line shape may be determined from the measurements. For localized modes the measured structure factor may be Fourier transformed to give information on the extent to which the mode wave function is localized in space. Measurements have recently been made of two well-defined branches of the spin-wave dispersion relation for the split-band systems $(Co,Mn)F_2$ and $K(Co,Mn)F_3$, while for the single-band system $(Mn,Zn)F_2$ the spin-wave branch falls in frequency and broadens as the zinc concentration is increased. Near the critical zinc concentration for percolation, p_c, where the long range magnetic order disappears the spin-waves are found to be overdamped. The coherent potential approximation has recently been extended to apply to antiferromagnets, and it is now possible to calculate the spin-wave line shape for a quasi-binary alloy of any composition. The theory includes part of the effect of the off-diagonal randomness arising from the transverse exchange interactions as well as the differing interactions with the various neighboring clusters. Detailed numerical calculations are found to yield correctly

---

[*]Research sponsored by the U. S. Atomic Energy Commission under contract with Union Carbide Corporation.

[†]Guest scientist from Atomic Energy of Canada Ltd., Chalk River, Ontario, Canada.
383

the split-band and single-band behavior observed experimentally, and good agreement is found for the frequencies and line shapes of the spin-waves.

The neutron experiments were performed at the NRU reactor, Chalk River, and are described in the review by E. C. Svensson, W. J. L. Buyers, T. M. Holden, R. A. Cowley and R. W. H. Stevenson (AIP Conference Proceedings 5, 1315, 1972), in the work of G. J. Coombs, T. M. Holden, E. C. Svensson, W. J. L. Buyers and D. A. Jones on the dilute antiferromagnet (to be published) and in the review by R. A. Cowley and W. J. L. Buyers (Rev. Mod. Phys. 44, 406, 1972) where the low concentration Green's function theory is also given. The theory for high concentrations has been developed by W. J. L. Buyers, D. E. Pepper, and R. J. Elliott (J. Phys. C., September, 1972) for mixed antiferromagnets and for the dilute antiferromagnet by the same authors in a forthcoming publication (J. Phys. C.).

DISCUSSION

S.M. Bose: I wonder, when you include the off-diagonal randomness, is the calculation self consistent?

W. Buyers: Yes, it is done self consistently. It is done by iterating a pair of equations and the transverse part of the self-energy is retained throughout.

S.M. Bose: You take the off-diagonal terms into the Green's functions?

W. Buyers: Yes. It turns out within this approximation you can do that.

R. Tahir-Kheli: I would like to go on record as saying that the manganese zinc fluoride system that Professor Buyers is looking at is the kind of system that I was trying to describe this morning in the first part of my talk. My results, which refer only to the single-site density of states, and therefore do not describe q-structure, but only the total density of states structure, would lead me to believe that I am in substantial agreement with what he has said. The only disagreement that I could have would be the critical concentration which I find is 33.3% and I would believe that it should never be less than that, whereas these systems have a slightly lower critical concentration.

W. Buyers: What I referred to were measurements of the critical concentrations. It is less than 33% apparently.

R. Tahir-Kheli: For the nearest neighbor interactions?

W. Buyers: These are measurements of where the long-range order disappears. It seems to disappear at about 72% of zinc which is 28% of manganese.

R. Tahir-Kheli: The other point of disagreement would be possibly, I would say, that you would never see the zone boundary magnons because (although I agree with you that they will be much stiffer than their width) they will have very little density of states available to them.

W. Buyers: In point of fact it turns out that the zone boundary ones are the easiest ones to see. They have a response which extends to the highest frequency as perhaps you saw from the last few slides. In fact they stick out as a shoulder and there is no doubt that there is a well-defined response from zone boundary magnons.

R. Tahir-Kheli: In other words you say that the partial density of states associated with them . . .

W. Buyers: No, I am not making any comment about the density of states. I am making a comment about whether we can see zone boundary magnons with neutron scattering.

R. Tahir-Kheli: No, I am saying that if by seeing them you make the statement they there is a whole lot of density of states available to them, then I would disagree with that but if by seeing you simply say, well, that they happen to be rather long lived and you can scatter from them, and therefore you do know something about them, then I would have nothing to disagree.

W. Buyers: The effect of the overdamping of the spin waves will certainly be to produce a large enhancement of the density of states at low frequencies. All of the magnons, including the zone boundary ones, have a finite response at zero frequency near the critical concentration. I think someone mentioned yesterday the importance of zero-frequency response for the appearance of peaks in the low frequency susceptibility and it certainly plays an important part in the critical behavior.

W.M. Hartmann: Do you make a spherical approximation when you do the averaging of the T-matrix, or do you put in lattice structure?

W. Buyers: The different configurations give potentials which depend only on the number of neighbors which are defect or host, so one doesn't have to worry about the different spatial configurations of defects. The discrete lattice structure is of course, included in all Fourier transforms and sums over the Brillouin zone.

W.M. Hartmann: Well, one should worry about them. What do you do about it?

W. Buyers: The different spatial configuration of defects around a central ion influence only the off-diagonal elements of the self-energy. The Goldstone approximation we have made for these off-diagonal terms is equivalent to retaining only the part of the self-energy having s-like symmetry. Σ^{12} is then the same for all nearest neighbors, 2, of 1, but Σ^{22} between neighbors 2 and 2' of 1 is zero.

COMMENTS ON SPIN-WAVE STATES IN SUBSTITUTIONALLY DISORDERED ANTIFERROMAGNETS[*]

D. C. Licciardello[†], E. N. Economou and
C. T. Papatriantafillou

Department of Physics, University of Virginia
Charlottesville, Virginia   22901

ABSTRACT

A recent theory by Economou of magnetic excitations in substitutionally disordered antiferromagnets is briefly presented and compared with the theories of Buyers et al and Lyo. Emphasis is given to the interrelation of the different approaches. Some further experimental work is proposed as well as theoretical improvements by incorporating off-diagonal randomness.

Considerable attention has been focused recently on the various excitations in substitutionally disordered crystals. To a large extent agreement between theory and experiment has been markedly qualitative due to both the crude nature of the model calculations and the rarity of simple measurable systems. These problems are circumvented, however, in the case of simple antiferromagnets which are not uncommon and have been widely investigated in their pure forms. Moreover, the exchange interactions are short range and well known which make them quite amendable to theoretical description.

Buyers et al[1] have recently studied the spin wave spectra in several single crystals of the systems $K(Co,Mn)F_3$

[*]Supported by the Center for Advanced Studies at the University of Virginia.
[†]NDEA Fellow.

and $(Co,Mn)F_2$ by neutron inelastic scattering. Their results provide well-defined magnetic dispersion branches at selected "impurity" concentrations which permit a comparison with existing theories of substitutional disorder where quantities like the density of states are available. Furthermore, the neutron data yields information about the nature of these states via the excitation widths in k-space.

The most successful method for treating disorder of the present type has been the coherent potential approximation[2] (CPA), a self consistent average medium theory, which allows an interpolation between the concentration extremes. Although most work has been done within the single-site approach, considerable attention has recently been given to effects due to fluctuations within a larger cluster[3] including potential variation associated with off-diagonal randomness[4].

Due to the presence of two magnetic ions per unit cell in the simple antiferromagnets under discussion, some extensions of the traditional approaches are in order. Economou[5] considered a two sublattice model Hamiltonian where each ion is coupled to only one set of neighbors on the opposite sublattice. Moreover, the square of the linearized Heisenberg Hamiltonian, $\hat{H}$, was examined rather than $\hat{H}$ itself. $\hat{H}^2$ can be separated into terms referring exclusively to each of the magnetic sublattices and one which weakly couples them through small matrix elements of random sign. In the periodic case this weak coupling vanishes exactly and in the present disordered case it is assumed zero for all concentrations as an approximation. It should be noted, however, that the two sublattices interact via the spin-coupling matrix elements which are present in H before it is squared. These coupling terms appear finally in the part of $\hat{H}^2$ which is retained.

The above procedure and approximations reduce the problem to a consideration of a Hamiltonian which refers to only one of the magnetic sublattices and whose matrix elements are of the form

$$<\underline{n}|\hat{H}_1|\underline{m}> = \varepsilon_{\underline{n}} \, \delta_{\underline{n}\underline{m}} + V_{\underline{n}\underline{m}} \qquad (1)$$

where $\underline{n}$, $\underline{m}$ refer to lattice sites in one of the magnetic sublattice and $V_{\underline{n}\underline{m}}$ is different from zero for both nearest and

next nearest neighbor sites. Both the diagonal and off-
diagonal matrix elements are random variables with probability distributions which can be easily expressed in terms
of the concentration and the spin-spin coupling matrix
elements.

To simply the calculations the probability distribution
of ε_n has been approximated by

$$P(\varepsilon_n) = x\delta[\varepsilon_n - \varepsilon_A(x)] + (1 - x)\delta[\varepsilon_n - \varepsilon_B(x)] \quad , \quad (2)$$

where $\varepsilon_A(x)$ and $\varepsilon_B(x)$ are the conditional average of ε_n when
when the site n is occupied by an atom A or B respectively.
Thus the problem becomes equivalent to the random binary
alloy $A_x B_{1-x}$ characterized by the Hamiltonian (1). A
serious approximation is introduced by neglecting randomness
in the off-diagonal matrix elements in order to simplify the
self-consistent CPA theory. To partly remedy this the
following considerations were taken into account: If the
eigenstates were Bloch-type extended functions the general
localization theory[6] suggests that the proper procedure is
to average over the logarithms of V_{nm}. On the other hand,
if the eigenstates are perfectly localized on a given site,
e.g. on an A site, then a conditional average V_{nm}^A would be
more appropriate. As the concentration varies from 0 to 1
we pass from the one extreme to the other. It was assumed,
arbitrarily, that a proper V_{nm} for any concentration, x, is
given by a linear combination of the two extreme cases, the
coefficients being 1-x and x respectively. Similar considerations were used for the other band. This approach
improves to a small degree the quantitative agreement
between theory and experiment especially for the position
of the lower band edge of the lower band (see Ref. 5).
Moreover, the serious underestimation, resulting from the
neglect of off-diagonal randomness, of the regions of
localized states is not affected by these approximations.

It is instructive to compare this theory with the
recent work of Buyers et al[7] who followed a different
approach in extending the single site CPA to simple antiferromagnetic systems. They considered $\hat{H}$ itself and
employed a matrix formulation to take into account the
presence of two sublattices. Moreover, using the basic
results of the CPA they calculated in detail the dispersion
relations, widths, relative intensities and line shapes of

the excitations, providing thus a good direct comparison with neutron scattering experiments.

Economou[5] considered the "propagator" $\hat{R}(E^2)$ defined as $\hat{R}(E^2) = 1/(E^2 - \hat{H}^2)$ which for numerical simplicity was taken to predict a semi-circular density of states for the periodic system. The transformation which relates the matrix Green's function, $g(\vec{k},\omega)$, of Buyers et al to the scalar $\langle\vec{k}|\hat{R}|\vec{k}\rangle$ is given by

$$\mathbb{1}\langle\vec{k}|\hat{R}(\omega^2)|\vec{k}\rangle = g(\vec{k},\omega)g(\vec{k}, -\omega) \quad , \tag{3}$$

where $\mathbb{1}$ is the unit 2 x 2 matrix. The approximation of ignoring the coupling between the two magnetic sublattices in $\hat{H}^2$ can be shown to be analogous to the Buyers et al neglect of the off-diagonal elements in the self-energy.

Economou's approximation[5] of neglecting randomness in the off-diagonal matrix elements of $\hat{H}^2$ is analogous to Buyers' et al approximation of neglecting randomness in the transverse interactions. It may be noted, however, that the two approximations are not equivalent since the diagonal matrix elements of $\hat{H}^2$ contain both the transverse and longitudinal interactions so that randomness in the transverse part is included in Economou's $\hat{H}^2$ approach. On the other hand, the simple "binary alloy" type probability distribution for these matrix elements employed by Economou probably compromises this advantage. Indeed Buyers et al has shown that more realistic probability distributions in their single site theory leads to better agreement with experiment especially near the zone center in the dispersion relations. Study is presently underway to incorporate the effects of randomness in the off-diagonal matrix elements using either the $\hat{H}^2$ approach or Buyers et al matrix formulation.

Even more interesting is the series of conjectures and theories developed to explain the nature of states within these magnon spectra. Buyers et al[1] interpreted their experimental result of localized eigenstates for finite x as a manifestation of randomness induced localization as was originally proposed by Anderson[8]. Economou and Cohen[9] developed a theory within the framework of Anderson's model, of which the basic result is the existence of a localization function $L(E)$ such that to every value of E satisfying $L(E) > 1$, there corresponds extended eigenstates. On the other hand, if $L(E) < 1$, either there exist no eigenstates of energy E or they are localized. At the critical energies E_c, termed mobility edges where $L(E_c) = 1$, the character of the states changes abruptly from localized to propagating. A very general but complicated expression for $L(E)$ has been obtained involving quantities like

$$\tilde{G} \equiv \left\langle \ell n \left| \frac{1}{E - \varepsilon_{n_i} - \Delta_{n_i}^{0,n_1,\ldots n_{i-1}}} \right| \right\rangle$$

where $\Delta_{n_i}^{0,n_1,\ldots n_{i-1}}$ are self energies (see Ref. 9). The most crude approximation one can make in calculating $L(E)$ is to ignore all the self energies, Δ. We then obtain a result for $L(E)$ which, although it contains some correct features, nevertheless disagrees with well known percolation theory limits[9]. Such a result was arrived at by Ziman[10] and recently rederived for the magnon case by Lyo[11]. A much better approximation is the $F(E)$ method[9,12] in which the CPA is employed to calculate G. It can be shown[9] that the $F(E)$ approximation is such that the effect of randomness on localization is always underestimated. What is more important, the $F(E)$-method predicts correctly[12] the percolation limits as well as the periodic limit. It should be pointed out that there is no contradiction in using an effective medium theory to give information about localization. The CPA is used to approximate a quantity which is known to predict the nature of the states, namely $L(E)$ (and not <G> which does not provide any information about localization).

One should keep in mind that only indirectly can one obtain some information about the localization of the states from neutron scattering data. As a result of that the available data are consistent with both the crude estimate of $L(E)$ resulting from neglecting Δ altogether and the more sophisticated $F(E)$-method. Nevertheless further experiments with $KCo_xMn_{1-x}F_3$ for $.4 < x < .6$ and for $x \sim .9$ could serve as a check of these two approximations. In addition the study of the upper mode in $Co_xMn_{1-x}F_2$ for x in the range $.4 < x < .7$ would be extremely interesting since a non-monotonic behavior of localization as a function of x is expected in this region.

REFERENCES

1. W.J.L. Buyers, T.M. Holden, E.C. Svensson, R.A. Cowley, and R.W.H. Stevenson, Phys. Rev. Lett. 27, 1442 (1971).
2. D.W. Taylor, Phys. Rev. 156, 1017 (1967); P. Soven, ibid. 156, 809 (1967); B. Velický, S. Kirkpartick, and H. Ehrenreich, Phys. Rev. 175, 747 (1968).
3. K.F. Freed and M.H. Cohen, Phys. Rev. B3, 3400 (1971).
4. E-Ni Foo, S.M. Bose and M. Ausloos, to be published and references therein.
5. E.N. Economou, Phys. Rev. Lett. 28, 1206 (1972).
6. E.N. Economou, Journal de Physique, to be published.
7. W.J.L. Buyers, D.E. Pepper, and R.J. Elliott, to be published.
8. P.W. Anderson, Phys. Rev. 109, 1492 (1958).
9. E.N. Economou and M.H. Cohen, Phys. Rev. B5, 2931 (1972).
10. J.M. Ziman, J. Phys. C: Proc. Phys. Soc., London 2, 1230 (1969).
11. S.K. Lyo, Phys. Rev. Lett. 28, 1192 (1972).
12. E.N. Economou, S. Kirkpartick, M.H. Cohen, and T.P. Eggarter, Phys. Rev. Lett. 25, 520 (1970).

DISCUSSION

S.C. Moss: Did you include manganese-cobalt interactions?

D.C. Licciardello: The formulation includes nearest neighbor spin-spin coupling between all possible constituent pairs, i.e. Co-Co, Mn-Mn, Co-Mn. The manganese-cobalt interaction can be determined by requiring the theory to predict a single defect localized mode which has been measured.

S.C. Moss: Is there evidence for any chemical short range order? Does the manganese want to surround itself with manganese?

W. Buyers: There is no evidence for chemical short range order. The manganese-cobalt interactions are probably on a magnetic scale of 10^{12}. Measurements show no evidence for chemical order in the diffraction patterns.

AMORPHOUS FERRO- AND ANTIFERROMAGNETISM WITHIN AN RPA-CPA

THEORY

Raza A. Tahir-Kheli

Department of Physics, Temple University

Philadelphia, Pennsylvania 19122

FORMALISM

We consider the system to be governed by a Heisenberg spin Hamiltonian with vanishing external field,i.e. $\mu \to 0$

$$H= A \sum_{f,p} J(f,p) \; S_f \cdot S_p \; - \; \mu \sum_f S_f^z \qquad (1)$$

Here A=-1 for the ferromagnetic case and +1 for the anti-ferromagnetic case. The exchange integral between any site p and its surrounding shell of atoms (which we take to consist of z atoms on the average) is to be random with a probability distribution P(J) such that P(J) is zero outside the range $J_2 > J > J_1$ and

$$\int_{J_1}^{J_2} P(J) \; dJ= 1 \; . \qquad (2)$$

(Note that it is the fluctuations in the neighboring environment, in general, and in the separation vectors (f-p) as compared with the corresponding vectors for a crystalline lattice of coordination z , in particular, which cause these fluctuations in the exchange integral J(f,p).

The central physical assumption of the present work is that (a) the long wave-length collective magnetic phenomena are governed primarily by the structure of the fluctuations in the exchange integral J(f,p) and (b) that these phenomena are not strongly dependent upon the structure of the fluctuations in the vectors(f-p) themselves. This is to say that a model assuming a crystalline lattice structure but

393

appropriately random exchange integrals should possess the
salient features of the amorphous system. We expect,however,
that in systems with strong spin-lattice coupling this assumption will be inaccurate.

For antiferromagnets an additional assumption about
the character of the ground state is needed. We assume that
if antiferromagnetic long range order (LRO) obtains, it
does so in an interpenetrating two sub-lattice pattern such
that all the atoms in the first shell around any given atom
(f,λ) belong to the sub-lattice $-\lambda$.

It is convenient to view the approximations necessary
to do the statistical mechanics of such systems as having
two features. Firstly, they truncate and linearize the
infinite hierarchy of coupled non-linear equations describing interactions and secondly they tell us how to handle the
randomness. While both these features are inextricably
mixed together, we can usefully talk about an RPA like decoupling[1] to handle the former and two site T-matrix truncation type of coherent potential approximation (CPA) to
deal with the latter[2]. The result of such a treatment is
that the retarded equilibrium self-Green's function, for
the amorphous ferromagnet, is given by the following relation

$$G_{g;g}(E) = <<S_g^+(t);\exp(aS_g^z(t'))S_g^-(t')>>=\Omega(a)\Gamma(E) \quad (3a)$$

where
$$\Omega(a) = <[S_g^+ , \exp(aS_g^z)S_g^-]> \quad (3b)$$

$$\Gamma(E) = (N)^{-1}\sum_k [E - 2Mzj(1-\gamma_k)]^{-1} \quad (3c)$$

$$M = <S_g^z> \quad ; \quad \gamma_k = (z)^{-1} \sum_f \exp[ik.(g-f)] \quad (3d)$$

For the amorphous antiferromagnet, the same Green's function
for the sub-lattice λ is (note that $<S_{g,\lambda}^z>=- <S_{g,-\lambda}^z>=M_\lambda$)

$$G_{g,\lambda;g,-\lambda}(E)=\Omega_\lambda (a) (N)^{-1} \sum_k \Gamma_k^{\lambda\lambda}(E) \quad (4a)$$

$$\Gamma^{\lambda\lambda}(E)=(E+2M_\lambda zj)(E^2-(2M_\lambda zj)^2(1-\gamma_k))^{-1}$$
$$\Gamma^{\lambda\lambda}(E) = (N)^{-1}\sum_k \Gamma_k^{\lambda\lambda}(E) \quad (4b)$$

Here the CPA like coherent exchange integral, j, is a
function of the frequency E and it is determined from the
following integral relations for the ferro- and antiferromagnetic cases respectively:

$$\int_{J_2}^{J_1} dJ (J-j) P(J) (zj-2(J-j)(E\Gamma(E)-1))^{-1}=0 \quad (5)$$

$$\int_{J_2}^{J_1} dJ \ (J-j)P(J) \ (zj-2(J-j)\sigma(E))^{-1} = 0 \qquad (6a)$$

where $\sigma(E) = -1 + E^2(N)^{-1} \sum_k \ (\ E^2 - (2M_\lambda zj)^2 (1-\gamma_k^2)\ )^{-1} \qquad (6b)$

In three dimensions Eqs. (5) and (6a,b) can only be solved numerically. For given E, one first computes the coherent exchange integral $j(E)$. Next, one determines the corresponding density of states from the relations:

$$\rho(\omega) = -(\pi)^{-1} Im(\Gamma(\omega+i\varepsilon)); \quad \rho_\lambda(\omega) = -\pi^{-1} Im(\Gamma^{\lambda\lambda}(\omega+i\varepsilon)) \ (7)$$

The ferromagnetic (or the λ-sub-lattice antiferromagnetic) magnetization is obtained from these density of states in the usual fashion[3] and we get

$$M = \frac{(S-\Phi\nu)(1+\Phi\nu)^{2S+1} + (S+1+\Phi\nu)\ (\Phi\nu)^{2S+1}}{(1+\Phi\nu)^{2S+1} - (\Phi\nu)^{2S+1}} \qquad (8)$$

Here S is the magnitude of the spin in Dirac's units and

$$\Phi\nu = \int_{-\infty}^{+\infty} d\omega \ \rho_\nu(\omega)(e^{\beta\omega}-1)^{-1}; \ \beta = (k_B T)^{-1} \qquad (9)$$

The index $\nu\equiv\lambda$ if we refer to the antiferromagnetic case for the λ sub-lattice. For the ferromagnetic case Eqs. (8) and (9) areto be written without the suffix ν.

THE APPROXIMATION

The treatment presented above is subject to several different types of errors. Firstly, the use of the RPA introduces errors into the description of the quantum statistical mechanics of the interacting spin system. Secondly, the assumption that the disorder dynamics of the many spin system can be approximated by the use of a two-site CPA[2] is only approximately correct and consequently its use is subject to error. Thirdly, the numerical computations to be presented below for the z=6 case are subject to numerical errors inherent in our computational scheme. Finally, the inadequacies of the model itself(i.e. the description of the amorphous system through the use of random exchange integral only) will be reflected in the results obtained.

Any discussion of the accuracy of the RPA is complicated because here the RPA is being used on a spatially disordered system. Even for the ordered system(i.e. where the Heisenberg spin Hamiltonian does not contain spatially random exchange integrals) the precise nature of the level

of approximation inherent in the use of the RPA for studying the spin dynamics is not clear. It has, however, been
noticed[1] that the non-dynamical, thermodynamic results obtained by the use of the RPA for such a system are much like
those following from the use of the spherical model[4]. Consequently, the usual statement that the thermostatics of
the spherical model approximates that of the Heisenberg
model to the two leading orders in an asymptotic large z
expansion, implies that at least non-dynamically the RPA
represents the next stage beyond the mean-field approximation.
In any event, we can show that our two-site T-matrix trunc-
tion scheme leads to a density of states which exactly
preserves all the frequency moments to the two leading orders
in z^{-1}.

The numerical accuracy of the present computations is
such that the density of states is computed to an accuracy
of about 2.5% except in the immediate vicinity of zero frequency. There the density of states is small(except for
situations close to the percolation limit) and the relative
percentage accuracy of our computations can be quite large.
The net result of this inaccuracy is that the computation
of the(sub-lattice)magnetization near the magnetic transition temperature can be in somewhat greater percentage error
than it is elsewhere in the interval $0<T<T_c$.

COMMENTS AND RESULTS

The renormalization of the long wave-length,i.e. $k \to 0$,
spin waves determines the critical temperature of the system[5]
For $k<0$, Eqs. (5) and (6a,b) yield the appropriate zero frequency coherent exchange j_o through the relation

$$\int_{J_1}^{J_2} dJ \, (J-j_o) \, P(J) \, (\, (z-2)j_o - J \,)^{-1} = 0 \qquad (10)$$

The interesting thing about this relation is that it is the
same for the ferr- and the antiferrmagnetic cases. (The
reason this happens lies in our approximation procedure,
namely the use of the RPA and the quasi-classical two sub-
lattice model of anti-ferromagnetism. Note that for the pure
non-random Heisenberg ferro- and antiferromagnets the Curie
and the Néel temperatures are the same if we use the RPA
and the spin-wave theory antiferro- ground state which
uses the quasi-classical model for its representation).

The magnetic transition temperature, T_c or T_N, is

$$T(\text{transition})=(2/3)S(S+1)zj_o/[(1/N)\sum_k(1-\gamma_k)^{-1}] \qquad (11)$$

Defining a virtual-crystal-RPA theory transition temperature T(vc) through the relation

$$[T(vc)/T(transition)] = \int_{J_1}^{J} 2P(J)JdJ/j_0 , \qquad (12)$$

one finds that for a wide variety of of probability distributions, P(J), the ratio of these temperatures given in
(12) is greater than unity. In other words,T(vc)>T(transition). These two temperatures are equal only for the pure
cases; otherwise, the amorphous sytem always seems to
lose magnetic LRO at a lower temperature than the corresponding pure system with exchange integral equal to the
average of that in the amorphous system. Another interesting
feature of the results given by Eq.(10) is that the inclu-
sion of small amounts of negative exchange integrals,i.e.
$J_1<0$, can cause a rapid breakdown in the magnetically ordered state. The magnitde of such negative J's depends upon the
relevant probabilities, P(J), and the coordination z. Details
of these results and the mathematics used here to derive
Eqs. (3a)-(12) will be presented elsewhere.

The density of states for the amorphous ferromagnet
have also been computed for a variety of distributions P(J).
The general feature of these results is a depletion of the
higher energy states and a corresponding increase in the
density at lower frequencies[6]. Near the points(in the J
space) for which the magnetic LRO breaks down the enhanc-
ment in the low frequency density of states becomes particularly pronounced.

For a very simple model of exchange randomness, i.e.
$$P(J) = c \, \delta(J - J_0) + \delta(1-c) \, \delta(J)$$
which corresponds to the case where those neighboring spins
which are closer than a certain distance interact via an
exchange integral equal to J_0 while others do not interact,
the properties of the system are particularly revealing.Here

$$T(transition)/T(vc)=c^{-1}(c-2/z)/(1-2/z) \qquad (13)$$

For concentrations $c \leq c_0$ where $c_0=2/z$, the system does not
possess magnetic LRO because here T(transition) is a negative quantity. This citical concentration c_0 defines the
percolation limit.

In the vicinity of the percolation concentration,i.e.
$c-c_0 <<c_0$, from equations (3a-d),(5),(8) and (9) one can
show that the tremendous enhancement in the density of states near small frequencies results in the appearance of
a critical mode of frquency ω_c (which may or may not be
a propagating mode) which one may assign a wave-vector
k_c. The dependence of these parameters on the density

reveals an interesting set of exponents, i.e.

$$\omega_c \simeq (c-c_o)^{5/3}; \rho(\omega_c) \simeq (c-c_o)^{-2/3}; k_c \simeq (c-c_o)^{1/3} \qquad (14)$$

It is interesting to inquire as to whether such a strong
enhancement in the density of states will not affect the
low temperature thermodynamic properties of the system in
a similarly dramatic fashion. In particular, one may expect
the specific heat to be dramatically enhanced due to the
availability of a large density of states at such low frequency.
One can show, however, that the form of the low
temperature magnetization as well as the specific heat remain
unaltered as $c \to c_o$. Only the coefficient of the $(T/T_c)^{3/2}$
term, and possibly also of the higher order temperature dependent
terms, gets modified. But this modification does not
occur like any divergent power of the factor $(c-c_o)$. Rather,
it is only a few times the corresponding factor for the c=1
case.

For the amorphous ferromagnetic we have computed the
magnetization using the extremely simple model given above.
The structure of the results at the low temperature end
agrees with the foregoing prediction and as we proceed from
the pure limit c=1 to the percolation limit c=0.33, the
coefficient of the temerature dependent term varies by about
an order of magnitude.

In conclusion, it should be stated that our model of
amorphous magnetism is essentially similar to that used previously
in the literature.[7] The only difference in the present
work and the previous works is that they specifically
assumed only symmetrical fluctuations of the exchange integral
around its mean. Secondly, our use of the two-site CPA
contrasts with a virtual-crystal or perturbation(2nd order)
approximation of the previous works. The assymptotic accuracy
to the two orders in (1/z) is also new. Therefore, the
previously not reported results relating to the critical
concetration region as well as the restatement of the previously
reported results relating to the lowering of T_c and
the computation of the magnetization that emerge from the
present analysis are based on at least a qualitatively sound
basis. The present analysis ,however, is incomplete in the
sense that the critical exponents of the various thermodynamic
funtions are determined only with respect to the concentration
c (that is, the system temperature was assumed to be
far from T_c). Future work, using the same physical assumptions
as this work, will investigate the region near T_c more
fully.

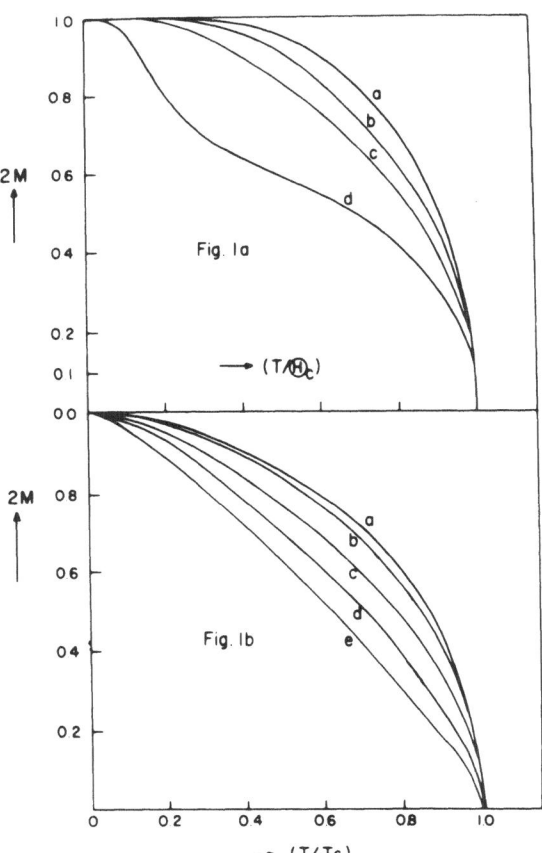

Figure 1: Magnetization curves for the model P(J) =
$c\delta(J-J_0) + (1-c)\delta(J)$. In Fig. 1a we give the
relative magnetization for S = 1/2. Θ_c
= $3J_0c$. Curves a,b,c, and d are given by
Kaneyoshi's theory[8] for concentrations 1, 0.6,
0.4, and 0.2 respectively. Our results are
given in Fig. 1b. Curves a, b, c, d, e, are
for c = 1, 0.8, 0.5, 0.4, 0.35.

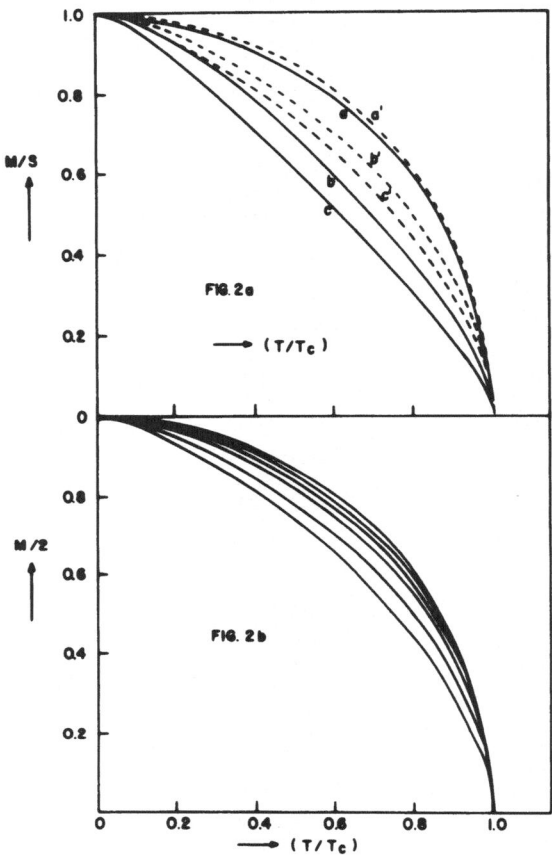

Figure 2: (M/S) for spins 1/2 and 2 are plotted in Fig.
2a against their relative temperatures T/T_c.
Here c = 1, 0.4 and 0.35 for curves (a, a'),
(b, b') and (c, c'). Broken curves are for
S = 2 and full ones for S = 1/2. For S = 2,
M/2 is plotted in Fig. 2b. The curves from
top to bottom respectively refer to concentrations 1.0, 0.7, 0.6, 0.5, 0.4, and 0.35.

REFERENCES

1. R. Tahir-Kheli and D. ter Haar, Phys. Rev. 127, 88 (1962).
2. R. Tahir-Kheli, Phys. Rev. B6, 2808, 2826, 2838 (1972).
3. H.B. Callen, Phys. Rev. 130, 890 (1963).
4. T.H. Berlin and M. Kac, Phys. Rev. 86, 821 (1952).
5. G.A. Murray, Proc. Phys. Soc. (London) 89, 87 (1966).
6. J.E. Gubernatis and P.L. Taylor (paper in this issue).
7. K. Handrich and S. Kobe, Acta Phys. Polon. A38, 819 (1970); C.G. Montgomery et al., Phys. Rev. Letters 25, 669 (1970); A.W. Simpson and D.R. Brambley, Phys. Stat. Sol. (b) 49, 685 (1972).
8. T. Kaneyoshi, Prog. Theoret. Phys. 44, 328 (1970).

DISCUSSION

M.E. Collins: You talked about the density of states in the specific heat. Would you like to say what would happen to the ferromagnet and antiferromagnet if you actually measured the spin wave dispersion curves?

R.A. Tahir-Kheli: Dispersion curves relate the spin wave energy E_k to k. On the other hand, the single site density of states, $\rho(\omega)$, is related to an integral over all the wave vectors. Therefore, the connection between $\rho(\omega)$ and the E_k is complicated. One can, however, look at $\rho(\omega)$ and make some qualitative statements about E_k. This we do by stating that when c is close the the percolation limit c_0, the wave vector k_c and the spin wave energy, ω_c, have certain peculiar c dependence. Now, the fact that $\rho(\omega_c)$ is large can figure into, say, neutron scattering measurements of the spin wave dispersion. Here, the inelastic neutron scattering will become large as long as (1) the system temperature is such that ω_c spin waves can be excited and (2) the wave vector k_c and ω_c satisfy the usual momentum-energy conservation requirement for neutron scattering.

M.E. Collins: By strong scattering you mean the energy is low?

R.A. Tahir-Kheli: What I mean is that near the percolation limit, there is a large density of states available in the neighborhood of ω_c.

M.E. Collins: Do all these spin waves have the same energy?

R.A. Tahir-Kheli: No. They have different energies. Because
there is a whole lot of density of states available for
$\omega \simeq \omega_c$ we expect that (if $k_B T > \omega_c$) there will be a whole
lot of spin-waves around for these energies. Now, whether
these spin waves are localized or propagating, one does not
know. I had a short conversation with S. Kirkpatrick regarding this point and my feeling is that these spin waves
are part localized and part propagating. In any event, the
statement that they refer to wave vectors $k \simeq k_c$ is not made
with any great confidence. Therefore, precisely how, i.e.,
for what range of wave vectors, they will show up in neutron
scattering experiments can also not be stated with confidence. What can, however, be said is that there is a whole
lot of density of states available to these spin waves and
given $k_B T > \omega_c$, they should be around, in large numbers.
Secondly, we stated that the low temperature thermodynamic
effects of these spin-waves, e.g. magnetic specific heat,
temperature dependence of the magnetization etc., will not
be as dramatic as their large density of states might otherwise suggest.

S.M. Bose: How is your calculation any different from Wu
and Foo's [Phys. Rev. B5, 98 (1972)]? Your approximations
and final equations are exactly the same as theirs (cf. Eq.
4 of Wu and Foo).

R.A. Tahir-Kheli: Fortunately, my approximations and final
equations are not exactly the same as Wu and Foo's. If they
were, they would give many misleading and wrong results!

S.M. Bose: When you say their calculation is wrong, do you
mean they have mistakes in the numerical part of their calculations?

R.A. Tahir-Kheli: What I mean by wrong is the following:
The Wu-Foo work deals with a binary ferromagnetic alloy.
They have three exchange parameters J^{AA}, J^{AB} and J^{BB}. To
take a particular instance consider the limit $J^{BB} = J^{AB} = 0$.
In this limit their alloy reduces to a dilute ferromagnet
with magnetic concentration c^A and exchange J^{AA}. Now for
this system their results are that the Curie temperature
scales quadratically with the magnetic concentration. Also,
they find the percolation limit to be about 58% for a simple
cubic lattice. These results are clearly wrong, for all the

existing theories of the dilute ferromagnet give a linear
dependence of the Curie temperature on c^A. Also the percolation limit is generally thought to be close to half of
what they predict. This is not all! One can show that
their theory gives a total single-site density of states
which exactly preserves only the zeroth and the first frequency moments. Note that this is as poor as a mean-field
approximation theory. However, in contrast with a correct
version of such a theory, and contrary to the statement of
the authors, the Wu-Foo theory is not correct in the mean-
field $z \rightarrow \infty$ limit (i.e., all the frequency moments are in
error in the leading order in z^{-1}). Now regarding their
computations, there is only one error. This is purely an
oversight and is not too significant. When they computed
the coherent exchange, $J^{(E)}$, at one point in the computation they introduced an erroneous factor $1/2$ in the second
term in the denominator of their Eq. (4). For the densities
of states computed in their paper, this omission causes
only $\simeq 2\%$ error which is insignificant. For the dilute
ferromagnetic case (for which they did not in their paper
supply the density of states) the omission of this factor
is serious especially near the percolation limit. Finally,
I might add that my theory for the dilute ferromagnet is
given in Phys. Rev. **B6**, 2808 (1972).

R. Hasagawa: In your last slide you showed the decrease of
T_N with respect to $\Delta J/J$. If you give the sign of ΔJ, what
do you expect? I ask this for in the molecular field theory,
T_N or T_c can go up or down depending on the sign of J.

R.A. Tahir-Kheli: There is a slight confusion. ΔJ is a
measure of the fluctuation around the mean J. What I am
saying is that T_N or T_c, measured in the units of T_N (pure)
or T_c (pure), where the latter corresponds to the pure case
with exchange equal to the mean J, behaves in a certain
fashion when ΔJ increases from zero to a certain maximum
value. At the maximum value the system undergoes a first
order (in many cases) or a second order (in some special
cases) phase transition and the state, with the original
type of magnetic LRO, breaks down. [Authors Note: this
slide was omitted from the manuscript for reasons of space].

R. Hasagawa: How about (T_N or T_c) going up?

R.A. Tahir-Kheli: For the P(J)'s that I have studied, it
has always gone down.

SPIN WAVES IN AN AMORPHOUS HEISENBERG FERROMAGNET*

J.E. Gubernatis[+] and P.L. Taylor

Case Western Reserve University
Physics Department
Cleveland, Ohio 44106

Theoretical investigations of the ferromagnetic properties of noncrystalline solids have taken two directions - mean-field theories and Green-function calculations. The Green-function calculations[1-3] have been applied only to disordered alloys or, at least, to systems in which a Bravais lattice exists; on the other hand, only mean-field theories have been applied to truly amorphous systems. As an approach to investigate the properties of an amorphous ferromagnet, the mean-field calculations[4-7] have been deficient in several respects. They have not taken into proper consideration the structure of the amorphous material. Furthermore, although experiments have indicated that structural order reduces the spontaneous magnetization σ and Curie temperature T_C of the amorphous material below that of the corresponding crystalline problem, some mean-field calculations[6,7] predict an increase in σ and T_c. We now present a Green-function calculation of the spectrum of an amorphous Heisenberg Ferromagnet which takes into consideration the characteristic structure of an amorphous material and which also predicts a reduction of σ and T_c. The central point of the method is the use of a transformation, suggested by Taylor and Wu[8] (TW) in a discussion of the dynamics of atomic motion in glasses, which allows the long-range correlations in the low-frequency modes to be accurately taken into account.

In the Green-function approach the frequency density of states per spin, $g(\omega)$, is found from

$$g(\omega) = -\frac{1}{\pi N} \text{ Im } \sum_{\vec{\ell}} D(\vec{\ell},\vec{\ell},\omega+i0) \tag{1}$$

where $D(\vec{\ell},\vec{\ell},\omega)$ is the appropriate Green function. For a system of N spin-1/2 particles we obtain in the RPA the following equation of motion of a spin wave with energy ω:

$$\sum_{\vec{\ell}'} [(\frac{\omega}{\sigma} - \sum_{\vec{\ell}''} J(\vec{\ell},\vec{\ell}'')) \delta_{\vec{\ell}\vec{\ell}'} + J(\vec{\ell},\vec{\ell}')] B(\vec{\ell}') = 0$$

where σ is the average magnetization per spin, $B(\vec{\ell})$ the Pauli spin-wave creation operator for site $\vec{\ell}$, and $J(\vec{\ell},\vec{\ell}')$ the exchange interaction between sites $\vec{\ell}$ and $\vec{\ell}'$. The Green function $D(\vec{\ell},\vec{\ell},\omega)$ is now defined from Eq. 1 by

$$\sum_{\vec{\ell}''} [(\omega+i0)\delta_{\vec{\ell}\vec{\ell}''} - V(\vec{\ell},\vec{\ell}'')] D(\vec{\ell}'',\vec{\ell}',\omega) = \delta_{\vec{\ell}\vec{\ell}'} \tag{2}$$

where

$$V(\vec{\ell},\vec{\ell}'') = \sum_{\vec{\ell}'} J(\vec{\ell},\vec{\ell}')\delta_{\vec{\ell}\vec{\ell}''} - J(\vec{\ell},\vec{\ell}'') .$$

Expressed more simply in operator form the equation for the Green function becomes

$$D = D_o + D_o VD \tag{3}$$

where $D_o = (\omega+i0)^{-1}I$ with I the identity operator.

In the treatment of a perfect lattice, since the trace in Eq. 1 is independent of representation, one would transform D to a wave-number representation by the operator which we will call S. The matrix elements of S are

$$S_{\vec{\ell}\vec{q}} = N^{-\frac{1}{2}} e^{-i\vec{q}\cdot\vec{\ell}} .$$

Since S would be unitary, its adjoint E would be its inverse, that is, ES = I. In the case of an amorphous solid, S is no longer unitary; however, following the suggestion of TW, we write

$$ES = I + R$$

from which it follows that

$$S^{-1} = (I+R)^{-1}E \tag{4}$$

where $E = S^{\dagger}$ and S is defined as in the case of a perfect crystal. The operator R represents the structural disorder

in the system. In the special case of a Bravais lattice
R = 0 and the transformation reduces to the usual traveling-
wave description of the normal modes of the system. Trans-
forming Eq. 3 to the wave-number representation defined by
S, we obtain

$$\tilde{D} = D_o + D_o S^{-1}VS \, \tilde{D} \tag{5}$$

where $\tilde{D} = S^{-1}VS$. Again following the suggestion of TW, we
decompose $S^{-1}VS$ by writing

$$S^{-1}VS = V_o + W \tag{6}$$

where V_0 is chosen to be diagonal in wave-number space and
W is therefore the off-diagonal part. Equation 5 can now be
written as

$$\tilde{D} = \tilde{D}_o + \tilde{D}_o W\tilde{D} \tag{7}$$

where $\tilde{D}_o = (\omega+i0 - V_o)^{-1}$.

The Green-function $\tilde{D}$ in Eq. 7 is thus expressed as a
perturbation-series expansion in terms of W, the off-diagonal
elements of the transformed interaction. An explicit form
for W can be obtained from TW's Eqs. 14-19. It is shown that

$$(S^{-1}VS)_{\vec{q}\vec{q}'} = N^{-1} \sum_{\vec{q}''} [I+R)^{-1}]_{\vec{q}\vec{q}''} \sum_{\vec{\ell}} e^{i(\vec{q}''-\vec{q}')\cdot\vec{\ell}}$$

$$\times \sum_{\vec{L}(\vec{\ell})} V(\vec{\ell},\vec{\ell}+\vec{L}) \, e^{-i\vec{q}\cdot\vec{L}(\vec{\ell})}$$

where $\vec{L}(\vec{\ell}) = \vec{\ell}'-\vec{\ell}$ and the set of lattice vectors $\{\vec{L}(\vec{\ell})\}$ represents the configuration of lattice sites for the solid
as seen from site $\vec{\ell}$. Next the sum over $\vec{L}$ is written as

$$\sum_{\vec{L}} V(\vec{\ell},\vec{\ell}+\vec{L}) \, e^{-i\vec{q}\cdot\vec{L}} = V_o(\vec{q}) + U(\vec{\ell},\vec{q}') \tag{8}$$

with

$$\sum_{\vec{\ell}} U(\vec{\ell},\vec{q}') = 0 \quad ,$$

and then from Eq. 8 it follows that

$$W(\vec{q},\vec{q}') = \sum_{\vec{q}''} [(I+R)^{-1}]_{\vec{q}\vec{q}''} \sum_{\vec{l}} N^{-1} e^{i(\vec{q}''-\vec{q}')\cdot\vec{l}} U(\vec{l},\vec{q}'). \qquad (9)$$

One sees from Eq. 9 that the TW transformation introduces the structural disorder factor as a series expansion of $(I+R)^{-1}$ and also enables one to expand $\check{D}$ in terms of a site-dependent scattering factor $U(\vec{l},\vec{q})$. The presence of the factors of R poses no limitation to the use of the transformation since it can be shown[8] that the absence of long-range order in the material leads to a rapidly converging perturbation series for $\check{D}$ as a function of R.

In a perfect Bravais lattice the sum over $\vec{L}$ in Eq. 8 would be independent of $\vec{l}$ and only depend on $\vec{q}'$. This will be approximately the case in many amorphous materials, especially when the interaction is short-ranged. For example, in an amorphous material each lattice site has approximately the same number of nearest neighbors. As one goes from one lattice site to another only the orientation of these nearest neighbors varies, and with an interaction extending only over nearest neighbors the sum over $\vec{L}$ may only weakly depend on the orientation about each site. Accordingly, Eq. 8 may be considered as being decomposed into its average over all $\vec{l}$ and its site-dependent deviation from that average $U(\vec{l},\vec{q})$. This decomposition is particularly convenient for studying spin-waves in an amorphous ferromagnet.

In the present calculation we consider a model in which each spin is assumed to be surrounded by an equal number z of nearest neighbors at an equal distance from it. When z = 4 for example, it is assumed that these neighbors are located to form a tetrahedron; only the orientations of the tetrahedra vary from site to site. The exchange interaction is limited to nearest neighbors and is assumed to be a constant and independent of position or orientation of a pair of spins. It is further assumed that both the density of the material and the relative orientations of nearest-neighbor vectors for an atom are the same in the amorphous states as in the crystalline material. Within this model

$$V_0(\vec{q}) = Jz[1 - j_0(qa)] \qquad (10)$$

while

$$U(\vec{l},\vec{q}') = Jzj_0(qa) - \sum_{\vec{a}(\vec{l})} e^{-i\vec{q}\cdot\vec{a}} \qquad (11)$$

where $j_0(qa)$ is the zero-order spherical Bessel function and $\{\vec{a}(\vec{\ell})\}$ the set of nearest-neighbor vectors about lattice site $\vec{\ell}$. The evaluation of $\tilde{D}$ is more conveniently carried out by use of Dyson's equation than the direct use of Eq. 7. Dyson's equation may be expressed as

$$\tilde{D}(\vec{q},\vec{q},\omega+i0) = [\omega+i0 - V_0(\vec{q}) - M(\vec{q},\omega)]^{-1} \quad ,$$

and the Dysonian $M(\vec{q},\omega)$ is defined as by TW. Of the many diagrams expressing the perturbation series for $M(\vec{q},\omega)$, we judged that those shown in Fig. 1 were the most significant. In Fig. 1 the dashed lines represent the scattering function

$$T(\vec{\ell},\vec{q},\vec{q}') = N^{-1}e^{i(\vec{q}-\vec{q}')\cdot\vec{\ell}} U(\vec{\ell},\vec{q}) \quad ;$$

the cross, the lattice site $\vec{\ell}'$; and the solid line, the propagator $[\omega+i0 - V_0(\vec{q})]^{-1}$. In the actual calculation the propagator in Fig. 1(b) was renormalized in the usual way to obtain the self-consistent set of equations for $\tilde{D}$ depicted in Fig. 2.

The characteristic presence of short-range order and absence of long-range order in an amorphous material plays a significant role in the evaluation of the diagrams in Fig. 1. For example, in the double-site scattering diagram shown in Fig. 1(b), if the distance between the scattering sites is greater than several times the nearest-neighbor distance, then in the absence of long-range ordering the positions of the scattering sites are effectively uncorrelated. But since

$$\sum_{\vec{\ell}} U(\vec{\ell},\vec{q}) = 0 \quad ,$$

the contribution to the diagram from such sites will be negligible, with the major contribution coming from those sites associated with the short-range order. Similar remarks apply to Fig. 1(a) since the function R is site-dependent. We also found that for higher-order diagrams than those in Fig. 1 the scattering sites become uncorrelated at smaller distances. As a consequence, the perturbation series expansion for $M(\vec{q},\omega)$ is a rapidly converging series not only of R but also of U.

The results of our calculation[9] of amorphous spin-wave spectra for different numbers of nearest neighbors indicate that when compared to the corresponding crystalline case the effect of the disorder is to deplete the high-frequency modes while enhancing those of lower frequency. The disorder,

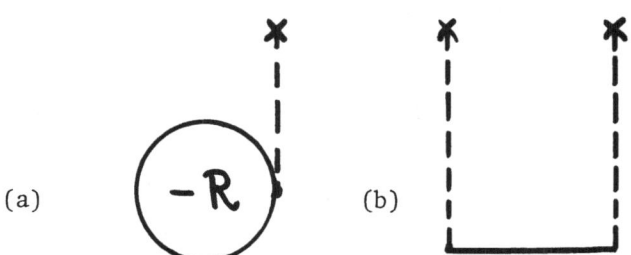

Fig. 1   Diagrams in the expansion of $M(\vec{q},\omega)$ that were evaluated.

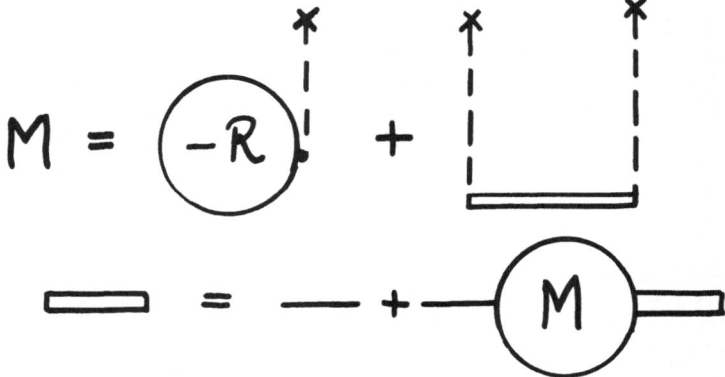

Fig. 2   Self-consistent set of equations for $\tilde{D}(\vec{q},\vec{q},\omega)$.

however, did not affect the modes of lowest frequency (the long-wavelength limit). The spin-wave spectrum for ω small in the amorphous case was found to be identical, within numerical error, to that for the corresponding crystalline case. When σ and T_c were calculated by a technique equivalent to that developed by Tyablikov[10] and later refined by Montgomery et al.,[1] it was found that both σ and T_c were reduced in comparison to the corresponding crystalline case; however, σ for both the crystalline and the amorphous case exhibited the same $T^{3/2}$ dependence at low temperatures.

*Work supported by the U.S. Atomic Energy Commission

†Present Address: Department of Physics, Ohio Northern
                  University, Ada, Ohio   45810

REFERENCES

1. C.G. Montgomery, J.I. Krugler, and R.M. Stubbs, Phys. Rev. Letters 25, 669 (1970).
2. E-Ni Foo and Der-Hsuch Wu, Phys. Rev. B5, 98 (1972).
3. R. Harris and M.J. Zuckerman, Phys. Rev. B5, 101 (1972).
4. K. Handrich, Phys. Stat. Sol.(b) 32, K55 (1969).
5. S. Kobe and K. Handrich, Phys. Stat. Sol.(b) 44, K33 (1971).
6. S. Kobe and K. Handrich, Fiz. Tverd. Tela. 13, 887 (1971) [Sov. Phys. - Solid State 13, 734 (1971)].
7. S. Kobe, Phys. Stat. Sol.(b) 41, K13 (1970).
8. P.L. Taylor and Shi-Yu Wu, Phys. Rev. B2, 1752 (1970).
9. Details of the calculation will be published elsewhere.
10. S.V. Tyablikov, Methods in the Quantum Theory of Magnetism (Plenum, New York, 1967).

DISCUSSION

R.M. Stubbs: For the Z=8 case, I was wondering if you saw any strange behavior in the disordered spin-wave spectrum at an energy corresponding to the infinity which exists in the spectrum of the perfect body-centered cubic case?

J.E. Gubernatis: We did not find any strange behavior. A common feature of all our calculations is the smoothing over of the singularities in the crystalline spectrum when disorder is added to the system.

DENSITY OF SPIN WAVE STATES FOR DISORDERED CUBIC FERROMAGNETS

R. M. Stubbs
NASA-Lewis, Cleveland, Ohio
and
C. G. Montgomery
University of Toledo, Toledo, Ohio

The interest in ferromagnetism in amorphous materials which first surfaced in a paper by Gubanov[1] was stimulated in the mid sixties by the discovery of several amorphous ferromagnets in various laboratories.[2] There has been some theoretical response to these experimental observations by Handrich[3] who used molecular field theory to show a decreased spontaneous magnetization arising from structure fluctuations and by Montgomery and coworkers[4] who applied a Green's function method to a disordered spin one-half Heisenberg ferromagnet of simple cubic topology. The present paper expands this latter work to show the effects of disorder in higher spin systems of single cubic (sc), body-centered cubic (bcc) and face-centered cubic (fcc) symmetry.

GREEN'S FUNCTION FOR DISORDERED FERROMAGNETS

In the absence of an external field a Heisenberg ferromagnet is described by the Hamiltonian

$$\mathcal{H} = -\tfrac{1}{2} \sum_{f,g} J(f,g)\, \vec{S}_f \cdot \vec{S}_g \tag{1}$$

where $J(f,g)$ is the exchange interaction between spins at sites f and g and $\vec{S}_f$ and $\vec{S}_g$ are the spin operators associated with these sites. In a perfect crystal all the exchange interactions between nearest neighbors are of equal magnitude; similarly all next nearest neighbors have equal interactions, and so on. When disorder is present in the lattice these exchange interactions are no longer constant and will, in general, differ from spin pair to spin pair. The model of a disordered ferromagnet treated here is one in which the exchange interactions vary randomly about a mean interaction while the topology remains unchanged in the sense that the number of near neighbors remains the same in both the ordered and disordered case.

Double-time, temperature-dependent Green's functions have proved to be a powerful method for treating problems in statistical physics. To treat disordered ferromagnets of any spin value we have employed a Green's function which was first used by Callen,[6]

$$G^a_{fg}(t) \equiv \left\langle\!\!\left\langle S^+_f(t); e^{aS^z_g} S^-_g \right\rangle\!\!\right\rangle$$

$$\equiv -i\ \theta(t)\left\langle\left[S^+_f(t),\ e^{aS^z_g} S^-_g\right]\right\rangle \qquad (2)$$

where

$$\theta(t) = \begin{cases} 1, & t > 0 \\ 0, & t < 0 \end{cases} \qquad (3)$$

The single angular brackets denote an average over a grand canonical ensemble and the square brackets are the usual commutator brackets.

When the equation of motion of this function is written terms like $[S^+_f, H]$ lead to higher order Green's functions. To decouple this hierarch of equations we use the approximation of Tyablikov,[7]

$$\left\langle\!\!\left\langle S^z_h(t)\, S^+_f(t); e^{aS^z_g} S^-_g \right\rangle\!\!\right\rangle \simeq \langle S^z \rangle\, G^a_{fg}(t) \qquad (4)$$

This allows the equation of motion for the time Fourier transform of the Green's function to be written as

$$\hat{A}\,\hat{G} = \hat{1} \qquad (5)$$

where

$$A_{fh} \equiv \frac{2\pi}{\Theta(a)}\left\{\delta_{fh}\left[\frac{E}{\hbar} - \langle S^z \rangle \sum_j J(f,j)\right] + \langle S^z \rangle\, J(f,h)\right\} \qquad (6)$$

and

$$\Theta(a) \equiv \left\langle\left[S^+_f, e^{aS^z_f} S^-_f\right]\right\rangle \qquad (7)$$

Note that for $a = 0$, $\Theta(a)$ is related to the magnetization. That is $\Theta(0) = 2\hbar\langle S^z \rangle$.

In a perfect crystal the spatial Fourier transforms of the indexed quantities can be written and the Green's function for such a case with zero disorder is

$$\Gamma_{\vec{k}}^{a}(E) = \frac{\hbar\Theta(a)}{2\pi} \frac{1}{E - E_{\vec{k}}} \tag{8}$$

where the eigenvalues $E_{\vec{k}}$ are

$$E_{k} = \hbar\langle S^z \rangle \left(\mathbf{J}(0) - \mathbf{J}(\vec{k}) \right) \tag{9}$$

and

$$\mathbf{J}(\vec{k}) = \sum_{f-g} e^{i(f-g)\cdot\vec{k}} \, J(f,g) \tag{10}$$

$\vec{k}$, of course, is a reciprocal lattice vector. In a disordered ferromagnet translational symmetry is absent and the Fourier transform technique is not applicable.

Using the technique employed earlier for a spin one-half system[4] the Green's function for a disordered ferromagnet with nearest neighbor interactions of any spin can be shown to be

$$\left\langle G^{a}(E) \right\rangle_{\vec{k}} = \frac{\hbar\Theta(a)}{2\pi} \frac{1}{E - E_{\vec{k}} - \rho E_{\vec{k}} \frac{1}{N}\sum_{\vec{k'}} \frac{E_{\vec{k'}}}{E - E_{\vec{k'}}}} \tag{11}$$

Here ρ is a disorder parameter related to the mean square deviation of the exchange interactions from the mean,

$$\rho \equiv \frac{2}{Z} \frac{\langle (J - J^o)^2 \rangle}{J^{o2}} \tag{12}$$

Z is the number of nearest neighbors and J^o is the mean exchange interaction.

DENSITY OF SPIN WAVE STATES

The density of spin wave states is related to the trace of the imaginary part of the Green's function.[4] We have normalized the spin wave energies by dividing by the energy of the highest energy spin wave in the corresponding perfect crystal.

$$x \equiv \frac{E}{\hbar \langle S^Z \rangle J'} \tag{13}$$

where J' has values of 12 J, 16 J and 16 J for the sc, bcc and fcc cases respectively. Here J is the nearest neighbor exchange interaction. The density of spin wave states is

$$g(x,\rho) = \lim_{\epsilon \to +0} \frac{i}{2\pi} \frac{1}{N} \sum_{\vec{k}} \left[\frac{1}{x + i\epsilon - x_{\vec{k}} - \rho x_{\vec{k}} \frac{1}{N} \sum_{\vec{k}'} \frac{x_{\vec{k}'}}{x + i\epsilon - x_{\vec{k}'}}} \right.$$
$$\left. - \frac{1}{x - i\epsilon - x_{\vec{k}} - \rho x_{\vec{k}} \frac{1}{N} \sum_{\vec{k}'} \frac{x_{\vec{k}'}}{x - i\epsilon - x_{\vec{k}'}}} \right] \tag{14}$$

Since the eigenvalues of the ordered crystal, $x_{\vec{k}}$, are known,[8] the density of states for disordered ferromagnets with nearest neighbor interactions can be computed from Eq. (14).

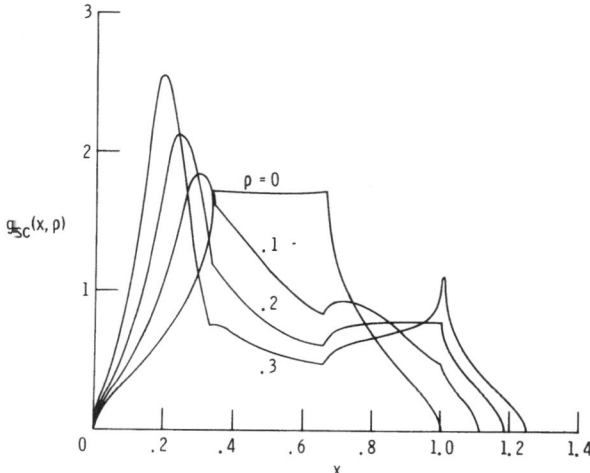

Figure 1. - Density of spin wave states for disordered simple cubic ferromagnets.

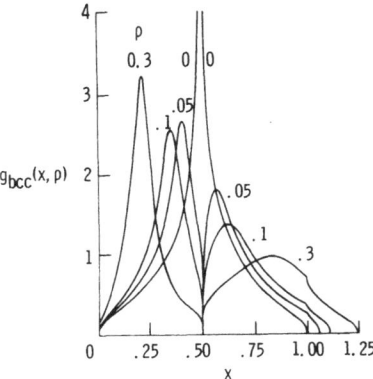

Figure 2. - Density of spin wave states for disordered body-centered cubic ferromagnets.

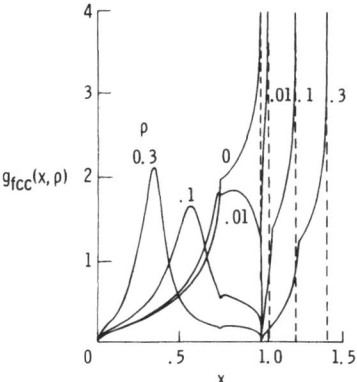

Figure 3. - Density of spin wave states for disordered face-centered cubic ferromagnets.

Figures 1, 2 and 3 show the state densities for disordered sc, bcc and fcc ferromagnets. In each figure the perfect crystal or $\rho = 0$ curve is presented. In these the density of states has nonzero values only in the range $0 < x < 1$ and in all three spectra singularities are present. For both the bcc and fcc cases the density of states has logarithmic singularities and there are discontinuities in the slope of the sc and fcc spectra. These are the familiar Van Hove singularities that appear as a consequence of lattice periodicity.

When disorder is introduced in a ferromagnet the density of states undergoes a more or less continuous change of shape so that at small values of the disorder parameter $g(x,\rho)$ is similar in appearance to $g(x,o)$. Exceptions to this continuous change of $g(x,\rho)$ with ρ occur where there are infinities in $g(x,o)$. These infinities become finite when ρ becomes nonzero. The most noticeable general effects of disorder on the density of states is to enlarge the number of spin wave states at lower energies, decrease the number in the upper portion of the perfect crystal energy band and then to extend the maximum energy to higher values.

An important part of the density of states is the low energy end of the spectrum. Since spin waves behave as bosons only the low energy states are occupied at low temperatures. In calculating low temperature magnetic properties, then, only the low energy end of the density of states curve is important. We have seen that disorder introduced more low energy states. More quantitatively

$$g(x,\rho) \underset{x \to o}{\simeq} (1 - \rho)^{-3/2} g(x,o) \tag{15}$$

Low temperature properties such as specific heat will show increases of the order $(1 - \rho)^{-3/2}$ over the corresponding perfect crystal as a result of Eq. (15).

The state densities that have been derived can be used to calculate the spontaneous magnetization of disordered ferromagnets. For systems of spin greater than one-half the technique of Callen[6] has been employed giving

$$\langle S^z \rangle = \hbar \frac{(S - \Phi)(1 + \Phi)^{2S+1} + (S + 1 + \Phi)\Phi^{2S+1}}{(1 + \Phi)^{2S+1} - \Phi^{2S+1}} \tag{16}$$

In perfect crystals

$$\Phi = \frac{1}{N} \sum_{\vec{k}} \left(\exp \frac{E_{\vec{k}}}{kT} - 1\right)^{-1} \tag{17}$$

The analogous quantity for disordered cases can be evaluated using $g(x,\rho)$.

$$\Phi = \int \frac{g(x,\rho)\, dx}{\exp\left(\frac{\hbar J' \langle S^z \rangle x}{kT}\right) - 1} \tag{18}$$

The spontaneous magnetization, $\sigma = \langle S^z \rangle / S$, can be solved self-consistently from Eqs. (16) and (18) for disordered magnets of any spin value.

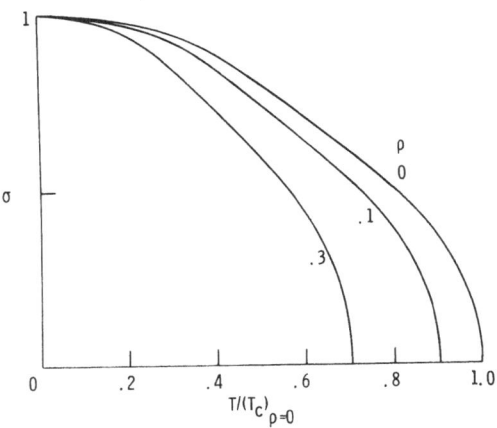

Figure 4. - Spontaneous magnetization for disordered fcc ferromagnets of spin 7/2.

Figure 4 shows the effect of disorder on the spontaneous magnetization of disordered fcc ferromagnets of spin seven-halves. The state densities of Fig. 3 were used to generate these curves. They show that the Curie temperature is depressed by disorder as $(1 - \rho)$. Similar results for sc and bcc systems[8] show that disorder also produces a slight flattening of σ versus T curves that decreases with increasing spin. Unlike the ferromagnetic Curie temperature the paramagnetic Curie point for disordered ferromagnets is identical to the perfect crystal case, indicating the effects of disorder are more pronounced in the lower temperature ferromagnetic phase.

REFERENCES

1. A. I. Gubanov, Fiz. Tver. Tela 2, 502 (1960). [Sov. Phys.-Solid State 2, 468 (1961).]

2. S. Mader and A.S. Norwick, Appl. Phys. Lett. 7, 57
    (1965); C.C. Tsuei and P. Duwez, J. Appl. Phys. 37,
    435 (1965); B. Elschner and H. Gartner, Z. Angew.
    Phys. 20, 342 (1966); P. Duwez and S.C.H. Lin, J.
    Appl. Phys. 38, 4096 (1967); K. Tamura and H. Endo,
    Phys. Lett. 29A, 52 (1969).
3. K. Handrich, Phys. Stat. Sol. 32, K55 (1969).
4. C.G. Montgomery, J.I. Krugler and R.M. Stubbs, Phys.
    Rev. Lett. 25, 669 (1970).
5. D.N. Zubarev, Sov. Phys. Usp. 3, 320 (1960).
6. H.B. Callen, Phys. Rev. 130, 890 (1963).
7. S.V. Tyablikov, Ukr. Mat. Zhur. 11, 287 (1959); N.N.
    Bogolyubov and S.V. Tyablikov, Sov. Phys. Doklady 4,
    589 (1959).
8. R.M. Stubbs and C.G. Montgomery, Phys. Rev. (to be
    published).

DISCUSSION

S.M. Bose: There is a kink in the high-energy region of
your disordered density of states for the simple cubic system, Figure 1. We have found from a CPA calculation that
the state densities of amorphous systems are smooth functions of energy, which may be because of the mean field
approximation.

R. Stubbs: Yes, to the order of this calculation the Van
Hove singularities of the corresponding perfect crystal are
preserved when disorder is introduced. I would not want to
defend the real existence of these features in the disordered curves as much as I would the more general and important
effects of disorder, namely the enhancement of the low-
energy region of the state density.

THEORY OF SPIN-WAVES IN A DISORDERED HEISENBERG FERROMAGNET

S. M. Bose,[1] K. Moorjani,[2] T. Tanaka,[3]
and M. M. Sokoloski[4]

[1]Department of Physics, Drexel University
Philadelphia, Pennsylvania 19104

[2]Applied Physics Laboratory
The Johns Hopkins University
Silver Spring, Maryland 20910

[3]Department of Physics
The Catholic University of America
Washington, D. C. 20017

[4]Harry Diamond Laboratories
Washington, D. C. 20438

ABSTRACT

The pair-theory of the coherent potential approximation
in the presence of off-diagonal randomness (CPA-ODR) has
been applied to a disordered spin-1/2 Heisenberg ferromagnetic binary alloy. The spin-wave Green's function has been
evaluated in the random phase approximation and the randomly varying exchange interaction has been treated in the CPA-
ODR. The density of spin-wave states (DSWS) has been calculated for a bcc lattice for the entire range of impurity
concentrations. It is found that the general effect of disorder is to remove the singularities, modify the band width
and introduce an additional structure in the DSWS. The
ferromagnetic Curie temperature T_c for all impurity concentrations has also been investigated. For impurities which
are ferromagnetic in nature, the results from the present
theory are found to be close to those obtained from the
mean field theory.

INTRODUCTION

Most of the theoretical studies[1,2] on the properties of disordered Heisenberg ferromagnets have so far been limited to either the dilute or the weak-scattering limits. Recently, however, noticing the fact that a disordered ferromagnet has much in common with a disordered electronic system, Foo and Wu[3] applied the coherent potential approximation (CPA), originally developed for the electronic case[4], to a random ferromagnetic alloy with arbitrary impurity concentrations and coupling strengths. Although this theory was able to explain qualitatively some experimental results, yet because of its basic single-site nature it was unable to reproduce some of the peaks in the density of spin-wave states (DSWS)[5] occurring from pairing or clustering effects. Furthermore, this theory is rather unrealistic because it fails to realize that, unlike the electronic case, the randomness in the diagonal term of the Heisenberg Hamiltonian would automatically mean randomness in the off-diagonal terms.

In an attempt to overcome the shortcomings of the Foo and Wu theory we have applied the pair-theory of the coherent potential approximation in the presence of off-diagonal randomness (CPA-ODR)[6] to the case of a disordered spin-1/2 Heisenberg ferromagnetic binary alloy. This theory was originally developed by the present authors in connection with their studies of the electronic properties of a random binary alloy and includes, self-consistently, the effects of off-diagonal randomness and scattering from two-site clusters. This theory is ideally suited for studying these effects in a disordered ferromagnetic alloy.

THEORY

In this paper the ferromagnetic system has been treated in the spin-wave approximation i.e. the Green's function of the system has been evaluated in the random phase approximation (RPA), and the randomly varying exchange interactions have been treated in the CPA-ODR. In the RPA, the Hamiltonian for a spin-1/2 Heisenberg ferromagnetic system can be expressed as

$$H = \frac{1}{2} \sum_{\ell,m} J(\ell,m) \{ |\ell><\ell| + |m><m| - |\ell><m| -$$

$$- |m><\ell| \} \tag{1}$$

where $|\ell>$ represents a single spin deviation state at site ℓ. For a disordered alloy of type $A_x B_{1-x}$, $J(\ell,m)$ is composition dependent and can take values J_{AA}, J_{BB} or J_{AB} depending on whether the nearest neighbor sites ℓ and m are occupied by two A atoms, two B atoms or one A and one B atoms respectively.

In the spirit of the coherent potential theory, a configuration independent effective Hamiltonian H_0 containing two coherent exchange interactions J_0 and J_1 is introduced as

$$H_o = \frac{1}{2} \{ J_o [\sum_\ell |\ell><\ell| + \sum_m |m><m|]$$

$$- J_1 \sum_{\ell,m} |\ell><m| + |m><\ell| \} \tag{2}$$

At this point we start deviating from the Foo and Wu theory or the original Soven theory where the effective Hamiltonian was defined in terms of a single coherent potential.

The exact Green's function and the coherent Green's function satisfy the equations

$$G = \frac{1}{E-H} \qquad \text{and} \qquad G_o = \frac{1}{E-H_o} \tag{3}$$

and are related to each other by

$$G = G_o + G_o T G_o \tag{4}$$

where the scattering matrix T is given by

$$T = \Gamma + \Gamma G_o T \tag{5}$$

Here Γ is the scattering potential and is obtained from

$$\Gamma = H - H_o \tag{6}$$

Taking configurational average of (4) we obtain

$$< T > = 0 \tag{7}$$

This equation which is so far an exact condition can be used to determine J_o and J_1. Since the exact evaluation of the T-matrix is almost impossible we evaluate it in the nearest neighbor pair approximation. Noting that the operator Γ has diagonal and off-diagonal components it is convenient to introduce two T-matrices by the following equation

$$T^{(i)} = \Gamma^{(i)} + \Gamma^{(i)} G_o T^{(i)}, \quad i = 1,2 \tag{8}$$

where

$$\Gamma^{(1)} = \sum_{\ell} \Gamma^{(1)}(\ell) = \sum_{\ell} |\ell > [\sum_m J(\ell,m) - J_o] < \ell| \tag{9a}$$

and

$$\Gamma^{(2)} = \sum_{\ell,m} \Gamma^{(2)}(\ell,m) = \sum_{\ell,m} |\ell > [J(\ell,m) - J_1] <m| \tag{9b}$$

The total scattering matrix T can then be expressed in terms of $T^{(1)}$ and $T^{(2)}$ as

$$T = [T^{(1)} + R] Q + [T^{(2)} + S] P \tag{10}$$

where

$$R = T^{(2)} G_o T^{(1)} \tag{11}$$

$$S = T^{(1)} G_o T^{(2)} \tag{12}$$

$$P = (1 - G_o S)^{-1} \tag{13}$$

$$Q = (1 - G_o R)^{-1} \tag{14}$$

The total T-matrix given by Eq. (10) includes independent scatterings by $\Gamma^{(1)}$ and $\Gamma^{(2)}$ plus all other correlations where scatterings occur alternatively by $\Gamma^{(1)}$ and $\Gamma^{(2)}$.

The two unknown coherent exchange interactions J_o and J_1 can be determined by using the self-consistency requirement that both the diagonal and the nearest neighbor off-diagonal matrix elements of the configurational averaged T-matrix be equal to zero i.e.

$$< T >_{\ell\ell} = 0 \qquad \text{and} \qquad < T >_{\ell m} = 0 \tag{15}$$

Since in the pair approximation T is completely specified in terms of $T^{(1)}$ and $T^{(2)}$, Eq. (15) implies that we must evaluate the diagonal and off-diagonal matrix elements of $T^{(1)}$ and $T^{(2)}$. Configurational average of these matrix elements are

$$< T^{(1)} >_{\ell\ell} = < t_\ell > + Z < t_\ell^2 t_m g_1^2/(1-t_\ell t_m g_1^2)> \tag{16}$$

$$< T^{(1)} >_{\ell m} = Z < t_\ell t_m g_1/(1 - t_\ell t_m g_1^2)> \tag{17}$$

where

$$t_\ell = \Gamma^{(1)}(\ell) + \Gamma^{(1)}(\ell) G_o t_\ell$$

and

$g_1 = < \ell| G_o |m >$ and Z denotes the number of nearest neighbors. It can also be shown that

$$< t_\ell > = \sum_{n=0}^{Z} \frac{Z!}{n!\,(Z-n)!}\ x^n (1-x)^{Z-n}$$

$$\succ [x\ \frac{\varepsilon_A - J_o}{1-(\varepsilon_A - J_o)g_o}\ +\ (1-x)\ \frac{\varepsilon_B - J_o}{1-(\varepsilon_B - J_o)g_o}\,] \tag{18}$$

where

$$\varepsilon_A = nJ_{AA} + (Z-n)J_{AB}, \qquad \varepsilon_B = (Z-n)J_{BB} + nJ_{AB}$$

and

$$g_o = <\ell|\ G_o\ |\ell>.$$

Eqs. (16) and (17) indicate that we must also evaluate configurational averages of the form

$$< \frac{Ft_\ell t_m}{1-Kt_\ell t_m} > \ = x^2\ \frac{Ft_A'^2}{1-Kt_A^2}\ +\ 2x(1-x)\ \frac{Ft_A''t_B''}{1-Kt_A''t_B''}$$

$$+\ (1-x)^2\ \frac{Ft_B'^2}{1-Kt_B^2} \tag{19}$$

where

$$t_A' = \sum_{n=0}^{Z-1}\ \frac{(Z-1)!}{n!\,(Z-1-n)!}\ x^n\ (1-x)^{Z-1-n}\ \frac{\varepsilon_A' - J_o}{1-(\varepsilon_A' - J_o)g_o} \tag{20}$$

with

$$\varepsilon_A' = (n+1)J_{AA} + (Z-1-n)J_{AB}$$

and

$$t_A'' = \sum_{n=0}^{Z-1}\ \frac{(Z-1)!}{n!\,(Z-1-n)!}\ x^n (1-x)^{Z-1-n)}\ \frac{\varepsilon_A'' - J_o}{1-(\varepsilon_A'' - J_o)g_o} \tag{21}$$

with

$$\varepsilon_A'' = nJ_{AA} + (Z-n)J_{AB}.$$

t_B' and t_B'' can be obtained from t_A' and t_A'' respectively by simply interchanging the subscripts A and B in Equations (20) and (21). Notice that these configurational averages are somewhat more involved than those in the electronic case.

The diagonal and off-diagonal matrix elements of $T^{(2)}$ are exactly the same as the electronic case and are given by[6]

$$< T^{(2)} >_{\ell\ell} = Z < \Gamma_{m\ell}^2 \, g_o / [\,(1-\Gamma_{m\ell}g_1)^2 - \Gamma_{m\ell}^2 \, g_o^2] > \qquad (22)$$

$$< T^{(2)} >_{\ell m} = Z < \Gamma_{\ell m}(1-\Gamma_{\ell m} \, g_1) / [\,(1-\Gamma_{m\ell} \, g_1)^2$$

$$- \Gamma_{m\ell}^2 \, g_o^2] > \qquad\qquad\qquad (23)$$

where

$$\Gamma_{m\ell} = \Gamma_{\ell m} = J_{\ell m} - J_1$$

Given the matrix elements of $T^{(1)}$ and $T^{(2)}$, we can evaluate the matrix elements of T and then using the self-consistency criteria (15) we obtain two non-linear coupled equations which can be solved numerically for J_o and J_1.

RESULTS AND DISCUSSION

The knowledge of J_o and J_1 allows one to compute the Green's function g_o. The density of spin-wave states (DSWS) is then obtained from

$$\rho(E) = - \frac{1}{\pi} \, \mathrm{Im} \, g_o(E)$$

In Fig. 1 we have plotted the DSWS $\rho(E)$ for an alloy of bcc lattice symmetry with $J_{AA} = 1$, $J_{BB} = 2$ and $J_{AB} = 1.5$ and impurity concentrations varying from 0 to 1. For the pure case ($x = 1$) $\rho(E)$ shows a singularity characteristic of a bcc lattice. It is found that the general effect of disorder is to remove the singularity and modify the band width. The band width is found to be wider than that of the mean field theory. As the impurity concentration increases, DSWS shows an extra peak corresponding to the Van Hove singularities of the two pure cases.

The Curie temperature T_c, which is defined as the temperature at which magnetization vanishes, has been evaluated from the well-known relation[2]

$$T_c \sim - \frac{N}{Tr \; g_o \; (E=0)}$$

In Fig. 2 we have plotted T_c as a function x and for $J_{AA} = 1$ and $J_{BB} = 2$ and $J_{AB} = 1.0$, 1.5, 2.0, and 2.5. A comparison of T_c obtained from our theory (solid lines) with that obtained in the mean-field theory (dashed curves) shows that these two results are in close agreement, although our result is always lower than the mean-field values. This simply indicates that for ferromagnetic impurities the mean field theory works quite well at least at the low energy regions ($E \sim 0$). However, in the presence of nonmagnetic or antiferromagnetic impurities the situation gets drastically changed. At present we are involved in calculations of T_c in the presence of such nonmagnetic or antiferromagnetic impurities. Details of these calculations along with a comparison of our results with those of Foo and Wu will be presented elsewhere.

REFERENCES

1.   D. Hone, H. Callen and L.R. Walker, Phys. Rev. 144, 283 (1966).
2.   C.G. Montgomery, J.I. Krugler and R.M. Stubbs, Phys. Rev. Letters, 25, 669 (1970).
3.   E-Ni Foo and D.H. Wu, Phys. Rev. B5, 98 (1972).
4.   P. Soven, Phys. Rev. 156, 1017 (1967).

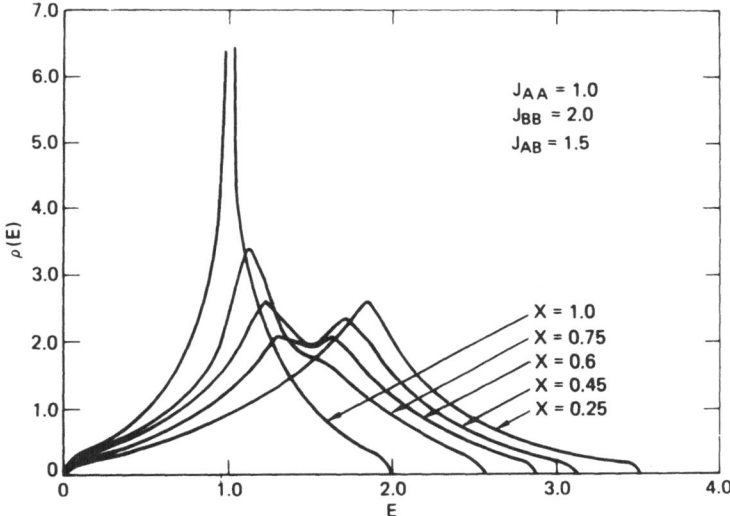

Figure 1   Density of spin-wave states for a disordered
           ferromagnetic alloy $A_x B_{1-x}$ with $J_{AA} = 1$, $J_{BB}=2$
           and $J_{AB} = 1.5$ for a bcc lattice for x = 1, 0.75,
           0.6, 0.45 and 0.25.

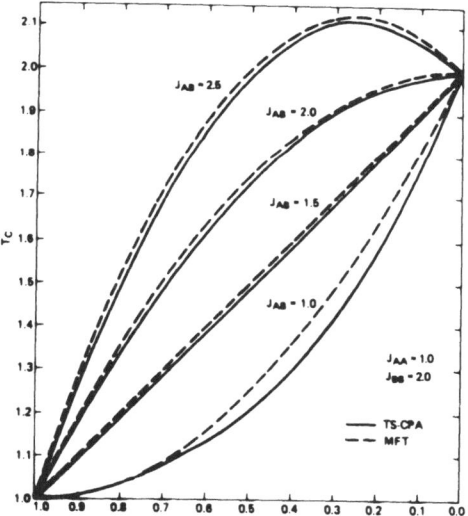

Figure 2   Curie temperature T_c for $A_x B_{1-x}$ alloy as a func-
           tion of impurity concentration with $J_{AA} = 1$,
           $J_{BB} = 2$ and J_{AB} varying between 1 and 2.5 as shown
           Solid curves are T_c in CPA-ODR and dashed curves
           represent the mean-field theory results.

5.   S.M. Bose and E-Ni Foo, J. Phys. C: Solid State Phys.
      5, 1082 (1972).
6.   T. Tanaka, M.M. Sokoloski, K. Moorjani and S.M. Bose,
      Jour. Non-Cryst. Solids 8-10, 155 (1972). Also to be
      published.

CATION DISORDER AND MAGNETIC PROPERTIES OF CERTAIN GARNET

TYPE CRYSTALS

R. Krishnan

Laboratoire de Magnétisme, C.N.R.S.

1, place Aristide Briand - 92190 BELLEVUE FRANCE

INTRODUCTION

The garnet of composition $Bi_{3-2x}Ca_{2x}Fe_{5-x}V_xO_{12}$ (BiCaVIG) is derived from the well known yttrium iron garnet (YIG) wherein Y^{3+} ions are completely replaced by a combination of Ca^{2+} and Bi^{3+} ions and Fe^{3+} ions are partly substituted with V^{5+} ions[1, 2]. Though BiCaVIG is isostructural to YIG one can expect a certain amount of difference between the crystal fields in the crystallographic sites in these two compounds. One way to study the crystal fields would be to study and compare the magnetic behaviour of a paramagnetic ion when present in these two crystals, as it is well known that the energy levels of an ion are perturbed by the crystal field and its symmetry. Further one can also expect that the presence of cations of different chemical nature as in the BiCaVIG lattice to modify the energy levels as it will be shown later. In this work we will describe the magnetic properties of some paramagnetic ions such as $Co^{2+}(3d^7)$, $Ru^{3+}(4d^5)$ and $Tb^{3+}(4f^8)$, when they are doped in BiCaVIG crystals and compare the results with those obtained when they are present in YIG crystal. Such differences in behaviour could tell us about the sites wherein are present the paramagnetic ions.

Let us briefly recall some structural properties of a garnet lattice. This structure belongs to the space group Ia 3d of the cubic system. The important aspects of this structure in relevance to the magnetic properties are to do with the cation sublattices. There are three types of crystallographic sites : tetrahedral 24d, octahedral 16a and dodecahedral 24c. If one considers an octant of the unit cell with a side of length a/2 (a being the lattice constant) it is a bcc structure built up of 16a sites. The 24d and 24c sites alternate along chains that are on the faces of the octant. The disposition of O^{2-} ions are such that each of them has four near neighbours ; two 24c sites and one of 24d and 16a each. Besides, none of the oxygen polyhedra are regular causing distortions in the site symmetry. For instance the 16a site has a trigonal distortion and the 24c an orthorhombic one. These properties go a long way in determining the magnetic properties of a particular paramagnetic ion.

EXPERIMENTAL

The crystals studied were prepared by the flux method[3]. The actual composition of the crystals was determined by spectrophotometric method and emission spectroscopy. The crystals were exempt from any rare-earth impurity and the silica content was less than 10 ppm.

Ferrimagnetic resonance was observed in these crystals in the X-band and in the temperature range 4.2 to 300°K. From the angular dependence of the field for resonance H_{res} in the (110) plane the anisotropy constants K_1 and K_2 were computed.

RESULTS AND DISCUSSIONS

The host lattice was kept the same for all dopings and corresponds to the chemical formula $Ca_{2.70}Bi_{0.30}Fe_{3.65}V_{1.35}O_{12}$. Thus the observed change in the properties ought to be attributed to the different paramagnetic ions added to the crystal. The doping levels for the different ions were as follows : $CoO = 0.1$, $Ru_2O_3 = 0.2$ and $Tb_2O_3 = 0.38$, all expressed in wt %. The results for each ion are presented separately.

$$Co^{2+} \text{ ION } 3d^7.$$

This ion is well known to contribute a large positive anisotropy in YIG [4] where it occupies 16a site. The ground state is an orbital triplet with a large spin orbit coupling. However the present results show that in BiCaVIG the contribution from Co^{2+} ion is negative (Fig. 1). This indicates that Co^{2+} ion is not present in the 16a site.

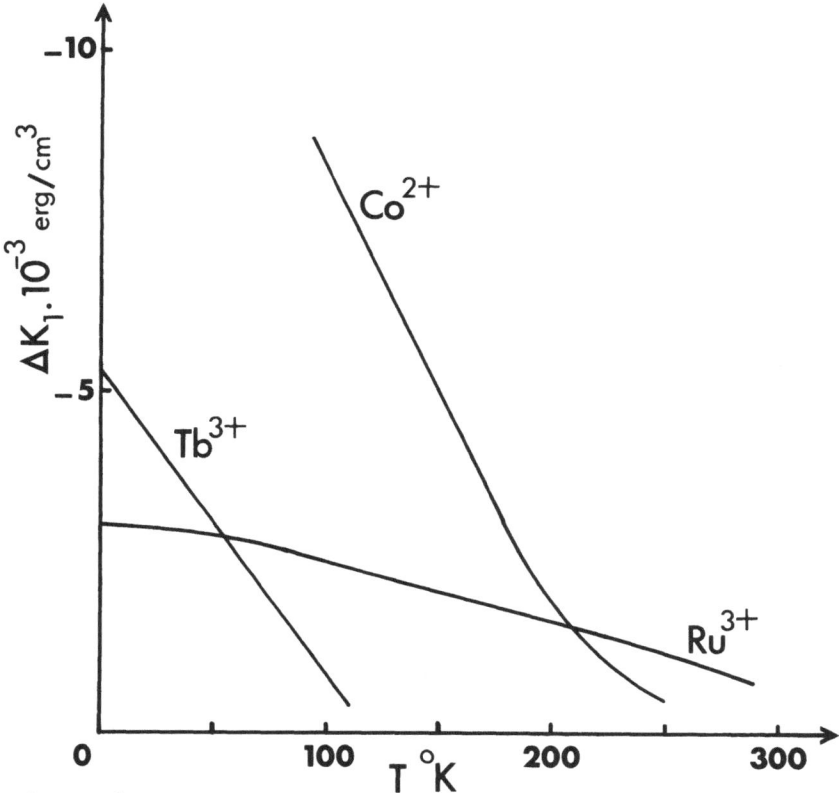

Figure 1.   Temperature dependence of anisotropy of the
paramagnetic ions.

Besides as shown in Fig. 2 the angular dependence of H_{res}
in the (110) plane shows at 4.2°K anomalous peaks. Such
anomalies are so far known to exist only for certain rare
earth ions [5]. They are attributed to the near cross over
of the energy levels in the exchange field [6]. Measurements
carried out on different crystallographic planes show that
the peaks could not be associated with the presence of
cobalt in either 16a sites of C_{3i} symmetry or 24d sites of
S_4 symmetry. Thus we are led to the conclusion that Co^{2+}
ion in BiCaVIG crystal enters the 24c site. This site with
its D_2 symmetry incidentally is quite favourable to pro-
voke near cross over of the energy levels. Thus the magnetic behaviour of cobalt ions in this crystal is indeed
original.

$$\text{Ru}^{3+} \text{ ION } 4d^5.$$

The properties of this ion doped in YIG have recently been studied [7, 8]. Though there is a certain amount of uncertainty as to whether ruthenium is present also as Ru^{4+} and contributing to the properties, the results obtained at the present stage on both magnetic anisotropy [8] and magnetostriction [9] in YIG : Ru seem to indicate that Ru^{3+} ion alone is operating. So we can safely suppose that in BiCaVIG crystals also ruthenium is present as Ru^{3+}. Now for Ru^{3+} present in 16a site, the cubic ground state is $^2T_{2g}$. The six fold degeneracy is removed by the Hamiltonian

$$\mathcal{H} = V_t(\vec{r}) + \lambda.\vec{L}.\vec{S}. + g.\mu_B.\vec{S}.\vec{H}_e$$

where $V_t(\vec{r})$ is the trigonal field, $\lambda.\vec{L}.\vec{S}.$ the spin orbit coupling and $g\mu_B\vec{S}\vec{H}_e$ the exchange field. Consideration of the two lowest lying energy levels explain well the observed K_1 and K_2. Thus in YIG, Ru^{3+} contributes to a large positive K_1 and a still larger negative K_2 and the ratio K_2/K_1 at 4K is of the order of -4. Also the angular dependence of H_{res} in the (110) plane shows a peak at the $|112|$ direction. But in the BiCaVIG crystal the contribution from Ru^{3+} to K_1 is not only negative but also small and K_2 is practically negligible (**Fig. 1**). This is at variance with the properties observed in YIG : Ru. Of course Ru^{3+} ion if present in 24d sites is expected to contribute to a negative K_1 but however the observed magnitude is far below that expected by the theory [8]. Also the angular dependence of H_{res} in the (110) plane is smooth down to 4.2K (**Fig.** 2). This result again shows that the 16a and the 24d sites in BiCaVIG have a very different property as compared to YIG.

$$\text{Tb}^{3+} \text{ ION } 4f^8.$$

Tb^{3+} ion has been studied in detail in YIG and the results have also been theoretically well interpreted [5]. The angular dependence of H_{res} in the (110) plane at 4.2 K shows several giant peaks and, as predicted by the theory the peaks occur at 19°, 32° and 77° respectively, the angles being measured from the (001) direction. As mentioned earlier these peaks are due to the near cross-over of the energy levels of the Tb^{3+} ion belonging to the ground state multiplet (J = 6). Fig. 2 shows the angular

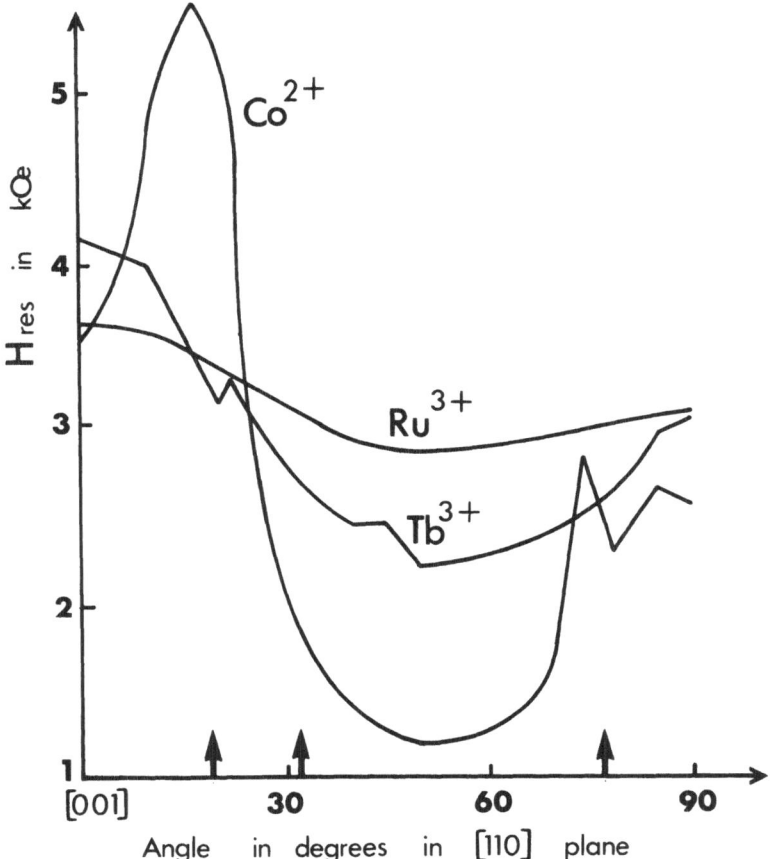

Figure 2. Angular dependence of H_{res} in the (110) plane at 4.2 K.

dependence of H_{res} at 4.2K in BiCaVIG : Tb. It is rather surprising that there are only two peaks and that too they are very small. The arrows indicate the peak positions in YIG : Tb. This result can be partly explained as follows. In BiCaVIG the partial substitution for Fe^{3+} by V^{5+} results in a reduction in the exchange field. It has been pointed out that the near cross over of energy levels $\emptyset_o$ must satisfy the relation [10] :

$$\cos^2\emptyset_o = \frac{E_1 E_2}{A_{12}^2 \mu_B^2 H_{ex}^2}$$

where $\emptyset_0$ is the angle the magnetization makes with one of the local axes of the 24c sites, E_1 and E_2 are the energies of the first and second excited levels of Tb^{3+} due to the crystal field splitting, H_{ex} is the exchange field from the ferric lattice acting on the spin of Tb^{3+}, A_{12} is a number and μ_B the Bohr magneton. Thus H_{ex} is seen to influence the near cross-over phenomenon. So though one can expect a shift in the peak position one cannot explain the great diminution of peak height in BiCaVIG : Tb. Thus it is evident that in order to explain the result one must also consider the second neighbours such as Fe^{3+} and V^{5+} of the 24c site and their possible effect on the crystal field. Of course one can also suppose that the cations in 24c sites that is Ca^{2+} and Bi^{3+} also can influence in the same manner.

To sum up, the above results on Co^{2+}, Ru^{3+} and Tb^{3+} clearly indicate the difference in the magnetic behaviour of these ions when present in BiCaVIG crystal. One possible approach to understand our results is to examine the crystallographical aspects of this crystal that could affect the crystal fields in the various sites. Detailed X-ray diffraction studies of this type of crystal [11] have yielded the oxygen co-ordinates from which one can deduce the interesting parameters such as the inter- ionic distances d-h, a-h and c-h, and the edges of the polyhedra (Table I) though the oxygen co-ordinates are very different in BiCaVIG from YIG the interionic distances BiCaVIG vary as follows. They are larger for a-h and c-h bonds while they are smaller for d-h as compared to YIG. This roughly indicates that in BiCaVIG one can expect a decrease in the crystal field strength for the a and c sites and an increase in the same for the d sites. Now it is obvious that such changes in the crystal field splitting cannot produce the anomalous properties that have been observed for the different ions. Further it also is seen from the calculated edges of the various polyhedra that such replacements of Fe^{3+} by V^{5+} leads to a reduction in the distortion of the oxygen tetrahedra and octahedra. So it is apparent that these factors are not responsible for the magnetic properties observed. Thus one is led to consider the next nearest neighbours for each site and their possible influence on the energy levels of the paramagnetic ion. In the garnet structure one can notice the configuration of the cation sites to be as follows. Each 24d site is surrounded by four of each 16a and 24c sites, each of 16a site by six of each 24c and 24d sites and finally each of 24c site by sixteen of each 24d

Table 1 : STRUCTURAL   PROPERTIES

of   BiCaVIG and   YIG

Compound	Bond length in Å		
	d – h	a – h	c – h
BiCaVIG for V= 1.46	1.799	2.039	2.521
YIG	1.887	2.005	2.411

and 16a sites. Now in YIG it is clear that 24d, 16a and 24c sites have a homogeneous cationic surrounding because there are only Fe^{3+} in the surrounding 16a and 24d sites and Y^{3+} in 24c sites being at a relatively greater distance. But in BiCaVIG the situation is quite complex. In 24d sites are present simultaneously both Fe^{3+} and V^{5+} ions each with a very different charge and having different oxygen affinity. This would perturb the crystal fields at all the sites along with the Coulomb interactions excited by these ions. Further the 24c sites are also occupied by both Ca^{2+} and Bi^{3+} ions causing the heterogeneity discussed above. They can also cause local distortion of the 24c sites due to their different ionic radii. A point charge calculation of the crystal field in BiCaVIG is quite complex and laborious and will not be yet complete in itself due to several approximations necessary, considering the different nature of cations in question.

In Conclusion, certain paramagnetic ions behave totally differently when they are doped in BiCaVIG crystal. Tb^{3+} ion which causes giant anisotropies in YIG does not posses any such property in BiCaVIG. Co^{2+} ion well known for its large positive anisotropy in YIG produces in BiCaVIG only smaller and negative anisotropy. Besides in the latter Co^{2+} causes anomalous peaks in H_{res} which is an original obser-vation. Finally the anisotropy from Ru^{3+} in BiCaVIG cannot be explained interms of existing theories which are quite valid for YIG. It is also shown that the cubic crystal field strengths in BiCaVIG are not responsible for such magnetic behaviour of the above ions but the cation dis-order in the BiCaVIG lattice and the heterogeneous nature in the cations present are causing such changes observed.

Though this compound is crystalline, the cation disorder alone is able to produce such tremendous changes in the properties of the crystal, and perhaps this is only one small step towards the amorphous state.

REFERENCES

1. G.A. Smolensky, V.P. Polyakov and V.M. Yudin, Izv. AN SSSR Ser. Fiz., 11, 1396 (1961).
2. S. Getter, G.P. Espinosa, H.J. Williams, R.C. Sherwood and E.A. Nesbitt, Appl. Phys. Letters 3, 60 (1963).
3. R. Krishnan, Phys. Stat. Solidi, 35, K63 (1969).
4. T. Okada, H. Sekizawa and S. Iida, J. Phys. Soc. Japan, 18, 981 (1963).
5. J.F. Dillon, Jr., and L.R. Walker, Phys. Rev. 124, 1401 (1961).
6. C. Kittel, Phys. Rev. 117, 681 (1960).
7. R. Krishnan, Phys. Stat. Solidi, 1, K17 (1970).
8. P. Hansen, Philips Res. Rept., suppl. No. 7 (1970).
9. R. Krishnan, V. Cagan and M. Rivoire, AIP Conf. Proc., No. 5, 704 (1972).
10. D.L. Huber, J. Appl. Phys. 36, 1005 (1965).
11. E.L. Dukhovskaya and Yu. G. Saksonov, Sov. Phys. Solid State, 10, 2613 (1969).

439